Millets Value Chain for Nutritional Security

A Replicable Success Model from India

FSC
www.fsc.org
MIX
Paper from responsible sources
FSC® C013604

Millets Value Chain for Nutritional Security

A Replicable Success Model from India

B. Dayakar Rao

Indian Institute of Millets Research (formerly DSR), Hyderabad, India

N.G. Malleshi

Ex-head, Grain Science Technology, Central Food Technological Research Institute (CFTRI), Mysore, India

George A. Annor

Department of Nutrition and Food Science, University of Ghana, Ghana

and

J.V. Patil

Indian Institute of Millets Research (formerly DSR), Hyderabad, India

CABI

CABI is a trading name of CAB International

CABI
Nosworthy Way
Wallingford
Oxfordshire OX10 8DE
UK

Tel: +44 (0)1491 832111
Fax: +44 (0)1491 833508
E-mail: info@cabi.org
Website: www.cabi.org

CABI
745 Atlantic Avenue
8th Floor
Boston, MA 02111
USA

Tel: +1 (0)617 682 9015
E-mail: cabi-nao@cabi.org

A catalogue record for this book is available from the British Library, London, UK.

Library of Congress Cataloging-in-Publication Data

Names: Dayakar Rao, B. (Benhur), editor.
Title: Millets value chain for nutritional security : a replicable success model from India / editors: B. Dayakar Rao, N.G. Malleshi, George A. Annor, J.V. Patil.
Description: Boston, MA : CABI, [2016] | Includes bibliographical references and index.
Identifiers: LCCN 2015048063 | ISBN 9781780648309 (hbk : alk. paper) | ISBN 9781780648323 (epub)
Subjects: LCSH: Millet industry--India. | Millets--Nutritional aspects--India.
Classification: LCC HD9049.M6 M355 2016 | DDC 338.1/731710954--dc23
LCrecord available at http://lccn.loc.gov/2015048063

ISBN-13: 978 1 78064 830 9

Commissioning editor: Rachael Russell
Editorial assistant: Emma McCann
Production editor: Tim Kapp

Typeset by SPi, Pondicherry, India
Printed and bound in the UK by CPI Group (UK) Ltd, Croydon, CR0 4YY

Contents

List of Figures

List of Tables

List of Appendices

List of Abbreviations

+iAUC	Positive incremental area under the curve
AAS	Amino acid score
AICPMIP	All India Coordinated Pearl Millet Improvement Project
AICSMIP	All India Coordinated Small Millet Improvement Project
ANGRAU	Acharya NG Ranga Agricultural University
APEDA	The Agricultural and Processed Food Products Export Development Authority
ATL	Above the line
BNI	Biological nitrification inhibition
BTL	Below the line
CAGR	Compound annual growth rate
CCSHAU	Chaudhary Charan Singh Haryana Agricultural University
CEO	Chief Executive Officer
CERES	Coalition for Environmentally Responsible Economies
CFTRI	Central Food Technological Research Institute
CGIAR	Consultative Group for International Agricultural Research
CGIR	Corporate Governance: an International Review
CIAE	Central Institute of Agricultural Engineering
CIPHET	Central Institute of Post-Harvest Engineering and Technology
CMC	Carboxymethyl cellulose
CoE	Centre of Excellence
CVD	Cardiovascular disease
DAC	Directorate of Agriculture and Cooperation
DES	Directorate of Economics and Statistics
DF	Dietary fibre
DFRL	Defense Food Research Laboratory
DNA	Deoxyribonucleic acid
DPPH	2,2-diphenyl-1-picrylhydrazyl
DSR	Directorate of Sorghum Research
EDP	Entrepreneurship Development Programme
FAO	Food and Agriculture Organization
FAQ	Fair average quality
FCI	Food Corporation of India

FLD	Front line demonstration
FP	Faecal protein
GI	Glycaemic index
GL	Glycaemic load
GSL	Grain sorghum lipid
HACCP	Hazard analysis and critical control points
HDL	High density lipoprotein
HYV	High-yielding variety
IBMA	Indian Biscuit Manufacturing Association
ICAR	Indian Council of Agricultural Research
ICMR	Indian Council of Medical Research
ICRISAT	International Crops Research Institute for the Semi-Arid Tropics
IIMR	Indian Institute of Millets Research
INR	Indian Rupee
INSIMP	Initiative for Nutritional Security through Intensive Millets Promotion
IRR	Internal rate of return
ITC (ABD)	Imperial Tobacco Company of India, Agri-Business Division
kcal	Kilocalories
KVK	Krishi Vigyan Kendra
LDL	Low-density lipoprotein
MFP	Metabolic faecal protein
MMA	Macro-management in agriculture
MoA	Memorandum of Agreement
MoU	Memorandum of Understanding
MPET	Metallized polyethylene terephthalate
MSP	Minimum Support Price
NA	Not applicable
NAAS	National Academy of Agricultural Sciences
NAIP	National Agricultural Innovation Project
NARS	National Agricultural Research System
NAY	National average yield
NCDC	National Co-operative Development and Corporation
ND	Not detectable
NGO	Non-governmental organization
NIDDM	Non-insulin dependent diabetes milletus
NIN	National Institute of Nutrition
NIRS	Near infra-red spectroscopy
NPV	Net present value
NSC	National Seed Corporation (India)
NVIF	Nutritive Value of Indian Foods
PCS	Production-to-consumption system
PDCAAS	Protein digestibility-corrected amino acid score
PDS	Public Distribution System
PI	Protein intake
PoP	Package of practices
ppm	Parts per million
PPP	Public–private partnership
PRSS	Post-rainy season sorghum
qt	Quintal
R&D	Research and development
RADP	Rainfed Area Development Programme

RBP	Retinol binding protein
RCS	Recommended cropping systems
RKVY	Rashtriya Krishi Vikas Yojana
RTC	Ready-to-cook
RTE	Ready-to-eat
SAU	State agricultural university
SDS	Slow digestible starch
SHG	Self-help group
SKRAU	Swami Keshwan and Rajasthan Agricultural University
SKU	Stock-keeping unit
SPC	Soya protein concentrate
SWOT	Strengths, weaknesses, opportunities and threats
TBHQ	Tertiary butylhydroquinone
TE	Triennium ending
UAS	University of Agricultural Sciences
UNU	United Nations University
UP	Uttar Pradesh
UT	Union Territory
WHO	World Health Organization

Note: crore, a unit of the Indian numbering system equal to 10,000,000; lakh, a unit of the Indian numbering system equal to 100,000.

List of Abbreviations (in Cluster Analysis)

CHPT_A	Appearance of *jowar chapathi/roti*
CHPT_CLR	Colour of *jowar chapathi/roti*
CHPT_OAL	Overall rating of *jowar chapathi/roti*
CHPT_SFT	Softness of *jowar chapathi/roti*
CHPT_SML	Smell of *jowar chapathi/roti*
CHPT_SWL	Ability of *jowar chapathi/roti* to swell
CHPT_TST	Swell test of *jowar chapathi/roti*
CHPT_TX	Texture of *jowar chapathi/roti*
COM_APR	Compared raw appearance
COM_TST	Compared taste
COM_PRP	Compared preparation process
JBS_A	Aroma of *jowar*-rich biscuit
JBS_CLR	Colour of *jowar*-rich biscuit
JBS_CRP	Crispiness of *jowar*-rich biscuit
JBS_FLT	Fullness of test after eating *jowar*-rich biscuit
JBS_MFL	Mouthful of *jowar*-rich biscuit
JBS_OAL	Overall rating of *jowar*-rich
JBS_SML	Biscuit smell of *jowar*-rich biscuit
JBS_SWT	Sweetness of *jowar*-rich biscuit
JBS_THK	Thickness of *jowar*-rich biscuit
JBS_TST	Test of *jowar*-rich biscuit
JFL_APR	Raw appearance of *jowar* flakes/*poha*
JFL_PRP	Preparation process of *jowar* flakes
JFL_TST	Test of *jowar* flakes/*poha*
JMAT_APR	Raw appearance of *jowar* multigrain *atta*
JMAT_PRP	Preparation process of *jowar* multigrain *atta*
JMAT_TST	Test of *jowar* multigrain *atta*
JOAT_CLR	Colour attribute of *jowar* multigrain *atta*
JOAT_ERL	Ease of rolling out *roti* attribute of *jowar* multigrain *atta*
JOAT_OAL	Overall attribute of *jowar* multigrain *atta*
JOAT_SFD	Softness of dough attribute of *jowar* multigrain *atta*
JOAT_SML	Smell attribute of *jowar* multigrain *atta*
JOAT_TX	Texture attribute of *jowar* multigrain *atta*

JOFL_A	Appearance of *jowar flakes/poha*
JOFL_CLR	Colour of *jowar* flakes/*poha*
JOFL_OAL	Overall rating of *jowar* flakes/*poha*
JOFL_SML	Smell of *jowar* flakes/*poha*
JOFL_THK	Thickness of *jowar* flakes/*poha*
JOSJ_A	Appearance of *jowar suji/rawa/semolina*
JOSJ_CLR	Colour attribute of *jowar suji/rawa/semolina*
JOSJ_OAL	Overall rating of *jowar suji/rawa/semolina*
JOSJ_ROS	Ease of roasting *jowar suji/rawa/semolina*
JOSJ_SML	Smell attribute of *jowar suji/rawa/semolina*
JOSJ_TX	Texture attribute of *jowar suji/rawa/semolina*
JOVM_A	Appearance attribute of *jowar* vermicelli
JOVM_CLR	Colour attribute of *jowar* vermicelli
JOVM_OAL	Overall rating of *jowar* vermicelli
JOVM_PRT	Preparation time of *jowar* vermicelli
JOVM_SML	Smell attribute of *jowar* vermicelli
JOVM_THK	Thickness attribute of *jowar* vermicelli
JSJ_APR	Raw appearance of *jowar suji/rawa/semolina*
JSJ_PRP	Preparation process of *jowar suji/rawa/semolina*
JSJ_TST	Test of *jowar suji/rawa/semolina*
JVM_APR	Raw appearance of *jowar* vermicelli
JVM_PRP	Preparation process of *jowar* vermicelli
JVM_TST	Test of *jowar* vermicelli
POHA_A	Appearance of cooked *jowar* flakes/*poha*
POHA_CLR	Colour of cooked *jowar* flakes/*poha*
POHA_OAL	Overall rating of cooked *jowar* flakes/*poha*
POHA_SFT	Softness of cooked *jowar poha*
POHA_SML	Smell of cooked *jowar* flakes/*poha*
POHA_STK	Stickiness/separateness of cooked *jowar poha*
POHA_TST	Test of cooked *jowar poha*
POHA_WAB	Water absorption *jowar* flakes/*poha*
SIRA_A	Appearance of (*sira*) sweet *jowar suji/rawa/semolina*
SIRA_CLR	Colour of (*sira*) sweet *jowar suji/rawa/semolina*
SIRA_OAL	Overall rating of (*sira*) sweet *jowar suji/rawa/semolina*
SIRA_SFT	Softness of (*sira*) sweet *jowar suji/rawa/semolina*
SIRA_SML	Smell of (*sira*) sweet *jowar suji/rawa/semolina*
SIRA_TST	Test of (*sira*) sweet *jowar suji/rawa/semolina*
SIRA_TX	Texture of (*sira*) sweet *jowar suji/rawa/semolina*
SIRA_WTA	Water absorption of (*sira*) sweet *jowar rawa*
UPIT_A	Appearance of (*uppit*) salty *jowar suji/rawa/semolina*
UPIT_CLR	Colour of (*uppit*) salty *jowar suji/rawa/semolina*
UPIT_OAL	Overall rating of (*uppit*) salty *jowar suji/rawa/semolina*
UPIT_SFT	Softness of (*uppit*) salty *jowar suji/rawa/semolina*
UPIT_SML	Smell of (*uppit*) salty *jowar suji/rawa/semolina*
UPIT_TST	Test of (*uppit*) salty *jowar suji/rawa/semolina*
UPIT_TX	Texture of (*uppit*) salty *jowar suji/rawa/semolina*
UPIT_WTA	Water absorption of (*uppit*) salty *jowar suji/rawa/semolina*
VMSW_A	Appearance of cooked *jowar* vermicelli
VMSW_CLR	Colour of cooked *jowar* vermicelli
VMSW_OAL	Overall rating of cooked *jowar* vermicelli
VMSW_SFT	Softness of cooked *jowar* vermicelli
VMSW_SML	Smell of cooked *jowar* vermicelli
VMSW_STK	Stickiness/separateness of cooked *jowar* vermicelli
VMSW_TST	Test of cooked *jowar* vermicelli

Foreword

Millets are a traditional staple food of the dryland regions of the world. In India, they are grown over about 20 million ha with annual production of 18 million tonnes and contribute 10% to the country's food grain basket. In the present scenario, demand for direct consumption of millets has been declining due to changes in food habits and the inconvenience attached to food preparation as compared to fine cereals. So it has become essential to direct R&D efforts towards millets to revive demand through diversification of processing technologies, nutritional evaluation and creation of awareness backed by backward integration. The subproject of NAIP, 'Millets Value Chain', focused such efforts on selected millet foods, particularly sorghum, with a view to extending it to other millets if successful.

This book reveals the various components in the value chain of sorghum and interventions attempted under the NAIP subproject. It describes in detail the sorghum product technologies that have been developed and standardized by the IIMR-led consortium. The nutritional values, methods of production and inputs incurred in researching these technologies reveal sustainability in the sorghum value chain, providing nutritional security to consumers and acceptable remuneration to farmers. Further, the Centres of Excellence established under the Indian government's INSIMP programme of the DAC to promote millet product technologies, have been developed due to efforts on the value chain of sorghum/millets. Plans have been made to replicate the already developed sorghum product technologies; this will exploit the processing capabilities of sorghum and millets to their fullest potential and could also enable them to attain export quality at feasible costs.

This book will be a valuable reference for planners, policymakers, processors, stakeholders, and food technologists alike, enabling them to advocate steps to promote the value chain of sorghum and other millets in India and other parts of the world. The value chain on sorghum attained national recognition and drew attention worldwide. This model is replicable to other crop commodities across the world as it has been sustainable beyond the project period. This is one of the few books on value chain concepts that is embedded with content that can be of use to many stakeholders, including the research and academic fraternity. I congratulate Dr. B. Dayakar Rao, PS, ICAR-IIMR and his co-authors for their achievements with this novel publication.

Dr. S. Ayyappan
Secretary DARE & Director General, ICAR
Government of India, New Delhi, India

Preface

Millets are important crops for dryland farmers; they are highly nutritious and are a climate-compliant crop. But overall millets consumption in India has declined over the years. In order to revive the demand for millets in India, the IIMR (formerly DSR) led a consortium under NAIP funding to undertake interventions to bridge the gaps identified in the millet value chain: at on-farm production, processing diversification, nutritional certification, promotion and marketing. It brought together all the stakeholders in the production-to-consumption system (PCS) value chain, linking them with poor dryland farmers. Before this project's implementation, the supply chain was disjointed and unorganized. But under this NAIP subproject, 'Millets Value Chain', an institutional mechanism was established to form a consortium of stakeholders in public–private partnerships, ensuring a win–win situation for all.

In this present era, consumers prefer high-quality foods with longer shelf life. They also look for products that are convenient to prepare. Moreover, people have increased their tendency to eat a greater variety of foods. Most people know little of the nutritious value of millets; despite the fact that they have a rich composition of nutrients and minerals, past health branding has not been exploited enough to commercialize millet foods. This has mainly been due to the absence of certification by competent national institutes and lack of awareness by the food industry. Professional market studies on consumer preferences and their potential in various segments of rural and urban markets tend not to focus on such small-scale products. Even some isolated efforts made to popularize them could not elevate millets' status due to paucity of funds and consistent public support for other grains. Further, there were no integrated efforts to link up all the stakeholders in the PCS value-chain with assurance of critical mass efforts for this service. Moreover, due to disincentives resulting from PDS for fine cereals and other practices, a sustainable value addition approach was necessary from producer to consumer to ensure nutritional security and simultaneously support the dryland farmers of the country.

This publication is a compilation of successful sorghum value chain interventions under the NAIP subproject. Each chapter contains original work produced by the IIMR-led consortium and derived from the experience and results or impact of IIMR-led interventions during and post-NAIP. It is hoped that this publication will provide a model on value chain of sorghum and other millets for many different users.

B. Dayakar Rao

N.G. Malleshi

George A. Annor

J.V. Patil

Executive Summary

Consumption of sorghum and other millets as direct food has significantly declined over the past three decades. The decrease in demand led to a decline in sorghum production from 9.86 million t in 1969–70 to 7.29 million t in 2009–10 (Dayakar *et al.*, 2011). That sorghum consumption has suffered is mainly a result of its laboriousness to prepare, in combination with the paucity of available processing technologies and lack of awareness of its nutritional benefits. Added to this are government policies which have tended to favour the supply of fine cereals at subsidized prices.

The IIMR (formerly DSR)-led consortium, under NAIP funding, was vested to undertake the challenge of reviving demand for millets in India through an innovative value chain approach. They have intervened to bridge the identified gaps on different aspects of on-farm production, processing diversification, nutritional certification, promotion and marketing of sorghum in the Indian market. This attempt has enabled all the stakeholders in the production-to-consumption system value chain to be brought to a common platform and link poor dryland farmers with markets as well as consumers. In this regard, the IIMR has made links with private partners such as ITC (ABD) Ltd and Britannia Industries Ltd, and with public institutes such as NIN, SAUs, DFRL, CFTRI, CIAE and CIPHET.

On-farm production interventions

The first intervention on the backward integration model of product specific on-farm production – covering 3000 acres in Parbhani (*rabi*) and Nanded (*kharif*) of Maharashtra and in Adilabad (*kharif* for two seasons) in Andhra Pradesh – was tested for four successful years under the *e-Choupal* market-assured model of ITC. The beneficiaries were technology-backstopped by IIMR product-specific cultivars (more than 12), bringing change in the mindset of farmers on the commercial aspects of sorghum cultivation. The recommended package of practices for receiving better yield and quality was extended in public-private partnership (PPP) mode of farm extension services. The impact has been visible through increased farm productivity and assured net incomes (as compared to the baseline survey), which in turn led to a shift in allocation of sorghum cultivation from marginal to better soil and water environments. This is an indicator of stabilization of acreage in the study area and has provided additional employment to the stakeholders. In fact, backward integration resulted in overall livelihood improvement, in addition to imparting commercial colour to the otherwise neglected, 'poor man's' crop of sorghum.

Value addition through processing interventions in millets

The second intervention was aimed at reversing the declining consumption of sorghum, which was primarily due to inconvenience in its preparation as direct food. Thus diversification of processing technologies related to sorghum were attempted, to remove the inconveniences and to develop, fine-tune and standardize sorghum product technologies. For this purpose, the IIMR has installed and retrofitted 30 machines. The primary processing and secondary processing methods have been developed and fine-tuned using this equipment resulting in 30 products, such as multigrain *atta*, semolina, flakes, extruded products (vermicelli and pasta) and biscuits. Similarly, our partner ANGRAU has come up with another set of ten pearl millet and sorghum products, of which nine from IIMR and five from ANGRAU have been commercialized. These products were made ready for commercialization with improved shelf life.

Processing interventions in sorghum improved nutritional value, convenience and shelf life. They are being continued at IIMR and target both niche markets such as gluten-free products and other generic sorghum products for mass-markets at the national level.

Nutritional evaluation and certification

The third intervention was an important milestone in the promotion of sorghum and other millets. The NIN undertook an organoleptic study of all the sorghum products developed by IIMR, finding that they are either superior than rice products or similar to wheat-based products. Another clinical assessment studied the effect of a sorghum diet in controlling diabetes. The third study analysed over 8 months the effect of a sorghum-based diet on schoolchildren's nutrition in terms of body mass, anthropometric and biochemical parameters. These studies established that sorghum has potential to offer better nutrition in general over the market-available products made from wheat, rice and maize. The amino acid profile of pulse (soy blend)-incorporated sorghum products contained greater amounts of lysine, which was the limiting factor in sorghum. The glycaemic index of sorghum foods was analysed to determine the mean glycaemic response for reference and test foods using international standards. The study reported that there was a decrease in the mean incremental area under glucose curve levels after consuming sorghum products. Thus, the nutritional data generated were beneficial for substantiating claims and labelling sorghum products as healthy and nutritional products.

Entrepreneurship development

Entrepreneurship Development Programmes were developed for sorghum cultivation, processing, and marketing of sorghum-based products. They were jointly organized by ITC and IIMR with active participation from institutes like IIMR, ITC, ANGRAU, NIN and the College of Home Science and MAU, both in-house and on-site. Processing machines for standardized sorghum products were demonstrated to farmers, and 2000 rural women and another 3000 self-help groups (SHGs), farmers and urban entrepreneurs were trained in sorghum food processing technologies.

Promotion and popularization

One of the most relevant millet interventions was promotion; creating awareness of the health and nutritional merits of millet products and reinventing them as popular, convenient and healthy foods. They have found a place on the shelves of retail markets in Hyderabad, and nutritionists/doctors/

dieticians were sensitized to IIMR's *eatrite* products. Event managers were outsourced to popularize sorghum products in urban markets and electronic media. Massive awareness was created on sorghum as a health food through roadshows and promotions in public parks, malls and institutes in Hyderabad as well as reaching 40,000 consumers through the *Jowar Rath* van in cities such as Pune, Bangalore, Jabalpur, Chennai, Coimbatore and New Delhi. Rural consumer drive was undertaken by ITC's rural *choupal haats*, to sensitize consumers to the convenience and nutritional aspects of the sorghum products.

Commercialization

The pilot commercialization of sorghum products in Hyderabad began with a landmark decision to launch its own brand, *eatrite*. The sorghum products were fine-tuned, standardized, labelled and branded as health foods based on nutritive values established by NIN, targeting separately the urban (middle and higher income classes) and rural markets. The range of sorghum products included sorghum-rich multigrain flour, semolina, pasta, vermicelli, flakes, roasted flakes and biscuits/cookies. Suitable packaging, labelling, marketing and pricing strategies were adopted for targeting them to urban markets (IIMR and ANGRAU). These commercialization interventions have provided convenient options for consumers (especially those ailing with lifestyle diseases) among sorghum foods.

Thus a model emerged for a producer-to-consumer-system value chain on sorghum/millet foods. This led to enhanced consumption levels of targeted groups, income and employment of stakeholders through value-addition and branding of sorghum and millets as health foods. The demonstration of market-linked production, procurement, primary processing and buy-back arrangements were done through ITC's market assurance model; all important in establishing the confidence of entrepreneurs for supply chain management.

Under the NAIP, the IIMR-led PPP consortium has explored various issues from farm to consumer. Through their synergic efforts, various nutritionally rich and convenient sorghum product technologies have been developed and successfully commercialized at the pilot scale in Hyderabad. Now recipes for items such as *idli* and *dosa* can be made with sorghum and other millets instead of rice and wheat. Implementation of effective promotional strategies and policy sensitization has attracted entrepreneurs and policymakers to consider sorghum as a priority.

Socio-economic impacts

Intensive sorghum cultivation through technological backstopping with end product-specific improved cultivars resulted in increased net incomes for the participating farmers over the baseline. In fact, backward integration resulted in an overall improvement, with better quality of produce (sorghum grain), better utilization of fallow land and commercial colour added to the crop through sustainable linkages among all the stakeholders in the value chain. This impacts our primary stakeholders, the poor dryland sorghum farmers. The linking of poor dryland farmers with markets as well as consumers in the value chain is a hallmark of on-farm intervention in sorghum.

Considering these successes, the government of India allocated INR 300 crores in 2011 under Rashtriya Krishi Vikas Yojana for promotion of millets as 'nutri-cereals' under INSIMP (now subsumed under the NFSM programme), a project-replication of NAIP's subproject. As such, more than 200 processing clusters have been set up across the country. Now, with the setting up of Centres of Excellence (CoEs), IIMR has been actively working on influencing consumer eating habits by producing, promoting and popularizing various millet-based products. Promotional activities and entrepreneurship development programmes have continued.

It is worth noting that 15 MoUs (Memoranda of Understanding) have been signed with the entrepreneurs for production and marketing of sorghum food products while many more are in line to sign MoU with IIMR for the same purpose. The entry of big players in the food processing business – such as Britannia Industries Ltd and ITC Ltd – in the nutri-choice sector, is expected to generate demand for sorghum cultivation in the long term due to the high volumes these companies require. Many private labels have also been established in the market to undertake business in millet value addition and the brand franchisee of *eatrite* could extend its operations across the country.

This indicates that there is a growing tendency among the food players to invest more in the millet food business, and consumption of millets will once again be revived in India. With regard to horizontal expansion, it is envisaged that markets will be targeted across the country, even in non-traditional sorghum-consuming urban metropolises, such as Delhi, Bangalore, Chennai and Pune in phase I, expanding to other areas in phase II.

Since the value chain model has proved to be successful and sustainable, it is ready for replication to the remaining millets and other crop commodities in India and across the globe. The constraints outlined in this book may be taken as lessons for replication; the reinvention of millets has commenced.

Reference

Dayakar Rao, B., Patil J.V., Nirmal Reddy, K., Sriharsha, B., Kachui, N. and Kiranmai, E. (2011) Millets value chain to pilot innovative agri-business approaches & farm entrepreneurial development - a success story. In: *International Conference on Innovative Approaches for Agricultural Knowledge Management*. NASC Complex, New Delhi, India 9–12 November.

Acknowledgements

At the very outset we would like to thank ICAR and its dignitaries for giving us opportunities and all the required encouragement. We acknowledge Dr D. Rama Rao, National Director of NAIP, and Dr R. Ezekiel, National Coordinator of NAIP Component-II, and former national directors and national coordinators (NC-II) of the PIU for their funding through World Bank assistance in the IIMR-led consortium on millets' value chain, and for their relentless support throughout the project period.

We are indebted to the directors and associates of IIMR (formerly DSR), ANGRAU, ITC (ABD) Ltd and NIN for their support and contributions in different capacities. Our deep appreciation to all the CoPIs, including Mr K. Nirmal Reddy, ITC (ABD) Ltd, Dr Rajendra Prasad, NIN, and Dr Hymavathi, ANGRAU, who joined the brigade as consortium partners and contributed immensely to the development of value chain on sorghum. My own colleagues from IIMR, Dr Ch. Shashidhar Reddy, Dr K.N. Ganapathy, P. Mukesh and Dr C.V. Ratnavathi, Dr H.S. Talwar, Dr S. Aruna, have contributed greatly to the development of value chain. Our CAC committee and CLs Drs N. Seetharama and S.V. Rao are gratefully acknowledged. We are grateful to the DAC, Ministry of Agriculture, for recognizing the subproject and for establishing the Centre of Excellence under INSIMP in IIMR, Hyderabad. The profound interest in millets, especially sorghum, by the multinational company Britannia Industries Ltd, and its continuing efforts through the joint project on development and commercialization of sorghum biscuits has been one of the motivational factors in our endeavours. The support of ITC (ABD) Ltd in backward linkages of sorghum farmers through the *e-Choupal* platform is also acknowledged.

We also acknowledge all the private firms, NGOs, farmers, SHGs, entrepreneurs, SAUs and government servants who have been involved directly or indirectly in on-farm technological backstopping, value addition, promotion, commercialization and sensitization of sorghum and other millet foods developed under the subproject that is expected to create nationwide demand in the long run. Not least, we are grateful to all the project staff, especially N. Kachui, Latika Devi, Arlene Christina and others, for their enormous contributions at various capacities to bring this book to fruition.

B. Dayakar Rao

N.G. Malleshi

George A. Annor

J.V. Patil

1

Production and Utilization of Millets

Millet crops comprise sorghum (*Sorghum bicolor*), pearl millet (*Pennisetum glaucum*), finger millet (*Eleusine coracana*) and small millets – namely foxtail millet (*Setaria italica*), proso millet (*Panicum miliaceum*), kodo millet (*Paspalum scrobiculatum*), little millet (*Panicum sumatrense*) and barnyard millet (*Echinochloa frumentacea*). These crops are grown on fragile marginal and drought-prone environments in the semi-arid tropics of Africa and Asia, cultivated in two seasons: *rabi* (post-rainy season) and *kharif* (rainy season). Millets are genetically suited to hot and dry agro-ecologies, where it is difficult to grow other food grains (ICRISAT, 2004). Around 90% of the sorghum area in the world and 95% of the total area under millet lies in developing countries, mainly in Africa and Asia. The majority of its growing area in Africa runs across the Sahara, West Africa, Ethiopia, Sudan and the upper regions of Somalia. It is also grown in the upper regions of Uganda, Burundi, Kenya, Egypt, Tanzania and Zambia. It is an important crop in Pakistan, India, northern China, Thailand, Australia, the drier areas of Brazil and Argentina, Venezuela, USA, France and Italy (Léder, 2004). Sorghum is a staple food for about 300 million people worldwide (Léder, 2004).

In these agrarian economies, livestock is the backbone of business. Millet crops provide food, feed, fodder and biofuel, and the dried vegetative parts are also used to make shelters; thus the crops are known for their whole plant utilization. Also, small millets are largely cultivated by tribal farmers for their sustenance. These crops are environmentally friendly because of their low water-input requirement. Cultivation of millets, particularly pearl millet, as a trap crop in oilseeds and pulses also helps in reducing pesticide consumption. Recent studies on mitigating the emission of greenhouse gases through biological nitrification inhibition (BNI) (Subbarao *et al.*, 2013) have also revealed that sorghum and pearl millet have a BNI ability in root exudates. This could play a vital role in mitigating the impact of global warming by regulating the emission of N_2O into the atmosphere. The majority of millet grains contain higher protein, fibre, calcium and mineral content than wheat and rice, and so are often hailed as 'noble' or 'miracle' grains (*adbhut anaj*) and 'nutri-cereals'.

1.1 Production of Millets

1.1.1 Global production

Sorghum

Sorghum is mostly grown in arid and semi-arid regions of the world, over an area of 45.33 million ha (triennium ending (TE) 1983). India has the largest coverage (16.47 million ha) followed by the USA (5.10 million ha), although the USA is the major sorghum producer followed by India, Argentina and China. The current productivity of sorghum is highest in Argentina (4085 kg/ha) followed by the USA (3739 kg/ha) and Mexico (3735 kg/ha). The area, production and productivity of sorghum in major producing countries are given in Table 1.1.

The data show that the area of sorghum has declined not only in India, but also in countries such as the USA, China and Argentina. However, the productivity of sorghum has improved enormously in all major countries except Sudan and Nigeria, where it has marginally declined. The factors affecting the increase in productivity are suitability of climate, crop duration (time from sowing to harvesting), the varieties/hybrids used and technology adopted (in China, Argentina, Mexico and USA). The area under sorghum has increased in Sudan (former), Nigeria, Mexico and Ethiopia, mainly because of the stable demand and suitable climatic conditions.

Pearl millet

Pearl millet is an important crop of the rainfed areas of the arid and semi-arid tropics, where its cultivation is confined to drier tracts of Southern Asia (mainly India) and Africa (Nigeria, Niger, Mali, Chad, Tanzania, Sudan and Senegal) for grain and fodder purposes. However, in the USA, Canada, Japan, Italy and Australia it is grown on a small scale, mainly as a fodder crop. Africa has been recorded as the primary centre of origin of pearl millet, from where it has spread to India and other countries. Pearl millet is a plant that does not resist drought but cuts short its life cycle and flowers early under such adverse conditions. Thus it can grow in areas that have extended dry periods and grows well in regions having a rainfall of less than 500 mm. The area, production and productivity of pearl millet in major producing countries are given in Table 1.2. These data indicate that the area under pearl millet has declined not only in India but also in China, despite having better productivity, whereas the area under production in countries like Nigeria and Niger has increased enormously mainly because of the climatic limitations to other crops and the increase in demand for consumption.

Small millets

The small millets comprise a group of crops that include finger millet, foxtail millet, proso millet, kodo millet, little millet and barnyard millet, mainly grown in Asia and Africa under various kinds of soils and adverse agro-ecological situations, including a wide range of temperatures. Some of them are most suited to contingency planning under late receipt of rain, failure of main *kharif* crops due to long dry spells or even under receding flood situations. These are extremely important commodities that are largely cultivated by tribes for their sustenance.

Finger millet is an important millet grown in Africa, Madagascar, Malaysia, China, Japan and India. De Candolle (1886) reported that finger millet probably originated in India, from where it has spread to Abyssinia and the rest of

Table 1.1. Area, production and productivity of sorghum in major producing countries. (Source: FAO Statistics, 2015.)

	TE 1983			TE 2013		
Countries	Area (million ha)	Production (million t)	Yield (kg/ha)	Area (million ha)	Production (million t)	Yield (kg/ha)
USA	5.10	18.61	3649.12	2.64	9.88	3739
Sudan (former)	3.72	2.41	647.39	7.14	4.52	634
China	2.70	7.34	2716.12	0.55	2.02	3664
Nigeria	2.22	3.60	1624.61	5.50	6.70	1218
Argentina	2.38	7.88	3316.97	0.89	3.64	4085
India	16.47	11.58	703.02	6.18	5.28	854
Mexico	1.50	5.22	3484.91	1.69	6.31	3735
Ethiopia	0.91	1.32	1462.21	1.85	4.34	2349
World	45.33	67.96	1499.15	42.12	61.38	1457

TE, triennium ending

Africa through Arabia, whereas, according to Vavilov (1926) the centre of origin is Abyssinia. Foxtail millet is also grown in China, India, Indonesia, the Korean peninsula, and some parts of southern Europe. Proso millet is widely cultivated in the Russian Federation, Kazakhstan, Ukraine, the USA, Australia and Argentina; it is a plant with open branches and drooping panicles and is tolerant of a wide range in temperature. Kodo millet is harvested as a wild cereal in western Africa and India, where it grows abundantly along paths, ditches and low spots. Little millet is widely grown in India, Nepal, Pakistan, Sri Lanka, eastern Indonesia and western Myanmar. Barnyard millet is important in the tropics and subtropics of India (FAO/ICRISAT, 1996). In India, these crops are widely grown in tropical as well as subtropical climates from Kanyakumari in the south to the Himalayan region in the north and north-east, Gujarat in the west and West Bengal in the east. These crops have different vernacular names in India (Table 1.3).

Table 1.2. Area, production and productivity of pearl millet in major producing countries. (Source: FAO Statistics, 2015.)

	TE 1983			TE 2013		
Countries	Area (million ha)	Production (million t)	Yield (kg/ha)	Area (million ha)	Production (million t)	Yield (kg/ha)
India	17.69	10.32	583.31	9.66	11.44	1184.27
China	4.01	6.63	1654.81	0.74	1.71	2316.74
Nigeria	1.73	2.71	1569.99	2.72	2.52	925.25
Niger	3.09	1.31	423.19	7.08	3.26	460.71
Mali	0.78	0.57	738.91	1.86	1.46	783.54
Chad	0.24	0.13	546.17	0.90	0.58	648.15
Tanzania	0.29	0.33	1127.54	0.31	0.31	989.36
World	36.86	28.06	761.21	32.18	27.92	867.72

TE, triennium ending

Table 1.3. Vernacular names of small millets in India.

Language	Names					
English	Finger millet	Foxtail millet	Common (proso) millet	Kodo millet	Little millet	Barnyard millet
Latin (Botanical name)	*Eleusine coracana*	*Setaria italica*	*Panicum miliaceum*	*Paspalum scrobiculatum*	*Panicum sumatrense*	*Echinochloa frumentacea*
			Indian vernacular name			
Assamiya	–	–	–	–	–	*Konidha*
Bangla	*Mandua*	*Kaon*	*Cheena*	*Kodo*	–	*Shyama*
Gujarati	*Nagli, Bavto*	*Kang*	*Cheno*	*Kodra*	*Gajrao/Kuri*	*Banti*
Hindi	*Mandua*	*Kakun, Kangni*	*Chena, Bari*	*Kodon*	*Kutki, Savan*	*Sanwa*
Kannada	*Ragi*	*Navane*	*Baragu*	*Harika*	*Same/Save*	*Oodalu*
Malayalam	*Moothari*	*Thena*	–	*Varaku*	*Sama*	–
Marathi	*Nachni*	*Kang, Rala*	*Vari*	*Kodra*	*Sava*	*Banti*
Oriya	*Mandia*	*Kangam, Kanghzu*	*Bachari*	*Kodua*	*Suan*	*Khira*
Punjabi	*Mandhuka, Mandhal*	*Kangani*	*Cheena*	*Kodra*	*Swank*	*Swank*
Telugu	*Ragulu*	*Korra*	*Variga*	*Arika*	*Samalu*	*Oodalu*
Tamil	*Kelvaragu*	*Tenai*	*Panivaragu*	*Varagu*	*Samai*	*Kutdiravalli*

1.1.2 Indian production

Cultivation of millets is of primary importance to the millet-growing rainfed areas in India; known for its whole plant utilization, it provides food, fodder, fuel and ferment. Millets are either grown as a pure crop or intercropped with pulses and oilseed, providing a composite, nutrient-packed farm. Despite these attributes, millet has declined in its importance in both production and consumption in India mainly as a result of the patronage given to rice and wheat and the growth in their production, procurement and distribution.

The shift towards fine cereal (rice and wheat) cultivation and consumption in India is a consequence both of supply-led factors (such as subsidized input supply) and demand-led factors (such as Indian government policy to supply fine cereals at subsidized prices, output incentives, etc.), which have resulted in the lowering of consumer demand and in turn have reduced the acreage under millet. Other factors, such as the convenience of cooking, difficulties in large-scale millet processing and relatively lower productivity giving economic advantage to the cultivation of fine cereals, are also responsible for the shift from millets to other cereals. Apart from the nutritional disadvantage of losing millet from the diet, extensive cultivation of cereals using groundwater in arid areas is threatening water security in the country.

Cultivation of millets in India covered an area of 15 million ha with an annual production of 16.0 million t in 2012–13 (Tables 1.4, 1.5). Of all the millet crops, sorghum is the most important, being grown across about 6.0 million ha and producing about 5.28 million t in India during TE 2013 (FAO Statistics, 2015). State-wise, Maharashtra has 46% of the total sorghum area of India, followed by Karnataka (18%). Meanwhile, Rajasthan has about 56% of the total area given over to pearl millet in the country, followed by Uttar Pradesh (11%) and Maharashtra (9%). Similarly, Karnataka has more than 59% of India's total area of finger millet and Madhya Pradesh has the greatest area (33%) of small millets of the country during 2013–14 (Appendix 1). As far as the biodiversity of millet crops across the states is concerned, Maharashtra and Tamil Nadu are the richest, with cultivation of all major and minor millets.

Table 1.4. Cultivated area of millet compared with other crops in India, pre- and post-Green Revolution. (Source: DES, 2014.)

	Pre-Green Revolution		Post-Green Revolution						
Crop/year	1955–56	1965–66	1975–76	1985–86	1995–96	2005–06	2008–09	2011–12	2012–13
Sorghum	17.36	17.68	16.09	16.10	11.33	8.67	7.53	6.25	6.21
Pearl millet	11.34	11.97	11.57	10.65	9.32	9.58	8.75	8.78	7.30
Finger millet	2.31	2.70	2.63	2.40	1.77	1.53	1.38	1.18	1.13
Small millet	5.34	4.56	4.67	3.16	1.66	1.06	0.91	0.80	0.75
Total millet	36.34	36.91	34.96	32.31	24.09	20.85	18.57	17.00	15.40
Rice	31.52	35.47	39.48	41.14	42.84	43.66	45.54	42.86	42.75
Wheat	12.37	12.57	20.45	23.00	25.01	26.48	27.75	29.87	30.00
Maize	3.70	4.80	6.03	5.80	5.98	7.59	8.17	8.78	8.67
Barley	3.42	2.64	2.80	1.37	0.82	0.63	0.71	0.64	0.70
Total cereals other than millets	51.01	55.48	68.76	71.31	74.65	78.36	82.17	82.15	82.12
Total cereals	87.35	92.39	103.72	103.62	98.74	99.21	100.74	99.15	97.52
Share of millets (%)	42.00	40.00	34.00	31.00	24.00	21.00	18.00	17.00	15.00
Soybean	0.00	0.00	0.09	1.34	5.04	7.71	9.51	10.11	10.84
Sunflower	0.00	0.00	0.32	0.75	2.12	2.34	1.81	0.73	0.83
Cotton	8.09	7.96	7.35	7.53	9.04	8.68	9.41	12.18	11.98

Note: Area in million ha.

Table 1.5. Production of millet compared with other crops in India, pre- and post-Green Revolution. (Source: DES, 2014.)

	Pre-Green Revolution		Post-Green Revolution						
Crop/year	1955–56	1965–66	1975–76	1985–86	1995–96	2005–06	2008–09	2011–12	2012–13
Sorghum	6.73	7.58	9.5	10.2	9.33	7.63	7.25	6.01	5.28
Pearl millet	3.43	3.75	5.74	3.66	5.38	7.68	8.89	10.28	8.74
Finger millet	1.85	1.33	2.8	2.52	2.5	2.35	2.04	1.93	1.57
Small millet	2.07	1.56	1.92	1.22	0.78	0.47	0.45	0.45	0.44
Total millet	14.07	14.22	19.96	17.6	17.99	18.14	18.62	18.66	16.03
Rice	41.34	45.88	73.35	95.82	115.57	91.79	99.18	105.31	105.23
Wheat	8.76	10.39	28.85	47.05	62.10	69.36	80.68	94.88	93.50
Maize	2.6	4.82	7.26	6.64	9.53	14.71	19.73	21.76	22.26
Barley	2.82	2.38	3.19	1.96	1.51	1.22	1.69	1.62	1.75
Total cereals other than millets	55.45	63.48	112.65	151.48	188.71	177.08	201.28	223.57	222.74
Total cereals	69.52	77.70	132.61	169.07	206.7	195.22	220	242.24	238.78
Share of millets (%)	20.00	18.00	15.00	10.00	9.00	9.00	8.00	8.00	6.70
Soybean	0.00	0.00	0.09	1.02	5.10	8.27	9.91	12.21	14.66
Sunflower	0.00	0.00	0.22	0.28	1.26	1.44	1.16	0.54	0.54
Cotton (million bales)	4.18	4.85	4.95	8.73	12.86	18.50	22.28	35.20	34.22

Note: Production in million t.

Trends in millet area and production compared to other crops

Despite the fact that the total area under cultivation has been drastically reduced over the years (Table 1.4), India is still one of the major global producers of millets. This is due to productivity gains in some varieties, with the production of millets showing some increase despite a shrinkage of area. Thanks to the Green Revolution in 1965–66, which led to an increase in per capita availability of food grains, cereals at the national level kept pace with population growth. However, since the mid-1990s Green Revolution fatigue has set in and production of cereals is tending to plateau. With population increase, the per capita availability of food grains, including cereals, has decreased over the years. Since cereal production is demanding in terms of water and fertilizers, millets that are hardy and can grow under less favourable conditions will have to meet the gap between demand and supply to ensure nutritional security. While the population pressure continues, the area under arable land is decreasing.

Table 1.4 shows that the area in India under millet cultivation has reduced drastically; during 1955–56 it was 36.34 million ha and by 2012–13 it had reduced to 15.40 million ha. Meanwhile the area under individual competing crops has increased (except with barley), with total area under these cereals increasing from 87.35 million ha in 1955–56 to 97.52 million ha in 2012–13. Millet covered 42% of the total cereal area in 1955–56, was reduced by half during 2005–06 (21%), and had only a 15% share by 2012–13. Crop-wise the greatest decline in area has been under sorghum (9 million ha), followed by small millets (4.5 million ha) and pearl millet (4 million ha) during the period under review.

The production of millets in India has risen marginally between 1955 and 2013 (Table 1.5); rising from 14.07 million t in 1955–56 to a peak at 19.96 million t in 1975–76, but subsequently reducing to 16.03 million t in 2012–13. This overall rise in millet production occurs despite a decline in area under millet cultivation from 36.34 million ha (1955–56) to 15.40 million ha (2012–13).

The increase in total millet production from a meagre productivity of 387 kg/ha (Table 1.6) during the 1960s (pre-Green Revolution period) is owed to the adoption of improved varieties, composite, hybrids, high-yielding varieties (HYVs) and improved technologies. The yield levels realized are no small achievement as there has been, over the past four decades, an increasing tendency to relegate the millet crops to marginal land with poor soil and water environments, vacating the best lands to competing crops.

However, the production of competing crops (e.g. rice, wheat, maize, cotton and sunflower) has risen tremendously when compared to millets. Millet production comprised 20% of total cereal production in the 1950s while it was around 7% during 2012–13. The yield of millet increased over the years, but the rate of increase is less than that of other competing crops (Table 1.6). The highest increase in yield among the millets was observed in pearl millet.

Decline/diversification of area under millets

The state- and crop-wise five-yearly mean area of millets shows a major decline (Appendix 1). The area under millet crops has declined in all the millet-growing states except Rajasthan, with the land being given over to competing crops. Since Rajasthan has no better crop option than pearl millet under low rainfall and sandy soils, the area under pearl millet has remained the same, apart from some fluctuation due to rain. However, the area under small millets has declined in Rajasthan. It would be difficult to locate the actual area shift to an individual crop in each state.

Competing crops

Almost 50% of the area under millet cultivation has been diverted largely to soybean (Madhya Pradesh, Maharashtra, Rajasthan, Andhra Pradesh and Karnataka); maize (Andhra Pradesh, Gujarat, Karnataka and Maharashtra); cotton (Andhra Pradesh, Gujarat and Haryana); sugarcane (Maharashtra and Uttar Pradesh); and sunflower (Karnataka) (NAAS, 2013). It has also been reported that some of the minor millets areas have been given over to housing plots and industries. These competing crops have proved to be more profitable because of their better market demand, production support and assured procurement through public agencies under mission mode programmes such as the Technology Mission on Oilseed and Pulses (later the Integrated Scheme of Oilseeds, Pulses, Oilpalm and Maize) and the government of India's Technology Mission on Cotton.

Table 1.6. Productivity of millets compared with other crops in India, pre- and post-Green Revolution. (Source: DES, 2014.)

	Pre-Green Revolution		Post-Green Revolution						
Crop/year	1955–56	1965–66	1975–76	1985–86	1995–96	2005–06	2008–09	2011–12	2012–13
Sorghum	387	429	591	715	823	797	962	962	850
Pearl millet	302	314	496	344	577	802	1015	1171	1198
Finger millet	800	492	1064	1049	1410	1534	1477	1641	1392
Small millet	388	341	412	386	469	443	445	565	578
Millets	387	385	571	545	747	870	1003	957	850
Rice	639	675	1041	1858	2329	2102	2178	2393	2461
Wheat	708	827	1410	2046	2483	2619	2907	3177	3117
Maize	704	704	1005	1146	1595	1938	2414	2478	2566
Barley	824	902	1139	1434	1834	1938	2394	2516	2500
Total cereals	796	841	1278	1632	2093	1968	2184	2641	2448
Soybean	0	0	975	764	1022	1073	1041	1208	1353
Sunflower	0	0	686	374	593	615	639	706	655
Cotton	88	104	138	197	242	362	403	491	486

Note: Productivity in kg/ha.

Reasons for the decline in area under millet

Demand-led factors:

1. Rapid urbanization.
2. Changing consumer tastes and preferences due to rising per capita income.
3. Government policies favouring other crops such as output price incentives and input subsidies.
4. Supply of Public Distribution System (PDS) rice and wheat at cheaper price introduced in non-traditional areas of fine cereals.
5. Poor social status and inconvenience in their preparation (especially sorghum).
6. Low shelf-life of grain and flour.

Supply-led factors:

1. Increasing marginalized cultivation.
2. Low profitability/low remuneration for millets vis-à-vis competing crops.
3. More remunerative crop alternatives in *kharif* competing with millets.
4. Decline in production and quality (as in *kharif* sorghum because of poor quality of grains due to blackening of grains, fetching low price for the farmers).
5. Lack of incentives for millet production.

Cropping seasons/sequence

In India, millets are predominantly grown during the *kharif* season. However, sorghum is grown on a larger area as a *rabi* crop in Maharashtra, Karnataka, Andhra Pradesh, Tamil Nadu and Gujarat; pearl millet as a *rabi* crop in Andhra Pradesh and a summer crop in Gujarat and south-western part of Uttar Pradesh; and finger millet is grown during the *rabi* season in Karnataka and Orissa. Millets have the best symbiotic relationship with oilseeds and pulses under a rainfed ecosystem. Pearl millet and sorghum are also raised as trap crops (i.e. those crops grown around a main crop so as to attract pests, which curb them without the use of pesticides) in oilseeds and pulses for integrated pest management.

An economic analysis undertaken by IIMR on *kharif* sorghum through on-farm trials over 3 years (Dayakar *et al.*, 2003) under the National Agriculture Technology Project (NATP) has concluded that:

1. The sole crop of sorghum was most profitable against groundnut in the Dharwad district and cotton in the Belgaum district of Karnataka.
2. Sorghum + pigeon pea (*arhar*) was most profitable in the Akola and Nanded districts of Maharashtra.
3. Sorghum + soybean intercropping was most profitable in Indore, where sorghum cultivation has diminished.
4. Sorghum + pigeon pea intercropping was most profitable in the Khargone district of Madhya Pradesh.
5. Sorghum + soybean intercropping was most profitable in the Adilabad district of Andhra Pradesh.

Interventions for improvement of millets

MILLET CROP IMPROVEMENT NETWORK. A well-established research network comprising the Indian Institute of Millets Research (IIMR), Hyderabad and the Project Coordinating Units for pearl millet (Mandore, Jodhpur) and small millets (University of Agricultural Sciences, Bengaluru), with a wide network of their centres located all over the country under the Indian Council of Agricultural Research (ICAR)/state agricultural universities (SAUs), has been entrusted with the improvement of these crops. CGIAR's International Crop Research Institute for Semi-Arid Tropics (ICRISAT) in Patancheru, Telangana state, is also associated with the improvement of pearl millet and sorghum and, more recently, finger millet. Several private companies are also associated with research and production of quality seed. They undertake R&D on millet in the country. In addition to the above network, SAUs proactively work for an individual state's crop improvement needs.

PROGRESS IN MILLETS IMPROVEMENT. CSH-1 and HB-1 were the first crop hybrid of sorghum and pearl millet developed in India in 1964 and 1965, respectively. More than 80 hybrids of sorghum (21) and pearl millet (60), more than 100 HYVs of millet crops, namely sorghum (41), pearl millet (15), finger millet (21) and small millets (34), have been developed over the last 15 years. However, half of them could not be utilized for commercial cultivation mainly because of meagre production of these seeds. The package of practices (PoP) of many states still includes age-old cultivars. The new hybrids/varieties have high genetic yield potential up to 5950 kg/ha (pearl millet), 5000 kg/ha (sorghum and finger millet) and 2648 kg/ha (small millets) as compared to the national average yields of 875 kg/ha (pearl millet), 796 kg/ha (sorghum), 1295 kg/ha (finger millet) and 440 kg/ha (small millets).

RESEARCH–EXTENSION–FARMER LINKAGES. The adoption of improved management/technologies is key to making agriculture more profitable. Therefore, the availability of improved technology – including seeds of new varieties/hybrids – is of paramount importance for the farmers. The IIMR (Hyderabad), the All India Coordinated Pearl Millet Improvement Project (AICPMIP) (Mandore, Jodhpur) and the All India Coordinated Small Millet Improvement Project (AICSMIP) (Bengaluru) are entrusted with the responsibility of improvement of these crops at the national level in coordination with SAUs under the aegis of ICAR. Similarly, the Directorate of Millets Development in Jaipur, a subordinate office of the Ministry of Agriculture, is entrusted with responsibility for the development of these crops at the national level in coordination with the research institution of ICAR/SAUs and the State Department of Agriculture. The state extension agencies work as a bridge between the research institution and farmers. The government of India has extensively taken up a number of initiatives to strengthen the research–extension–farmer (R-E-F) linkages.

In pursuance of an effective linkage between research and technology transfer, the Ministry of Agriculture launched a project called Front Line Demonstration (FLD) in 1995–96. FLDs on millets comprise recently released cultivars and improved technologies that are carried out on farmers' fields by agricultural scientists from ICAR/SAUs/KVKs. The financial support for these FLDs are provided to the crop specialized institutions of ICAR, namely IIMR, Hyderabad (for sorghum), AICPMIP, Mandore, Jodhpur (for pearl millet), and AICSMIP, Benguluru (for small millet), under direct funding from the Department of Agriculture and Co-operation's Macro-Management in Agriculture, which also provides assistance for organizing the national interface between research, extension, industry and trade activities on millets.

Over the last 5 years, the FLD programme, organized by the agricultural scientists of ICAR/SAUs over an area of more than 5000 ha across various states, indicates a yield gap of 6% to 236% over local practices and provides substantial scope for improved productivity of these crops (Singh and Dayakar, 2010), which have proved their responsiveness to high input application and crop management. Examples include:

- An average yield of 6500 kg/ha of sorghum has been recorded in rice fallows of the Guntur district of Andhra Pradesh over an area of 24,000 ha during *rabi* (2011–12), with the highest FLD yield recorded at 8500 kg/ha (Chapke *et al.*, 2011).
- Haryana has recorded an average yield of 1843 kg/ha of pearl millet over an area of 630,000 ha during *kharif* (2007) as against the national average yield (NAY) of 1042 kg/ha.
- Gujarat has recorded an average yield of 2434 kg/ha of pearl millet over an area of 1.75 lakh ha during summer (2009).
- Karnataka has recorded an average yield of 2250 kg/ha of finger millet over an area of 18,000 ha during summer (2006) as against NAY of 1534 kg/ha.
- An average yield of 1627 kg/ha of barnyard millet has been recorded in Bageshwar district of Uttarakhand over an area of 1180 ha during *kharif* (2001).
- Tamil Nadu has recorded an average yield of 1302 kg/ha of kodo millet over an area of 9000 ha as against NAY of 312 kg/ha (2001–06).

Millet production potential

A study by ICRISAT using the sorghum and pearl millet crop growth models of the Coalition for Environmentally Responsible Economies (CERES) and historical weather data, estimated the potential rainfed yields of pearl millet and sorghum (*kharif* and *rabi*) and the water balance in different zones of production for selected locations in India. Total yield gap (the difference between simulated potential rainfed yield and farmer's yield) ranged from 2130 to 2560 kg/ha for *kharif* sorghum, 280 to 830 kg/ha for *rabi* sorghum and 680 to 1040 kg/ha for pearl millet. This indicates that the productivity of *rabi* sorghum can potentially be increased by 1.4 to 2.7 times, that of *kharif* sorghum can be increased by 3.0 to 4.0 times and that of pearl millet by 1.8 to 2.3 times (Murty *et al.*, 2007).

These data suggest that for millet as a whole, average potential production is 2.5 to 3 times current production and achievable production could be 45–55 million t per annum. This production potential is achievable by bridging the yield gaps, not only through supply-side factors (high yielding cultivars, nutrient management and integrated pest and disease management) but also addressing demand-side issues by value addition processing through nutritional labelling, alternative industrial utilization and policy measures so as to make millet cultivation more remunerative. Accordingly, the breeding strategies should be redesigned to orient product-specific cultivars. The demand for millet can be increased by shifting its cultivation to better lands and in the long run this should realize an increase in area under millets through crop replacement from less profitable crops. The future sustainability of sorghum is described in more detail later in this chapter.

1.2 Utilization of Millets

Millet is a versatile crop providing food, feed, fodder and fuel. The entire plant is utilized and hence recognized as a high biomass crop. It has been an important staple food for millions of poor people in the dry and rainfed regions of the world.

1.2.1 Global utilization of millets

The pattern of utilization for millets varies across countries/regions. In most of the developing countries such as in Africa and Asia, these crops are used primarily for human consumption while in developed countries such as the USA, millets are primarily used for fodder and bird feed. Nearly 10% of total millet production is used in industries to produce various products, while livestock and feed consumption accounts for the lion's share of almost 42.5% (27 million t). The seed demand of millet grains accounts for 6.33% and food consumption accounts for 41.02%. The various industrial products of millets are given in Table 1.7.

1.2.2 Utilization of millets in India

Millets are used as human food and animal feed in India. They are traditionally used for the preparation of *roti, mudde*, fermented foods and beverages. Grains with high starch content (65–72%) and low protein content (8–12%) may be used to produce alcohol. The utilization of *kharif* grain in various industries is increasing, given the limited prospects of the rainy-season sorghum for human consumption. Important millet-derived traditional and fast foods are given in Table 1.8.

The main industries currently using millets in India include poultry feed, animal feed and potable alcohol distilleries. At present, the poultry feed sector is using approximately 2.0 million t sorghum annually; the animal feed sector uses about 0.60 million t sorghum and alcohol distillers use about 0.49 million t annually (Tonapi *et al.*, 2011).

Table 1.7. Utilization of millets in industries.

Millet	Industrial products
Sorghum	Malting, high fructose syrup, starch, jaggery (unrefined sugar), bakery, value-added products for diabetics, poultry and animal feed.
Pearl millet	Malting and brewing, starch, bakery, poultry and animal feed.
Finger millet	Malting and brewing, baby foods, bakery and food for diabetics.
Other millets	Value-added food for devotees (barnyard millet), feed, value-added food products for diabetics and bird feeds.

Table 1.8. Important millet-derived food products in India.

Millets	Food products
Sorghum	*Roti, ugali*, popped sorghum, malt food, snack/roasted mix grains.
Pearl millet	*Roti, ugali*, fermented food products, pizza, roasted mix grains.
Finger millet	*Roti*, dumpling, popped millet, malt food.
Other millets	*Roti*, cooked cereals.

Uses of millets in India in addition to the major food products are:

1. Poultry/animal feed. Poultry feed industry in the country is progressing well. Several independent feeding trials conducted on layer and broiler birds conclusively demonstrated that sorghum and pearl millet are as good as maize although the latter is marginally superior in total energy value. Millet-based feed in layer birds has to be supplemented with carotene to ensure a yellow egg yolk. In the case of cattle feed, sorghum-based feed is found to be satisfactory. Moulded *kharif* grain is also found to be an acceptable grain raw material in goat and swine feed. Sorghum is used when maize is in short supply and highly priced (up to 20% higher than sorghum).

2. Starch. Sorghum is rarely used by the starch industry due to processing problems and a lower starch yield as the extraction machinery is adapted to bigger size grains such as maize. While there is scope to improve starch recovery from sorghum by modified extraction, selective breeding and use of specifically suited machinery, starch production from the present sorghum cultivars will be economically viable only if the grain price is 20% lower than that of maize. Sorghum grain is found to be more suitable for production of glucose and liquid glucose.

3. Potable alcohol. The sorghum grain offers an excellent ethanol and feed stock with a high alcohol recovery of 422–448 l/t. This provides scope for utilization of rain-damaged, blackened and low market-valued sorghum grain for use as a raw material for malting and the production of quality potable alcohol. Currently, more than ten distilleries are using sorghum grain (mainly blackened) as a raw material for potable alcohol. The government of Maharashtra has announced incentives for those sorghum grain-based distilleries that are established as alternatives to sugarcane molasses; these industries could potentially benefit the farmers in assuring a market with remunerative prices through contract farming. Like sorghum, pearl millet is also being used for the production of potable alcohol.

4. Malt. The use of finger millet and sorghum malt as an adjunct in the brewing industry also has high potential. In Ghana and Nigeria, malted sorghum is used for the brewing of beer. Sorghum's comparative advantage would be its low output price, especially in the regions where *kharif* sorghum is produced in India.

5. Bioethanol. Sweet stalked and high energy sorghum has the potential to emerge as a major bioenergy crop for the production of industrial alcohol, gasohol and even electricity. While national trials on sweet sorghum showed a potential productivity of 50–60 t stalk/ha and 2–3 t grain/ha in 125–130 days, reports from China indicate a far higher productivity of 90 t stalk/ha and 6 t grain/ha from the crop in 140–150 days. This productivity potential places sorghum at an advantage over sugarcane or molasses. Sweet sorghum incorporation could monitor the ruling prices for ethanol production. Sweet sorghum as a bioenergy crop is more efficient than sugarcane because of sweet sorghum's high productivity combined with relatively low levels of fertilization and irrigation. IIMR, ICRISAT and many R&D organizations were actively involved in fine-tuning technologies related to sweet sorghum to project it as an alternative feed stock for bioethanol production. However, government policy support in declaring high ethanol pricing will determine its current utilization and future demand.

6. Other uses. Production of jaggery, a brown/colourless, high fructose syrup, could be carried out at farm/village level and by small-scale enterprises.

Trends in millet consumption in India

Sorghum and other millets are the staple food in some parts of India, especially in western and central regions of Maharashtra and the northern area of Karnataka and Andhra Pradesh. The annual per capita consumption of sorghum declined by 68% in urban areas and 73% in rural areas during 1973–74 to 2004–05 (Table 1.9). However, in inland regions of eastern, central and western Maharashtra and northern Karnataka, sorghum is still an important crop. Sorghum's annual per capita consumption in rural areas ranges from 31.8 kg to 54.2 kg and in urban areas ranges from 9.9 kg to 34.0 kg (Basavaraj and Rao, 2011). In the last two to three decades, sorghum grain, especially that sourced from the *kharif* season, has been diverted to mainly industrial uses such as livestock and poultry feed, starch, potable alcohol and ethanol production due to poor quality and mould affected grains. The declining trend in sorghum consumption and other nutritious cereals in general is attributed to the shift in dietary

Table 1.9. Trends in direct consumption of sorghum vs major cereals in India. (Sources: various NSSO reports; GOI; Parthasarathy *et al.*, 2010.)

Commodity	1973–74	1983–84	1993–94	2004–05	2011–12
Rural					
Rice	84.0	80.7	85.4	78	71.7
Wheat	42.8	54.3	53.5	51	51.5
Sorghum	19.0	12.5	9.7	5.16	2.4
Urban					
Rice	65.5	64.7	64.2	57	53.8
Wheat	52.6	58.6	57.4	53	48.1
Sorghum	11.0	6.0	4.9	2.7	1.56

Note: values in kg/person/year.

patterns of consumption towards a more ready-to-eat food, which is mainly driven by an increase in income and urbanization wherein people are too busy to spend much time preparing their daily diet. Secondly, the shift is due to the consumption of fine cereals that are supplied through the Public Distribution System (PDS) at subsidized prices. The other reason that led to a decrease in consumption is the reduced area/production of millets, a consequence of their lack of competitiveness with other more profitable crops.

Table 1.9 shows that the decline in sorghum consumption in India from 1973–74 to 2011–12, was more drastic than in the case of wheat and rice, both in rural and urban households. However, the prospects for sorghum in direct food consumption tend to rise as interventions in its food processing and promotion have been successfully taken out by the IIMR-led consortium in collaboration with the Directorate of Agriculture and Cooperation (DAC) and other millets research institutes.

Human consumption of the grain is traditionally recognized as the primary cause for sorghum's demand, though the use of its stover for fodder of livestock animals is known as a secondary cause for its demand. However, the latter has become a lifeline for the continuation of sorghum farming in the dry-land regions despite the reduction of consumption demand for the grain. Thus, utilization of *kharif* sorghum grain is a constraint in the event of deterioration of the sorghum grain quality (caught up in October rains at the time of harvest) making it unfit for human consumption. Over a period of time, the scale of sorghum farming is linked to the extent of domestic fodder requirements in the *kharif* season, while the *kharif* sorghum grain evolved as a regular source of raw material for alternative uses such as poultry and cattle feed, alcohol and starch production.

However, due to the smallness in quantity of marketed surplus, supply chain management has become a major constraint thus making it unsustainable. Dayakar *et al.* (2003) reveals that decrease in consumption of sorghum was found to be proportional to the increase in consumer expenditure. Increased income is accompanied by increased consumption of wheat and rice, as products made from these cereals are easy to prepare and have better quality of keeping.

Similarly, pearl millet consumption also declined by 67% in urban areas and 59% in rural areas during 1972–73 to 2004–05 (Basavaraj *et al.*, 2010). The food use of pearl millet has declined sharply at the country level, but it is still an important crop in the major producing states. The increase in pearl millet production juxtaposed with its declining food use implies that its alternative uses have been increasing – mainly from animal feed (especially dairy, and to some extent in poultry), the alcohol, starch and processed food industries, and export demand. The potential demand for food processing, though at an early stage, presents an encouraging prospect for value addition.

Another study conducted by the IIMR, Hyderabad, *Macro and Micro Level Changes in Consumption of Nutritious Cereals in India* (Dayakar *et al.*, 2004), also indicated that sorghum in all the states – and the country as a whole – showed a decline in consumption among the urban and rural masses. Whereas pearl millet consumption in major producing states like Gujarat and Rajasthan exhibited increased growth in urban areas (urban Gujarat, 129%), in rural Gujarat it has declined (−18.16%) and in rural Rajasthan it has increased (39.1%). The magnitude of reduction in pearl millet consumption was highest in rural Tamil Nadu (−95.5%), followed by rural Andhra Pradesh (−80.95%). In finger millet, there was a slight decline of 9% in consumption in urban Karnataka, whereas in rural Karnataka the decline in consumption was 36%, i.e. four times that of the urban consumption decline. Rural consumption of finger millet in Andhra Pradesh has also fallen drastically (−67.27%).

Causal factors for the decline in millet consumption in India:

1. Easy availability of fine cereals like rice and wheat at a cheaper price to that of nutritious cereals ('nutri-cereals') under PDS.
2. Rapid urbanization.
3. Nutri-cereals are socially less valued, which necessitated the decline in their consumption due to the rise in per capita income.
4. Tedious and time-consuming preparation of nutri-cereal foods.
5. Preference for cash payment by labourers in lieu of taking food grain for work performed.
6. Change in food habits, particularly the preference for fast food by the younger generation.

Measures for increasing the consumption/ demand of millet:

1. Creating awareness in public about the nutritional superiority of millets.
2. Supply of nutri-cereals through PDS.
3. Value addition in nutri-cereals.
4. Provisions for supply of millet foods under the Mid-Day Meal Scheme or other food supplement programmes arranged by various states.
5. Utilization of millet foods under larger schemes like that of the Nutrition Programme for Adolescent Girls, set up under the Ministry of Women and Child Welfare.

1.3 Government Policies that Determine Millets Production in India

1.3.1 Minimum Support Price

Sorghum, pearl millet and finger millet are covered under the Indian government's Minimum Support Price (MSP), a form of market intervention whereby a minimum price is fixed; as yet small millets are not covered. An analysis of MSP from 1977–78 to 2012–13 reveals an increase in MSP of millet on a par with that of wheat and rice and of late even higher. However, MSP cannot help in sustainability of millets because government agencies have not put a system of procurement in place. Also, unlike wheat and rice, millets are not included under the PDS, which distributes subsidized food staples to the poor. Millet is included in the government's new Food Security Bill and so it is expected that the millets are going to be procured, but its operation will require proper monitoring.

The government of India has increased the MSP of pearl millet of fair average quality (FAQ) (i.e. of good enough quality to be sold) from 980 INR/quintal (qt) to 1175 INR/qt for the crop year 2012–13 and marketing season 2013–14; this is an increase of 195 INR/qt or 19.89% over the previous year's MSP fixed for pearl millet.

1.3.2 Public Distribution System

PDS provided a substantive captive market for rice and wheat produced in those areas associated with the Green Revolution. PDS has had another significant effect on the dietary habits of Indians; effectively, it made rice and wheat the staple diet across the country, replacing millets and other cereals.

The government of India's production policy for coarse grains/millets is similar to that for wheat and rice, but there is less focus on the former. Also, the government's support for coarse grain via procurement and PDS is very limited and ineffective compared with that for rice and wheat. Unlike for wheat and rice, the government does not typically maintain a buffer stock of coarse grains to keep the process in check. Decentralized procurement, storage and distribution of millets by elected bodies such as *panchayats* (a village council in India) or even state governments, remains a dream despite the obvious economic benefits. Assuming that decentralized procurement would provide communities with the choice to obtain produce of their liking, it would then be possible for coarse grains and cereals to find a way back into the dietary habits of people. However, political support is lacking. The obvious choice, then, is to hope that the markets suffice. Millets could be targeted in publicly funded welfare programmes such as the Mid-day Meal Scheme in those states where millets are traditionally grown and consumed, initially at least twice a week; targeting the preference of children in their midday meal and their options such as noodles.

1.4 Development Programmes for Millets in India

The government of India initiated and implemented various programmes for the enhancement of millet cultivation, production and consumption through centrally sponsored schemes. The programmes are outlined as follows.

1.4.1 Macro Management in Agriculture

The centrally sponsored Macro Management in Agriculture (MMA) scheme, operational from 2000–01 and covering millet crops, gives flexibility to the Indian states to develop and pursue activities on the basis of their location-specific needs, with the freedom to include new/innovative interventions, even if they are not covered under any other scheme of the central government or are not part of any ongoing state schemes. MMA also provides at least 33% of funds to small, marginal and female farmers, who tend to have a larger stake in millet cultivation. Assistance to the states/Union Territories (UTs) is provided as 100% grant-in-aid. All states are eligible to receive assistance under the MMA scheme.

1.4.2 Initiative for Nutrition Security through Intensive Millets Promotion

The Initiative for Nutritional Security through Intensive Millets Promotion (INSIMP) was a sub-scheme of the Indian government's Rashtriya Krishi Vikas Yojana (RKVY) that was allocated INR 3000 million in 2011–12 to promote millets as nutri-cereals. The scheme aimed to increase production of millets by developing post-harvest technologies in an integrated manner, as well as generating consumer demand for millet-based food products through the establishment of processing and value-addition techniques. The initiative supported on-farm production, seed production, establishment of processing clusters, the popularization and promotion of millets, and marketing linkages. Through this scheme, three Centres of Excellence (CoEs) were established: (i) at IIMR, Hyderabad, for sorghum processing, (ii) at CCSHAU, Hisar, for pearl millet and (iii) at UAS, Bengaluru, for small millet processing. The major objectives of these CoEs included the refinement of technology, retrofitting of machinery and demonstration of its use, and providing entrepreneurship development and training on sorghum post-harvest technologies and value addition.

1.4.3 Rainfed Area Development Programme

The Rainfed Area Development Programme (RADP) was also launched in 2011–12 as a sub-scheme of RKVY to address the needs of rainfed areas (it existed as a separate programme for several years prior to this). Its holistic approach to development in these areas is to support rainfed farming systems by focusing on the needs of small and marginal farmers. It promotes integrated farming practices, favours mixed farming systems, minimum soil disturbance, utilization of crop residues and crop rotation. It assists farmers in improving the productivity of existing cropping patterns and in diversifying production. Amongst the programme components is support for Recommended Cropping Systems (RCS), which includes millets.

1.4.4 National Food Security Mission's coarse cereals component

The INSIMP initiative was subsumed under the 12th Five-Year Plan (2012–17) of the National Food Security Mission (NFSM). The focus on promotion of millet has been lost, as millets became grouped together with maize under the same component of coarse cereals. In most cases, the funding allocation goes to maize. The value chain concept, which was a central scheme of INSIMP, is now absent. This plan is being implemented in all the coarse cereal-growing states in India. The result of these changes is that small millets will eventually be completely masked, a predicament that deserves greater policy attention to prevent their becoming 'museum crops' of the future.

1.5 The Need for a Value Chain Model of Millets in India

As discussed above, millets are used as food and feed in general in various countries. Currently, millets are utilized in industries to produce various value-added products. But over the years the area under millets has decreased in almost all the millet-growing countries. Similarly in India, over the past three decades, the area under millets has fallen due to the higher demand and profitability of competing crops (Seetharama *et al.*, 2007). Although sorghum is nutritionally rich, its consumption is declining significantly (Dayakar *et al.*, 2010). To augment the production and consumption of millets, a value chain model is needed.

MacKay *et al.* (1997) defined value chain analysis as the study of the full range of activities that are required to bring the product or services from production to final consumers. But according to Sturgeon (2001), value chains have three dimensions, namely organizational, spatial and the type of actors involved. From the organizational aspect, value chains are both complex and dynamic or simple, depending on their sustained supply of a variety of critical inputs, i.e. human resource requirements, capital equipment and services. The second dimension is spatial, which involves some value chains that are global and are sometimes referred to as 'global commodity chains' (Daviron and Gibbon, 2002) because they operate at international levels. The third dimension is the production actors in the value chain; these involve the firms that participate in the chain. The actors can either be producers (in the case of the agricultural production value chain), suppliers, retailers/wholesalers or lead firms (Sturgeon, 2001).

So, the value chain model essentially involves bringing all the stakeholders involved together on a common platform, with significant value creation at the end of each level. The farmers (our primary stakeholder) are ensured of higher income and benefit in the long run, as there is the creation of demand for millet cultivation through value addition and commercialization of millet-based convenience products, with the flagging up of nutritional benefits. This could involve the private sectors, public research institutes, policymakers, entrepreneurs and consumers. Subsequent chapters cover in more detail the value chain development model on millets to revive their demand in India through the ICAR-NAIP project.

1.6 Overview of Millets Value Chain Development in Various Countries

Millets are important crops grown in various rainfed/dryland regions of the world, thus an overview of value chain development of millet in pre- and post-harvest conditions is useful to set the Indian situation in context. The value chain development of various agricultural commodities such as wheat, rice, vegetable, fruits and meat are both well developed and well run in various countries. But the value chain of millets is very new and value addition is limited to the harvesting and marketing stages of whole sorghum grain; the value addition of sorghum grain processing at the large scale is at the initial stage and still limited.

Ogidi (2014) studied the sorghum value chain and enterprise development in Nigeria and found that the lack of initial processing facilities has undermined the marketability of sorghum for both local consumption and for exports. The linkages developed between policymakers and stakeholders give a better position to consider specific interventions that can raise standards of living and reduce poverty, generate employment, involve private/public investment and encourage entrepreneurial participation. It is important for governments to create a business environment that encourages the initiatives of rural, young and male/female entrepreneurs, and to enhance the human and institutional capacities required to foster entrepreneurial dynamism and increase productivity (Ogidi and Abah, 2012). Ogidi (2014) suggested that enterprise development with the right approach should be paramount;

where commercial sorghum farms, commercial sorghum food processing industries and industrial processing of sorghum will determine employment generation success. Private/public investors will only come into the chain if government policies favour them and constraints are reduced to the most minimum. Poverty reduction should be a vital tool in enterprise development so that more women can be involved in the chain activities and thus increase the earnings of individual households; this can be achieved through the provision of incentives to the womenfolk. The author concluded that an increase in food security and competitiveness of Nigerian sorghum requires improvements in technology, on-farm practices and supply chain linkages.

The project Harnessing Opportunities for Productivity Enhancement in Sub-Saharan Africa and South Asia, funded by the Bill and Melinda Gates Foundation, was undertaken by ICRISAT with the objective of discovering and developing new market opportunities for post-rainy season sorghum (PRSS) in the Maharashtra state of India (Basavaraj *et al.*,2014). Value chain mapping highlighted the limiting factors of supply, trade and demand and identified opportunities along the chain, from which upgrading options to enhance the market demand and value chain functioning for PRSS were proposed. Process and product upgrading were found to be two critical areas that were addressed as priority areas for PRSS value chain interventions. Choice of product was an important determinant of consumption and its acceptance. On the post-harvest front, farmers are constrained by lack of grading and cleaning facilities for the produce that help the realization of higher prices. Value addition in the existing value chain of sorghum is limited to physical processing; processing as a specialized activity for product development is still minimal. Therefore, growing the value chain by upgrading production to ready-to-use products will benefit different stakeholders along the chain and, most importantly, will potentially benefit the small and marginal farmers growing sorghum.

Rooney (2003) observed that processing technology is not the major obstacle to successful production of millet food products in various countries but consistent supply of a modest quality grain is the major constraint. The major limitation is the lack of high quality grain in sufficient quantities for processing. More efficient methods of threshing and cleaning the grain to remove sand and other impurities are essential. Millet grains in existing markets are extremely variable in kernel size, colour and cleanliness. The best strategy for developing convenient, shelf-stable millet foods is to use identity-preserved grains (those varieties/cultivars produced with a special end product in mind) to produce high-value products that can be priced slightly lower than imported products. The image of millets as a poor man's food can be overcome by developing highly improved products that have attractive, more socially acceptable names that appeal to wealthy consumers. The new name along with identity-preserved production schemes would lead to improved acceptability, linking farmers and other chain actors to the market, improvement of technology used in pearl millet processing, increasing production yield of the crop, provision of credit/financial support to chain actors, encouragement of the private sector to participate in the crop chain, provision of motivation and morale to chain players and promotion of the crop products.

Kaminski *et al.* (2013) studied the implications for competitive and inclusive value chains for millet in Mali. Here, like much of the Sahel, the value chain of millets remain relatively underdeveloped and weakly integrated, and only a small share (30%) of production is marketed despite a significantly larger urban demand in Mali. The decrease in demand for millet was due to urbanization, shifting to ready-to-eat food and the increase in imported food staples. Because of the limited marketability of millet and lack of value chain linkages within agro-industry, millets usually have no access to interlinked agreements such as outgrower schemes/contract farming that would facilitate input access. There were three main problems with the quality of the raw material originating from millet, in descending order: (i) a low degree of cleanliness; (ii) heterogeneity in grains; and (iii) unstable/insufficient quality. For traditional processors, problems of grain quality are even more important than their heterogeneity. A high percentage of impurities can be attributed to a lack of appropriate post-harvest handling techniques in storage and conservation, as well as to a lack of quality certification systems in domestic markets. Consistent product quality is necessary to develop customer loyalty. One of the main processing difficulties is consistency in quality over time, since quality is affected by a lack of established standards and non-standardized processing techniques, such as a lack of control over choice of ingredients, the heterogeneity in raw material and a lack of measuring instruments.

Thus, globally the constraints and opportunities of value chain development of millet are similar, although some factors are particular to a country. Thus far, value chain development in millet is restricted to the whole millet grain only. Most research proposes the development of processed food (ready-to-eat/ready-to-cook) using millet as the way forward, so efforts should be directed towards overcoming the constraints and developing new technologies that enhance the smooth running of the value chain of millets in the long run. This is of particular pertinence when one considers the emerging importance of millets in the context of future climate change.

References

Basavaraj, G. and Rao, P.P. (2011) *Regional Analysis of Household Consumption of Sorghum in Major Sorghum-Producing and Consuming States in India*. Working Paper Series no. 28. International Crops Research Institute for the Semi-Arid Tropics, Patancheru, India.

Basavaraj, G., Rao, P. P., Bhagavatula, S. and Ahmed, W. (2010) Availability and utilization of pearl millet in India. *Journal of SAT Agricultural Research* 1–8.

Basavaraj, G., Parthasarathy Rao, P., Pokharkar, V.G. and More, S. (2014) *Post-Rainy Season Sorghum: Value Addition and Value Chain Development.* Paper presented in workshop 'Global value chains for food and nutrition security' Roma, September 25–26 2014. International Crops Research Institute for the Semi-Arid Tropics, Patancheru, India.

Chapke, R.R., Mondal, B. and Mishra, J.S. (2011) Resource-use efficiency of sorghum (*Sorghum bicolor*) production in rice (*Oryza sativa*)-fallows in Andhra Pradesh, India. *Journal of Human Ecology* 34(2), 87–90.

Daviron, B. and Gibbon, G. (2002) Global commodity chains and African export agriculture. *Journal of Agrarian Change* 2(2),137–161.

Dayakar, R.B., Matthew, B., Kumar, B.K.A., Karthikeyan, K., Hyma Jyothi, S., Parwez, S., Rathnavathi, C.V. and Seetharama, N. (2003) *Industrial Utilization of Sorghum in India – Status and Prospects.* NATP (NRCS) Series no. 2. National Research Centre for Sorghum, Rajendranagar, India.

Dayakar, R.B., Karthikeyan, K., Seetharama, N. and Hyma Jyothi, S. (2004) *Macro and Micro Level changes in Consumption of Nutritious Cereals in India.* National Research Centre for Sorghum, Hyderabad, India.

Dayakar, R.B., Patil, J.V., Rajendra Prasad, M.P., Nirmal Reddy, K., Kamini Devi, Sriharsha, B. and Kachui, N. (2010) Impact of innovations in value chain on sorghum farmers. *Agricultural Economics Research Review* 23 (conf. no.), 419–426.

De Candolle, A. (1886) *Origin of Cultivated Plants*. Hafner Publishing Co., New York.

DES (2014) Directorate of Economics and Statistics, Department of Agriculture and Cooperation, Ministry of Agriculture, Government of India.

FAO/ICRISAT (1996) *The World Sorghum and Millet Economies Facts, Trends and Outlook*. FAO/ICRISAT, Rome.

FAO Statistics (2015) Food and Agriculture Organization of the United Nations [Internet] FAOSTAT Database. FAO, Rome. Available at faostat.fao.org/ (accessed 25 July 2015).

ICRISAT (2004) *Sorghum, a Crop of Substance*. International Crops Research Institute for the Semi-Arid Tropics, Patancheru, India.

Kaminski, J., Elbehri, A. and Samake, M. (2013) An assessment of sorghum and millet in Mali and implications for competitive and inclusive value chains. In: Elbehri, A. (ed.) *Rebuilding West Africa's Food Potential*. FAO/IFAD, Rome, pp. 479–502.

Léder, I. (2004) Sorghum and millets. In: Fuleky, G. (ed.) *Cultivated Plants, Primarily as Food Sources*. Encyclopedia of Life Support Systems (EOLSS), developed under the auspices of UNESCO. EOLSS Publishers, Oxford, pp. 250–252.

MacKay, A., Morrissey, O. and Vaillant, C. (1997) Trade liberalization and agriculture supply response: issue and some lessons. *European Journal of Development Research* 9(2), 129–147.
Murty, M.V.R., Singh, P., Wani, S.P., Khairwal, I.S. and Srinivas, K. (2007) Yield gap analysis of sorghum and pearl millet in India using simulation modeling. *Global Theme on Agro Ecosystems Report No. 37*. International Crops Research Institute for the Semi-Arid Tropics, Patancheru, India.
NAAS (National Academy of Agricultural Sciences) (2013) *Role of Millets in Nutritional Security of India*. Policy Paper No. 66. National Academy of Agricultural Sciences, New Delhi.
Ogidi, A.E. (2014) Sorghum value chain and enterprise development in Nigeria. *SCSR Journal of Agricultural Sciences* 1(1), 1–11.
Ogidi, A.E. and Abah, D.A. (2012) The impact of sorghum value chain on enterprise development: a holistic diagnosis of some actors in Benue State, Nigeria. *International Journal of Agriculture* 4(4), 79–92.
Parthasarathy, R.P., Basavaraj, G., Ahmad, W. and Bhagavatula S. (2010) An analysis of availability and utilization of sorghum grain in India. *Journal of SAT Agricultural Research* 8.
Rooney, L.W. (2003) *Overview: Sorghum and Millet Food Research Failures and Successes.* AFRIPRO, 2 April 2003. Available at: www.afripro.org.uk/papers/Paper23summary.pdf (accessed 12 February 2016).
Seetharama, N., Dayakar Rao, B., Elangovan, M., Vilas, A, Tonapi and Ratnavathi, C.V. (2007) *Changing Scenario of Millets Cultivation in India*. Souvenir, Farm Fest, 23–25 February 2007, Uppalam, Puducheri.
Singh, J.P. and Dayakar, B.R. (2010) Status paper on millets in India, 2010. Background paper prepared for *National Seminar on Millets*, organized by Directorate of Millets Development, Jaipur, and Directorate of Sorghum Research, Hyderabad, India.
Sturgeon, T.J. (2001) How do we define value chains and production networks? *IDS Bulletin* 32(3), 9–18.
Subbarao, G.V., Rao, I.M., Nakahara, K., Sahrawat, K.L., Ando, Y. and Kawashima, T. (2013) Potential for biological nitrification inhibition to reduce nitrification and N_2O emissions in pasture crop–livestock systems. *Animal* 2, 322–32.
Tonapi, V.A., Patil, J.V., Dayakar, R.B., Elangovan, M., Venkatesh Bhat, B. and Raghavendra Rao, K.V. (2011) *Sorghum: Vision 2030*. Directorate of Sorghum Research, Rajendranagar, Hyderabad 500 030 (AP), India. 38 pp.
Vavilov, N.I. (1926) Studies on the origin of cultivated plants. [Russian]. *Bulletin of Applied Botany and Plant Breeding* 14, 1–245.

2

Gaps in and Scope of the Value Chain in Millets

As discussed in Chapter 1, the direct consumption of millet as food has significantly declined over the past three decades in India. The declining trend in production and consumption of millets has been perceived as critical for nutritional and food security of the country in the long term, especially to the economy of dryland farmers where millet cropping is more suitable than any other cereal crops. The causes of this downward trend are demand-led, as discussed previously. An understanding of the gaps in millet production, utilization and marketing is essential for the development of the production-to-consumption system (PCS) value chain and to plan interventions to revive crop demand in the long term.

2.1 Gaps in Production, Utilization and Marketing of Millets

The gaps in millets production, utilization and marketing, core to addressing the revival of demand, were analysed as a prelude to developing an integrated value chain. They are as follows:

1. Inconvenience in preparation of millets especially *roti* (tortilla)-making due to lack of gluten formation (drudgery for modern working homemakers). It requires skill and is time-consuming.
2. Lack of diversification of processing technologies and unavailability of processing equipment specifically for millets. Except for some milling, no processing diversification technologies are commercially available.
3. Unexploited nutritional merits of millets. Very limited studies empirically established the health benefits and bioavailability of minerals in millets.
4. Inconsistent quantity due to smallholdings cultivation leading to low marketed surplus and unassured supply throughout the year may pose supply chain issues for those who want to commercialize them. The quality of produce is not uniform to address safety and quality assurance issues.
5. Economic theory suggests that changes in income, urbanization etc., would lead to a shift in consumption from coarse cereals to fine cereals and from a carbohydrate-rich to a protein-rich diet. Further, there would be changes in consumers' tastes and preferences that would lead them away from millets.
6. Neglected policy front: policy favours rice and wheat as evidenced from the lack of both input subsidies (fertilizer, electricity and irrigation) and output price incentives for millets. The Public Distribution System (PDS) supply of only fine cereal grains at subsidized cost disincentivizes the consumption of millets.
7. Lack of a procurement system for millets through public agencies as implemented in the case of rice and wheat as a part of the government of India's policy since independence.

It became imperative to reorient the R&D efforts on these crops to revive demand through value-added processing, with diversification of processing technologies, nutritional evaluation and creation of awareness linked up with backward integration (i.e. miller/processor of millets buys grains directly from the farmer, eliminating the intermediaries). To augment such efforts, a PCS model for specific products and markets was needed.

2.2 An Innovative Value Chain Model: the NAIP's 'Millets Value Chain Project'

Under the aegis of the Indian Council of Agricultural Research (ICAR), the National Agricultural Innovation Project (NAIP) was initiated in 1991 – essentially a public–private partnership (PPP) model that deals with the value chain of various commodities (59 in total) through active funding support from the World Bank. The subproject on millet (the 'Millets Value Chain' project) was initiated in 2008 with the then Directorate of Sorghum Research (DSR) (now the Indian Institute of Millets Research, IIMR) as the consortium leader, and other partners, namely the Indian conglomerate ITC (Agri-Business Division, ABD) Ltd, the National Institute of Nutrition (NIN) and College of Home Science, Acharya NG Ranga Agricultural University (ANGRAU) Hyderabad. The subproject focused efforts on selected millet foods with special emphasis on sorghum, with the intention of replicating with other millets if the pilot project proved successful.

2.2.1 Objectives of the project

In the wake of disincentives resulting from PDS for fine cereals at very low cost to consumers and other practices, it became necessary for R&D efforts on millet to re-orientate towards generating demand through value addition of processed foods (Dayakar *et al.*, 2014). To augment such efforts, PCS models for specific products and markets were needed. The NAIP subproject focused these efforts on selected millet foods, led by IIMR (under the Indian Council of Agricultural Research, ICAR). The objectives of the project were:

1. To enable market-driven millet production for specific end-uses, procurement and primary processing for continuous supply chain management.
2. Fine-tuning of the technologies for development of millet food products and up-scaling.
3. Testing for nutritional evaluation and safety of selected millet foods.
4. Assessing consumer acceptability, price and market strategies, and social and policy imperatives.
5. Developing entrepreneurship and appropriate strategies to promote and popularize millets for commercialization through value addition and branding as health foods.

2.2.2 Role of partners

A comprehensive methodology was worked out to achieve all the above-mentioned objectives. The lead centre, IIMR, apart from planning, co-ordinating and monitoring all project activities, undertook research on identification of suitable genotypes, R&D on enhancing the shelf life of grain as well as processed products, antioxidants, development and fine-tuning of sorghum food products, promotion and sensitization to consumers and policy makers, commercialization and, not least, entrepreneurship development in millet. IIMR facilitated coalition-building and establishing linkages among the various stakeholders and institutions for enabling PCS value-chain on millet foods.

ITC (ABD) Ltd, a private partner specializing in supply chain management in agri-enterprises, facilitated on-farm production of sorghum offering buy-back assurance. Along with IIMR, they also undertook integrated extension services and the procurement and aggregation of farm produce. They conducted an extensive market survey (through marketing consultants) for assessing the market potential and consumer acceptability of the on-shelf processed millet food products.

The NIN – a globally recognized institute – under the Indian Council of Medical Research (ICMR), evaluated the nutrient composition of millet foods based on metabolic and organoleptic (that is, its qualities to people in terms of taste etc.) studies. They conducted clinical trials and assessed the effect of the dietary intake of millet foods on diabetics and school children. ANGRAU,

Hyderabad, worked on the development of food products of pearl millet and their popularization and marketing.

2.2.3 Interventions

In order to meet the objectives, the following innovative interventions were envisaged:

1. Technological innovations:
   i. Identification of product-specific varieties in millets.
   ii. Nutritional evaluation of millet foods and using data for labelling which highlights nutritional features.
   iii. Shelf-life enhancement through improved packaging to meet modern food habits by providing balanced nutrition.
   iv. Standardization of multi-grain flour and food products.
   v. Fine-tuning of technologies in millet foods.
2. Change from commodity-based and on-farm production to market-driven end product-specific farm production with market assurance to the millet farmers through buy-back facilitation.
3. PPP model of holistic farm-extension services on seed supply, fertilizers, plant protection chemicals, crop management and training.
4. Ready-to-eat (RTE) and convenient foods to overcome cumbersome and time-consuming food preparation of millets.
5. Branding millets as health foods through value-addition, nutritional evaluation, certification and implementation of hazard analysis and critical control points (HACCP).
6. Innovative approaches for popularization, such as roadshows, wet sampling (where the customer is invited to test the product), and millet tableaux for Republic Day or Independence Day, or parades at important public places and functions.
7. Coalition building with other research and private groups working on millet elsewhere in the world through exchange visits, national seminars, workshops and further global meetings on millet.
8. Recognition of the value-addition of landraces that have been nursed by farmers over many years; possibly also securing intellectual property rights for local innovations.
9. 'Model of support' to dryland farmers by linking with government schemes such as the Mid-day Meal Scheme at a pilot scale in one district each of Andhra Pradesh, Maharashtra and Karnataka under INSIMP, so as to extract value, and to make the public welfare system more efficient and locally relevant by educating on the nutritious use of millet formulations.

The IIMR-led consortium was successful in pilot-testing the integrated millet value chain project on sorghum, with product development, nutritional studies, promotion and marketing. For this, the IIMR built links not only with NIN, SAUs and private partners such as ITC but also with publicly funded national institutions such as DFRL, CFTRI, CIAE and CIPHET. The backward integration model was tested with the help of ITC in Maharashtra.

Backward integration was provided by ITC leveraging the *e-Choupal* platform (an ITC initiative that supplies internet access to farmers for agricultural procurement) providing end-to-end service linkages with access to markets. It was implemented in both *kharif*-and *rabi*-cropped areas: in the Adilabad district in Andhra Pradesh and in the Parbhani district in Maharashtra, respectively, and involved about 3000 farmers in total. Later, intensive *kharif* cultivation trials were shifted to the Nanded district of Maharashtra. The farm-level product-specific interventions with buy-back assurance and backstopping by product-specific cultivars and procurement by ITC has practically impacted the mindset of sorghum farmers so that it is possible to reap higher net income through this crop. This backward integration has provided the industry with a consistent supply and quality of grain. The processing interventions backed by retrofitting of the sorghum processing machinery (leading to diversification in primary and secondary processing) made technologies available under a single roof, offering convenient sorghum products with improved shelf life and providing the potential for commercialization.

The value chain interventions were realized through realistic reassessment of crop research needs in terms of current and future demand, resolving specific production constraints, development of post-harvest processing, value addition technologies, and marketing strategies and policies that may result in additional income and employment without sacrificing the overall goal of attaining sustainable food and nutritional security, especially of the urban poor and sorghum farmers in dry regions.

2.2.4 Technical profile-contours of the value chain

Objectives

Objective 1: To enable on-farm production, procurement and primary processing for supply-chain management of specific end uses.
Activities covered:

1. Enabling on-farm production of specific end use through technological back stopping.
2. Providing integrated farm extension services and farmer training for intensive and profitable cultivation.
3. Facilitating grain procurement aggregation and storage function in target regions.

Objective 2: Fine-tuning the technologies for development of millet food products and upscaling.
Activities covered:

1. Fine-tune semi-processed and processed technologies and retrofitting machinery for the development of millet foods.
2. Upscaling the identified product(s) for widespread markets and niche markets for promotion, popularization and marketing (through newspapers and audio-visual advertisement).
3. Standardize combinations of different millets for making multi-grain *atta* that is gluten free for supply on a mass scale.

Objective 3: To test for nutritional evaluation and safety of selected millet foods.
Activities covered:

1. Assessment of nutrient composition of selected millet foods.
2. Nutritional evaluation of the millet foods for diabetic and obese patients.
3. Evaluation of semi-processed millets for their nutritional characteristics in the Mid-day Meal Scheme.
4. Implementation of HACCP.

Objective 4: To assess consumer acceptability, price and market strategies, and social and policy imperatives.
Activities covered:

1. Baseline survey.
2. Identification of the number of potential food products for wide-ranging and niche markets.
3. Economic feasibility of products developed and studies on consumer acceptability and pricing strategies of millet foods.
4. Sensitizing relevant governments departments for enhancing millet consumption.

Objective 5: To develop appropriate strategies to promote and popularize millets and innovative approaches for commercialization through value-addition and branding as health foods.
Activities covered:

1. Entrepreneurship development of stakeholders for intensive cultivation, product development and mechanization (on site and through demo units).
2. Value addition through branding of millet-based recipes as health foods by training and popularization.
3. Innovative approaches for popularization of millets.
4. Assess socio-economic and environmental impacts of the interventions for plan uptake.

The major outputs

Finished products, processes and protocols developed and/or adopted under NAIP, validated in the lab, on-farm, and through third-party validation, and licensed for commercialization, are summarized as follows:

1. On-farm technological backstopping with end product-specific sorghum cultivars/hybrids – such as CSH-14, CSH-16, CSH-23, CSV-20 and SPH-1148 in *kharif* and M35-1, CSV 216 R (Phule Yashoda) and SPV 1411 (Parbhani Moti), Phule Vasudha and Phule Revati in *rabi* – provided higher yields and production. Facilitation of procurement in PPP mode with ITC (ABD) Ltd has brought about a change in the mindset of sorghum farmers as, after 4 years of on-farm trials, they opted to allocate better lands for sorghum cultivation owing to higher returns from its cultivation. It has resulted in an increase in average income above the baseline in *kharif* and *rabi* seasons of 222% and 69%, respectively.

2. More than 30 kinds of processing equipment have been retrofitted to make them suitable for sorghum processing, such as: destoner and grader, dehuller, hammer mill, *chakki* mill, roaster, edge runner, parboiling unit, roller flaking unit, semolina making machine, cold extruder, dough kneader, ribbon blender, sifter, homogenizer, convection oven, planetary mixer, automatic cookie cutting machine, popping machine, grinder, automatic sealing machine, nitrogen flux packaging machine, sealing machine, foot-operated *roti*-making machine, automatic *roti*-making machine, etc.
3. More than 50 processing technologies (for ready-to-eat (RTE) and ready-to-cook (RTC) sorghum foods) have been developed, with sorghum-based products such as: coarse, medium and fine semolina; parboiled semolina; designer semolina; multigrain flour (four types); multigrain *roti* (two types); flakes; vermicelli; pasta; gluten-free pasta (three varieties: soy, caseinates and flax seed incorporated); biscuits (17 varieties: salty, sweet, groundnut, coconut, gluten-free salty and sweet biscuits (two types), transfat-free salty, sweet (two types), groundnut and coconut and low calorie salty, sweet, groundnut and coconut); RTE extruded snacks; gluten-free puffed snacks (two types); *lassi* (sweet and salty); instant foods such as *idli, dhokla, upma, dosa, pongal,* vermicelli *kheer* mix, flakes *kheer* mix and *masala* sorghum flakes; pops (two types: *aam chat masala* and *pudina masala*); *peda*; bran soup; bran *peda* were developed. More than 30 sorghum recipes have been developed and standardized, e.g. *idli, dosa, pongal, upma, roti,* multigrain *roti, bisibela bath, sankati, ambali, samosa, khichidi,* etc.
4. Nutritional evaluation of sorghum foods by NIN showed that sorghum is rich in complex carbohydrates, dietary fibre, folic acid, iron, calcium, zinc and magnesium, and can be consumed by all age groups.
5. Nine sorghum-based products (pure sorghum flour, sorghum-based multigrain flour, coarse, medium and fine semolina, flakes, pasta, vermicelli and biscuits) were successfully commercialized at the pilot scale under IIMR's *eatrite* brand, which is now marketed through retail outlets in Hyderabad.
6. 'ANGRAU foods', a trade mark, was also registered by the NAIP project partner ANGRAU to commercialize five products: sorghum snacks, multigrain biscuits, gluten-free pasta, vermicelli and hot extruder snacks.
7. Through entrepreneurial development programmes, more than 4000 farmers, self-help groups, small-scale processors, women groups and rural entrepreneurs were trained in sorghum processing, which resulted in the establishment of many rural industries (more than 100 processing clusters).
8. Sensitization of policymakers was an important output, achieved through implementation of roadshows in public parks and malls by food retailers such as M/s Reliance Fresh, M/s Food Bazaar, M/s *Choupal* Fresh, and also use of the *Jowar Rath* mobile van, creating awareness of sorghum's nutritional superiority. This resulted in policymakers directing INR 3000 million of INSIMP initiatives through the Directorate of Agriculture and Cooperation (DAC). Great emphasis is given to DSR's millet promotion by DAC under INSIMP, which is primarily due to the successful NAIP model tested at Hyderabad to create demand for sorghum through processing and commercialization.
9. IIMR built affiliations through Memoranda of Understanding (MoUs) with 15 potential entrepreneurs for technology transfer and marketing of sorghum foods and three Memoranda of Agreement (MoAs) for marketing of products under the project.

Creating demand for millets and enhancing mainstream dryland agriculture is important for India's sustainability, in terms of food, nutrition and livelihood security. To achieve this, the major challenge is to deliver millet-based technologies that are sustainable and market-oriented, aided by research that develops value-added products, new technologies, marketing strategies and policy measures in millet. Thus, a collaborative effort is needed to reduce those gaps that would help the millet ecosystem and all its stakeholders.

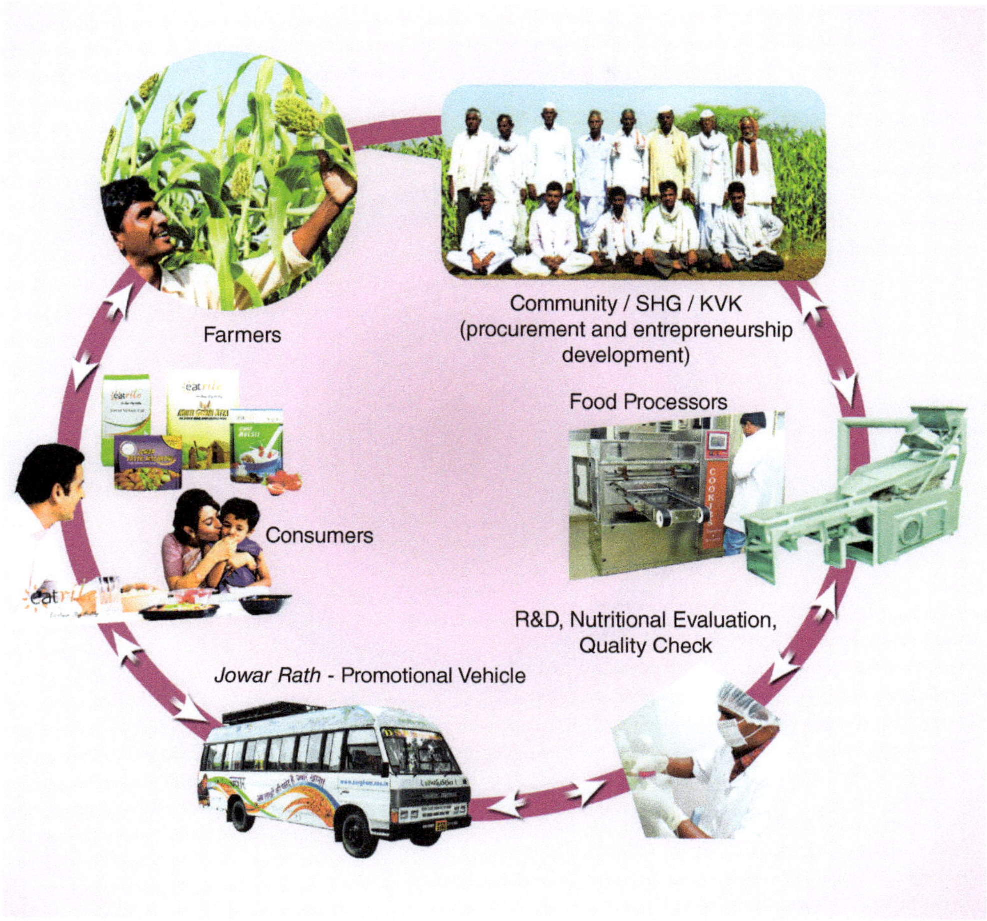

Fig. 2.1. Successful value chain of millets.

Reference

Dayakar, R.B., Patil, J.V., Hymavathi, T.V., Nirmal Reddy, K. and Rajendra Prasad, M.P. (2014) *Creation of Demand for Millet Foods through PCS Value Chain*. Final report of NAIP (ICCR). Directorate of Sorghum Research, Rajendranagar, India.

3

Nutritional and Health Benefits of Millets

Millets – the 'noble grains' – comprise sorghum, pearl millet, finger millet and five small millets. Among the millets, sorghum used to be the most important in terms of its cropped area and food use, though its position has been overtaken in recent years by pearl millet. Pearl millet and finger millet are gaining regional importance as food staples, though their distribution across many Indian states is not as pronounced as that of sorghum. Direct consumption of millets has reduced over the past three decades because of inconvenient, cumbersome and time-consuming preparation, lack of processing technologies, and the lack of awareness of its nutritional merits. This trend in consumption can be manipulated by creating value addition to the millet crops through post-harvest processing, and can be enhanced through diversification of processing technologies as well as nutritional evaluation.

In view of this, here we examine the nutritional and health implications of millet, both in terms of the whole grain and in terms of processed products – specifically sorghum products – which have been studied in detail under the auspices of the Indian government's National Agricultural Innovation Project (NAIP).

3.1 Nutritional Overview: Millet vs Major Cereals

The nutritional importance of sorghum and other millets in many populations cannot be underestimated. They contribute significantly to the energy requirements of low- to middle-income populations in most regions of Africa, Asia and Europe. In North America, however, they are used mainly for animal feed. Their gluten-free nature makes them reasonable alternatives for people suffering from coeliac disease and allergies that result from the consumption of wheat-based products (Taylor and Emmambux, 2008). Compared to major cereals crops such as maize, wheat and rice, pearl millet can be considered as superior in its nutrient composition. Ratnavathi *et al.* (2003) reported a starch content of 63.4–72.5%, a protein content of 7.9–11.5%, a fat content of 1.9–3.0% and a fibre content of 1.6–2.4% for various cultivars of sorghum. Values of 9.2–13.6% protein, 3.4–7.1% fat, 61.0–70.3% starch and 1.1–2.4% ash have been reported for pearl millet (Hadimani *et al.*, 1995; Malleshi and Klopfenstein, 1998). Vasan *et al.* (2008) also reported 9.46% and 10.95% protein for sorghum and pearl millet, respectively. Fibre and minerals are mostly concentrated in the pericarp while the germ contains mostly crude protein and fat (Rai *et al.*, 2008). With over 95% of the non-starch polysaccharides in millets, they are fibre rich. These polysaccharides help in preventing constipation and lowering the blood cholesterol level and also facilitate the glucose to be released slowly into the bloodstream during digestion, thus imparting hypoglycaemic properties. Regular millet consumption reduces the incidence of cardiovascular, gastrointestinal and lifestyle (e.g. diabetes) diseases. Pearl millet and sorghum especially are very good sources of fibre (Rai *et al.*, 2008). Sorghum also contains considerable amounts of tannins in addition to phenolic compounds in its pericarp. Pearl millet's pericarp contains polyphenols such as phenolic acids. Phenolic compounds such as gallic, protocatechuic, p-hydroxybenzoic, vanillic, caffeic-coumaric, ferulic, cinnamic, gentisic, salicylic, sinapic and syringic acids have been reported in sorghum (Dykes and Rooney, 2006, 2007). Pearl millet's germ, on the other hand, contains caffeic, cinnamic, gentisic, p-hydroxybenzoic, protocatechuic, syringic, vanillic, p-coumaric, ferulic and sinapic acids (Dykes and Rooney, 2006, 2007). As expected, both cereals and millets are deficient in lysine. They are, however, good sources of other amino acids especially the sulfur-containing amino acids. The deficiency in cereals is normally dealt with by adding legumes that are rich in lysine. The amino acid composition of pearl millet has been reported to be better than that of sorghum and maize (Abdalla *et al.*, 1998; Malleshi and Klopfenstein, 1998).

Millets are comparable to major cereals with respect to their broad nutritional features (Table 3.1) and are good sources of carbohydrates (60–73%), proteins (6–13%), fat (1–5%), crude fibre (1–10%), and phytochemicals that have nutraceutical properties. Pearl millet is rich in proteins (11–13%) and lipids (4–6%); finger millet, on the other hand, contains proportionately less protein (6–8%) and fat (1.5–2%) (Dayakar *et al.*, 2004). Crude fibre content is 10–50-fold more in millets when compared to fine cereals, the highest being in barnyard millet (Table 3.1).

Millet proteins contain high proportions of essential amino acids, being 1.2–1.5-fold higher in essential amino acid content than rice (particularly finger millet). When compared to rice, the essential amino acid histidine content is 1.2-fold higher in sorghum, and the same in pearl millet and foxtail millet. Compared to rice, phenylalanine levels are 1.3-fold higher in pearl millet and methionine levels are 1.4-fold higher in finger millet. Cystine levels are 1.2–1.5-fold higher (or the same as) in rice for all the millets. Isoleucine levels are 1.3-fold higher in proso millet and finger millet. Lysine was always found to be the most limiting amino acid in millets as well as fine cereals. The amino acid presence in all the millets are 1.2–1.9-fold higher than in wheat (Table 3.2).

Millets also contain important vitamins such as riboflavin, thiamine, folic acid and niacin. When compared to rice, the vitamin riboflavin levels are 2.5–5.2-fold higher and folic acid levels are 1.1–5.6-fold higher in all the millets, the highest being in pearl millet. Compared to rice, thiamine content is 1.4-fold higher in foxtail millet and similar in finger millet; niacin content in proso millet is similar to that in rice. All the vitamins are similar to the content present in wheat (Table 3.3).

Table 3.1. Nutrient composition of millets compared to fine cereals (per 100 g). (Source: NIN, 2007.)

Millets/cereals	Carbo-hydrates (g)	Protein (g)	Fat (g)	Energy (kcal)	Crude fibre (g)	Mineral matter (g)	Ca (mg)	P (mg)	Fe (mg)
Sorghum	72.6	10.4	1.9	349	1.6	1.6	25	222	4.1
Pearl millet	67.5	11.6	5	361	1.2	2.3	42	296	8
Finger millet	72	7.3	1.3	328	3.6	2.7	344	283	3.9
Foxtail millet	60.9	12.3	4.3	331	8	3.3	31	290	2.8
Proso millet	70.4	12.5	1.1	341	2.2	1.9	14	206	0.8
Kodo millet	65.9	8.3	1.4	309	9	2.6	27	188	0.5
Little millet	67	7.7	4.7	341	7.6	1.5	17	220	9.3
Barnyard millet	65.5	6.2	2.2	307	9.8	4.4	20	280	5
Rice (raw, milled)	78.2	6.8	0.5	345	0.2	0.6	10	160	0.7
Wheat (whole)	71.2	11.8	1.5	346	1.2	1.5	41	306	5.3

Note: Ca, calcium; Fe, iron; P, phosphorus.

Table 3.2. Essential amino acid profile of millets compared to fine cereals (mg/g of N). (Source: NIN, 2007.)

Millets/ cereals	Arginine	Histi- dine	Lysine	Trypto- phan	Phenyl- alanine	Tyro- sine	Methi- onine	Cyst- ine	Threo- nine	Leu- cine	Isoleu- cine	Valine
Sorghum	240	160	150	70	300	180	100	90	210	880	270	340
Pearl millet	300	140	190	110	290	200	150	110	140	750	260	330
Finger millet	300	130	220	100	310	220	210	140	240	690	400	480
Foxtail millet	220	130	140	60	420	–	180	100	190	1040	480	430
Proso millet	290	110	190	50	310	–	160	–	150	760	410	410
Little millet	250	120	110	60	330	–	180	90	190	760	370	350
Barnyard millet	270	120	150	50	430	–	180	110	200	650	360	410
Rice	480	130	230	80	280	290	150	90	230	500	300	380
Wheat	290	130	170	70	280	180	90	140	180	410	220	280

Table 3.3. Vitamin contents of millets and major cereals (mg/100 g). (Sources: NIN, 2007; NABARD, undated.)

Millets/cereals	Thiamin	Niacin	Riboflavin	Vit. A (carotene)	Vit. B6	Folic Acid	Vit. B5	Vit. E
Sorghum	0.38	4.3	0.15	47	0.21	20	1.25	12
Pearl millet	0.38	2.8	0.21	132	–	45.5	1.09	19
Finger millet	0.42	1.1	0.19	42	–	18.3	–	22
Foxtail millet	0.59	3.2	0.11	32	–	15	0.82	31
Proso millet	0.41	4.5	0.28	0	–	–	1.2	–
Kodo millet	0.15	2	0.09	0	–	23.1	–	–
Little millet	0.3	3.2	0.09	0	–	9	–	–
Barnyard millet	0.33	4.2	0.1	0	–	–	–	–
Rice	0.41	4.3	0.04	0	–	8	–	–
Wheat	0.41	5.1	0.1	64	0.57	36.6	–	–

The micronutrient compositions of different millets are presented in Table 3.4. When compared to fine cereals the copper content is 11-fold higher in proso and kodo millet and 7-fold higher in pearl millet, magnesium content is 1.4–1.9-fold higher in millets and highest in sorghum, manganese is 9.3-fold higher and zinc is 2.2-fold higher in finger millet. Millet lipids are also a good source of unsaturated fatty acid content (Table 3.5). Millets vary largely with respect to starch composition, measured as the proportion of amylose and amylopectin content (16–28% and 76–84%, respectively) (Table 3.6); this varies not only among different millets but also among different cultivars of the same millets. Amylose content in sorghum is 1.2-fold higher than that of fine cereals and pearl millet is 1.1-fold higher; all the other millets have amylose and amylopectin values similar to those of fine cereals. Due to these additional nutritional benefits, millets, when compared with fine cereals, are more nutritious.

3.2 Nutritional Profiles of Millets

The commercial processing of millet grain is an important driver for economic development and yet, to popularize millet consumption, it is important to understand its nutritional and health benefits. However, data on the nutritional composition of

Table 3.4. Micronutrient contents of millets compared to fine cereals (mg/100 g). (Sources: NIN, 2007; NABARD, undated.)

Millets/cereals	Mg	Na	K	Cu	Mn	Mb	Zn	Cr	Si	Cl
Sorghum	171	7.3	131	0.46	0.78	0.039	1.6	0.008	54	44
Pearl millet	137	10.9	307	1.06	1.15	0.069	3.1	0.023	147	39
Finger millet	137	11	408	0.47	5.49	0.102	2.3	0.028	160	44
Foxtail millet	81	4.6	250	1.4	0.6	0.07	2.4	0.03	171	37
Proso millet	153	8.2	113	1.6	0.6	–	1.4	0.02	157	19
Kodo millet	147	4.6	144	1.6	1.1	–	0.7	0.02	136	11
Little millet	133	8.1	129	1	0.68	0.016	3.7	0.18	149	13
Barnyard millet	82	–	–	0.6	0.96	–	3	0.09	–	–
Rice	90	–	–	0.14	0.59	0.058	1.4	0.004	–	–
Wheat	138	17.1	284	0.68	2.29	0.051	2.7	0.012	128	47

Notes: Cr, chromium; Cl, chlorine; Cu, copper; K, potassium; Mb, Molybdenum; Mg, magnesium; Mn, manganese; Na, sodium; Si, silicon; Zn, zinc.

Table 3.5. Fatty acid composition of millet lipids compared to fine cereals (mg/100 g). (Sources: NIN, 2007; NABARD, undated.)

Millets/cereals	Palmitic	Palmoleic	Stearic	Oleic	Linoleic	Linolenic
Sorghum	14	–	2.1	31	49	2.7
Pearl millet	20.85	–	–	25.4	46	4.1
Finger millet	–	–	–	–	–	–
Foxtail millet	6.4	–	6.3	13	66.5	–
Proso millet	–	10.8	–	53.8	34.9	–
Little millet	–	–	–	–	–	–
Rice	15	–	1.9	42.5	39.1	1.1
Wheat	24.5	0.8	1	11.5	56.3	3.7

Table 3.6. Amylose and amylopectin content of millet starches. (Source: NABARD, undated.)

Millets/cereals	Amylose (%)	Amylopectin (%)
Sorghum	24	76
Pearl millet	21.1	78.9
Finger millet	16	84
Foxtail millet	17.5	82.5
Proso millet	28.2	71.8
Kodo millet	24	76
Short grain rice	12–19	88–81
Wheat	25	75

millets are scant. Here, an attempt is made to compile a nutritional profile of different millets.

3.2.1 Nutritional profile of sorghum

The sorghum grain has certain nutritional qualities that have been suggested to help prevent and control lifestyle diseases and disorders (Dayakar *et al.*, 2014). The protein content and composition of sorghum varies from cultivar to cultivar due to agronomic conditions (water availability, soil fertility, temperatures and environmental conditions during grain development) and genotype. Sorghum proteins are mainly located in the endosperm (80%), germ (16%) and pericarp (3%) (Taylor and Schussler, 1986). Kafirins or prolamins and glutelins comprise the major fractions in sorghum proteins (Warsi and Wright, 1973) and are located in the endosperm, while germ is rich in albumins and globulins. Sorghum carbohydrate is composed of starch, soluble sugar and non-starchy polysaccharides or dietary fibre and the carbohydrates can be classified into non-structural (sugars, starch and fructosans) and structural (cellulose, hemicelluloses and pectin substances).

Sorghum contains good amounts of slow digestible starch (SDS), which prolongs digestion and absorption of carbohydrates in the intestine.

This SDS is favourable for dietary management and also for dealing with metabolic disorders such as diabetes and hyperlipidemia (Asp, 1994; Wursch, 1997). Sorghum contains a high amount of dietary fibre (9.7–14.3 g) which plays the role of a bulking agent, is a binding agent for cholesterol, increases transit time and retards carbohydrate absorption (Narasinga Rao, 2003). This has a significant positive effect on preventing and managing conditions like constipation, irritable bowel syndrome and obesity.

Lipid content of the sorghum grain is minor and located in the scutellar area of the germ. Sorghum lipid has the potential to lower cholesterol. A study examining the effect of grain sorghum lipid extract (GSL) on cholesterol absorption and plasma non-HDL (non-high-density lipoprotein) cholesterol concentration in hamsters (Carr *et al.*, 2005) found that liver cholesterol ester concentration was significantly reduced in hamsters fed with GSL. GSL diet lowers non-HDL cholesterol, by inhibiting cholesterol absorption; GSL extract consists of plant sterols, which reduce cholesterol absorption efficiency significantly, and policosanol, which is known to inhibit endogenous cholesterol synthesis. These findings indicate that grain sorghum contains components that are used as food ingredients and dietary supplements for managing cholesterol levels in humans (Carr *et al.*, 2005).

Sorghum is nutritionally rich and contains large amounts of minerals and vitamins, mostly located in the aleurone layer and germ. It is a major source of B vitamins except for vitamin B12 (Gazzaz *et al.*, 1989). Yellow coloured sorghum grain is rich in beta-carotene, leutin and zeaxanthin. However, content varies with environment and genes. It is poor in vitamin C, but can be synthesized with soaking and germination. Detectable amounts of vitamins E, K and D are found in sorghum grain. Decortications and de-germination reduce water- and fat-soluble vitamins since the micronutrients are located primarily in the aleurone and germ portions of the caryopsis. Grain sorghum is a good source of potassium and an adequate source of magnesium, iron, zinc and copper, but a poor source of calcium and sodium.

Sorghum is an excellent source of polyphenols, flavonoids and condensed tannins, which are potential antioxidants. The phenolic acids that are hydroxylbenzoic acids (galliac, vanillic, syringic, p-hydroxylbenzoic and protocatechic acids) and hydroxycinnamic acids (coumaric, caffeic, ferulic and sinapinic acids) are located in the sorghum pericarp, endosperm and aleurone layer (McDonough *et al.*, 1986). Sorghum contains anthocyanins (a major class of flavonoids) that lack the hydroxyl group in the 3-position of the C-ring, the 3-deoxyanthocyanins. These anthocyanins are responsible for the stability of the grain at a higher pH and also its colour; sorghum 3-deoxyanthocyanins are apigeninidin (yellow) and luteolinidin (orange) (Awika *et al.*, 2004). The sorghum pericarp contains the highest amounts of 3-deoxyanthocyanins, and flavonoids such as luteoforol and apiforol; these flavan-4-ols play an important role in mould resistance. Other identified flavonoids in sorghum grains are flavones (apigenin and luteolin), flavanones (riodictyol and eriodictyol 5-glucoside) and flavonol (kaempferol and taxifolin). Condensed tannins (the proanthocyanidins or procyanidins) are high molecular-weight polyphenols that are also present in sorghum (Dykes and Rooney, 2006).

3.2.2 Nutritional profile of pearl millet

Pearl millet is the second most widely grown millet in Africa and India. There is a variability across genotypes in pearl millet with respect to starch (62.8–70.5%), soluble sugars (1.2–2.6%) and amylose (21.9–28.8%) (Jambunathan and Subramanian, 1988). In high-yielding varieties of Indian pearl millet, the starch (56.3–63.7%) and amylose contents (18.3–24.6%) are lower than those of the traditional pearl millet varieties (Singh and Popli, 1973). Jambunathan and Subramanian (1988) found that the predominant component of total soluble sugars (2.16–2.78%) was sucrose (66%), followed by raffinose (28%). Other sugars that were found to be in measurable amounts were fructose, stachyose and glucose. The proportion of sucrose in the total sugar content was observed to be lower in pearl millet than in sorghum.

Pearl millet, like sorghum, contains about 9–13% protein, but large variations in protein content, from 6% to 21%, have also been reported (Serna-Saldivar, *et al.*, 1994). Lysine is a limiting amino acid and its level in the grain is inversely correlated with that of protein (Deosthale *et al.*, 1971). The essential amino acid profile shows more threonine, methionine and cystine in pearl millet protein than in the proteins of sorghum and other millets. It contains higher proportions of tryptophan (Table 3.2).

There were observed differences in fatty acid contents of pearl millet (Jellum and Powell, 1971), which related to the different lipid extraction procedures and the genetic variability. The principal fatty acids in both free and bound fatty acids are oleic, linoleic and palmitic acids. The fatty acid composition differed with the phospholipid, neutral lipid and glycolipid fractions (Osagie and Kates, 1984) due to the difference in estimation and genetic variety. Neutral lipids were found to be highest in linoleic acid and lowest in palmitic acid content; phospholipids were found to be lowest in oleic acid and highest in palmitic acid and glycolipids were highest in linolenic acid. The estimated total dietary fibre content of pearl millet (20.4%) was higher than that of sorghum (14.2%), wheat (17.2%) and rice (8.3%) (Kamath and Belavady, 1980).

3.2.3 Nutritional profile of finger millet

Finger millet is widely grown as a cereal in the arid areas of Africa and Asia. It is commonly known as *ragi* in India. It remains the staple diet in Karnataka in India. Nutritionally, finger millet is a good source of nutrients (e.g. calcium, iron, magnesium) and dietary fibre. Its total carbohydrate content has been reported to be in the range of 72–79.5% (Bhatt *et al.*, 2003), of which starch is the main constituent, at 59.4–70.2% of total carbohydrates (Mittal, 2002). Finger millet exhibits polygonal rhombic shape starch granules. About 80–85% of the finger millet starch is amylopectin and the remaining 15–20% is amylose. The non-starch polysaccharide accounts for 20–30% of the total carbohydrates. It contains around 1.5% reducing sugar and 0.03% non-reducing sugar (Bhatt *et al.*, 2003). The white varieties have higher protein content than the brown varieties. Finger millet contains 44.7% essential amino acids in its protein (Mbithi-Mwikya *et al.*, 2000), which is higher (33.9% more) than those of other cereal proteins.

The insoluble dietary fibre, total dietary fibre, and soluble dietary fibre content in finger millet is 12%, 11% and 2%, respectively (Ramulu and Udaysekhara Rao, 1997). Finger millet is a good source of energy for people suffering from lifestyle diseases, as it contains the tryptophan (amino acid) that reduces appetite. It has very good antioxidant capacity and has high amounts of calcium (344 mg) and potassium (408 mg). It has a higher content of dietary fibre, minerals, and sulfur-containing amino acids compared to white rice, the current major staple in India (Shobana *et al.*, 2009). However, finger millet also contains phytates (0.48%), polyphenols and tannins (0.61%), which are trypsin inhibitory factors. These were once considered as 'anti-nutrients' due to their metal chelating and enzyme inhibition activities (Thompson, 1993). However, nowadays they are seen as nutraceuticals (providing health and medical benefits) as they contribute to antioxidant activity in millet foods. Antioxidants play an important role in health, ageing and metabolic diseases; they are also a relaxant, serving to battle fidgetiness, sleeping disorders and depression. Additionally, antioxidants assist in treating headaches and cerebral pains. The antioxidant effects of finger millet were studied on the dermal wound-healing process in diabetes-induced rats (Rajasekaran *et al.*, 2004), finding that finger millet enhanced the antioxidant status, nerve growth factor production and wound healing parameters in early diabetic rats.

3.2.4 Nutritional profile of foxtail millet

Foxtail millet (Italian millet) is grown in northern China where it is widely consumed as soup or gruel and is seen to have many nutritious and medical functions (Prashant *et al.*, 2005). Yellow-seeded cultivars of foxtail are medicinally used as an astringent, digestive, emollient and stomach relaxant. It is used in dyspepsia, poor digestion and food stagnancy in abdomen treatment (Yeung, 1985). White-seeded cultivars act as a refrigerant and are used for cholera and fever treatment while the green-seeded cultivars are a diuretic and help to strengthen virility (Duke and Ayensu, 1985).

This millet contains 12.3% crude protein, 3.3% minerals (Vithal and Machewad, 2006) and the highest amounts of crude protein among all the millets. Foxtail millet protein characterization shows that its protein concentrate is a potential functional food ingredient and the essential amino acid pattern suggests that it can be used as a supplementary protein source for most cereal foods, as it is rich in lysine (Ravindran, 1992). The general properties of foxtail millet oil and the fatty acid profile suggest that it is a good source of natural oil (Liang *et al.*, 2010).

Foxtail millet exhibits significant antioxidant activity determined by the DPPH (2,2-diphenyl-1-picrylhydrazyl) method (Devi *et al.*, 2014). It is rich in minerals (copper and iron) and releases glucose steadily and thus does not affect the body's metabolism; diabetes incidence is rare among the population that consumes foxtail millet. This millet produced a significant fall (70%) in blood glucose in diabetic rats and had no such effect in normal rats (Devi *et al.*, 2014).

3.2.5 Nutritional profile of proso millet

Proso millet is the best alternative crop for diversifying and intensifying winter wheat-based dryland production systems. The protein content in proso millet (11.6%) is similar to that of wheat, but it contains no gluten, and its protein is significantly richer in essential amino acids (leucine, isoleucine and methionine) than that of wheat (Kalinova and Moudry, 2006). Proso millet is vitamin- and mineral-rich (copper and magnesium). Proso millet protein concentrate has protective effects against D-galactosamine-induced liver injury in rats (Ito *et al.*, 2008), while Park *et al.* (2008) and Choi *et al.* (2005) concluded that proso millet protein could be a potential therapeutic intervention in type-2 diabetes. It is also reported to improve glycaemic responses and plasma levels (Park *et al.*, 2008). Devi *et al.* (2014) reviewed the nature and resultant health benefits of polyphenols and dietary fibre in proso millet. The lipid composition in proso millet flours and bran were analysed and it was found that the free lipids, hydrocarbons, sterol esters, triacylglycerols, diacylglycerols and free fatty acids were present. The predominant fatty acids present were oleic, linoleic and palmitic acids, though in the bound lipids, monogalactosyl diacylglycerols, digalactosyl diacylglycerols, phosphatidyl ethanolamine, phosphatidyl serine, and phosphatidyl choline were tentatively identified.

3.2.6 Nutritional profile of kodo millet

Kodo millet is widely distributed in damp habitats across the tropics and subtropics. It is grown primarily in India, mostly as a minor grain crop (House, 1995) but is a staple in Uttar Pradesh, Kerala and Tamil Nadu. The grain has a hard husk, making de-branning difficult. Kodo millet grain consumption is reported to be poisonous after the rain, perhaps due to fungal infection (Doggett, 1989), but clean healthy grain seems to pose no health problem.

Kodo millet has a high protein (11%), medium fat (4.2%) and high fibre content (14.3%). It is rich in B vitamins (niacin, B6 and folic acid) and minerals (calcium, iron, potassium, magnesium and zinc). Regular consumption of kodo millet is very beneficial for postmenopausal women suffering from signs of cardiovascular disease (CVD) such as high blood pressure and high cholesterol levels. The nutritional value of the kodo millet protein is slightly better than that of foxtail millet but is comparable to that of other small millets. As with other food grains, the nutritive value of kodo millet protein could be improved by supplementation with legume protein. Kodo millet flour is used to make pudding in India. In Africa it is cooked like rice. It is also a good choice for animal fodder for cattle, goats, pigs, sheep and poultry.

3.2.7 Nutritional profile of barnyard millet

Barnyard millet (Japanese barnyard millet, *ooda, oodalu, sawan, sanwa, sanwank*) is a multi-purpose crop, cultivated for food and fodder. Nutritionally, barnyard millet contains high amounts of protein, which is biologically available. It is highly digestible and an excellent source of dietary fibre. The carbohydrate present in barnyard millet is very low and is digested slowly. The major fatty acids present are linoleic acid, palmitic and oleic acid. This millet contains a high degree of amylase that is more susceptible to retrogradation (changing from soluble to insoluble form on cooking), which helps in resistant starch formation. Hence, it is recommended for CVD and diabetes mellitus patients. Barnyard millet also helps in reducing blood glucose and serum lipid levels.

3.2.8 Nutritional profile of little millet

Little millet contains about 7% protein and, like other millets, is also a good source of slow digesting carbohydrates, dietary fibre, micronutrients and phenolic compounds. Total lipid content ranges from 5.1% to 8.3% in the little, kodo and barnyard millets (Saldivar, 2003). The flavonoids are present in small quantities which act as an antioxidant and play many roles in the body's immune defence system (Devi *et al.*, 2014).

3.3 Health Benefits of Millets

The regular consumption of millets has been reported to be associated with a myriad of health benefits. The health benefits are, in part, due to the presence of several phenolic phytochemicals, a group of compounds comprising phenolic acids, flavonoid-type and tannins. These are mainly present in the seed coat. It is important therefore to consume these grains whole, as any attempt to rid millet grains of their seed coat will result in significant loss of these important phenolic compounds. These compounds are responsible for the prevention and reduction of oxidative stress, having anti-cancer, anti-diabetic and anti-hypertensive properties, and are also important in the prevention of CVD (Taylor *et al.*, 2014).

The various health benefits of millets are:

- All millets are gluten-free – recommended for the gluten-intolerant and coeliac patients.
- They have relatively low glycaemic index (GI) and low glycaemic load (GL) – and so can reduce the risk of diabetes mellitus.
- They contain invisible fat (which lowers low-density lipoprotein (LDL) and improves high-density lipoprotein (HDL) cholesterol) and also have high antioxidant activity and help to control blood sugar.
- High fibre and low calorie food – beneficial for obesity, diabetes, CVD and inflammatory bowel disease. Acts as a detoxifying agent.
- Millets are rich in photochemicals, phytic acid (lowers cholesterol) and phytate (helps in reduction of cancer risks).
- Millets form a complete protein source and when combined with legumes are ideal for a vegetarian diet.

Various studies conducted with rats and with the supplementation of the human diet have shown these health benefits of millets. The specific health benefits are discussed in more detail below.

3.3.1 Anti-diabetic properties

As early as 1957, the importance of millets in the management of type-2 diabetes and its complications was established (Ramananthan and Gopalan, 1957). Since then, other studies have confirmed their hypoglycaemic properties. Chhavi and Sarita (2012) baked bread from wheat flour substituted with 30–60% millet flour and determined sensory preference and the GI of the bread samples. Bread baked from 30% millet substituted for wheat flour was most preferred. Bread from millet had a significantly lowered area under the blood glucose response curve as compared to that from refined wheat. The GI of bread containing millet flour was 49.5 while that for refined wheat flour was 67.8.

Anju and Sarita (2010) reported a GI of 50.8 for biscuits prepared from refined wheat flour substituted with 45% millet, versus 68 for biscuits from refined wheat flour only. Geetha and Parvathi (1990) showed a significant decrease in postprandial glucose after feeding type-2 diabetes subjects for a month with breakfast foods incorporated with 20% finger millet. A lower GI was reported for the consumption of 20% millet-substituted noodles versus that from 100% refined wheat flour noodles (Shukla and Srivastava, 2011). Abdelgadir *et al.* (2005) showed that the consumption of millet *acida* (porridge), followed by wheat *gorasa* (pancakes) resulted in significant lowering of postprandial glucose and insulin responses compared to the consumption of maize *acida*. In a recent study, feeding genetically obese type-2 diabetic mice under high-fat feeding conditions with millet showed improved glycaemic responses (Park *et al.*, 2008). The reason that millets have a lower glycaemic response compared to other cereals is very poorly understood but a better understanding would be beneficial in order to maintain the hypoglycaemic properties during processing.

In 2010, the GIs of sorghum-based foods were assessed by the National Institute of Nutrition (NIN) in Hyderabad, India, in collaboration with the Indian Institute of Millets Research (IIMR), Hyderabad, under the NAIP. The results revealed that sorghum-based foods have a low GI and reduce the postprandial blood glucose level and glycosylated haemoglobin. Another study also suggests that blood glucose levels of non-obese patients with non-insulin-dependent diabetes mellitus (NIDDM) was considerably reduced in those who consumed sorghum bran *papadi* (Shinde, 2004).

Dietary fibre in millet has also been mentioned as a contributor to its hypoglycaemic effects and, at about 3% and 20% soluble and insoluble fibre content, respectively, should not be overlooked. Fibre reduces the glycaemic response of foods by increasing their viscosity, especially the soluble fibre. A considerable increase in soluble fibre content of finger millet is observed after decortication, though total dietary fibre decreases. This increase in the soluble fibre content has special nutritional significance due to its physiological advantages in terms of hypoglycaemic and hypocholesterolemic characteristics. Some have attributed the low glycaemic property of millets to their rigid starch granule architecture (Mohan *et al.*, 2005), which slows down the rate at which the granules are hydrolyzed by starch digestive enzymes. The effects of the protein encapsulated granule matrix have been suggested as part of the reason for the hypoglycaemic property of millet (Annor, 2013); some protein fractions glue the protein bodies into a matrix surrounding starch granules, which then acts as a barrier to amylases. The removal of protein from millet flour clearly results in a significant increase in starch hydrolysis. Whole grain sorghum-based products have a lower release of glucose compared to those prepared from wheat (Lakshmi and Vimala, 1996). High tannin sorghums are slow to digest; hence they may be preferred since they contribute to a longer period of satiety (fullness in the stomach) as compared to other cereals (Kaur *et al.*, 2014).

3.3.2 Reduction of oxidative stress

Free radicals, which are by-products of cellular redox reactions, may be harmful to cell structures when produced at high concentrations in the human body. They are usually in the form of reactive oxygen or nitrogen species. Maintaining the balance between the formation and removal of these free radicals is crucial in preventing the development of chronic and degenerative diseases such as cancer, CVD and diabetes. When the rate of formation of free radicals is higher than the rate of removal from the body, it results in oxidative stress. The radical-scavenging properties of the phenolic compounds (Wursch, 1997) present in millet grains help in the reduction/prevention of the formation of free radicals in the human body, thus keeping them to a minimum.

3.3.3 Anti-cancer properties

The anti-cancer properties of millets are demonstrated by their ability to reduce oxidative stress, which is related to their free-radical scavenging property. It has been demonstrated that millet extracts have anti-prolific effects on the cancer cell line, inhibit DNA damage and induce the production of phase-2 detoxifying enzymes (Chen *et al.*, 1993). Regular and consistent consumption of sorghum is correlated with low incidence of oesophageal cancer in various parts of the world (Chen *et al.*, 1993). The phenol content of sorghum is correlated with its antioxidant activity (Dicko *et al.*, 2005).

Anti-carcinogenic properties of sorghum have been well documented. *In vivo* and *in vitro* studies have shown consumption of sorghum may help towards preventing cancer. The polyphenols and tannins present in sorghum have anti-mutagenic and anti-carcinogenic properties (Grimmer *et al.*, 1992) and can act against human melanoma cells, as well as positive melanogenic activity (Gómez-Cordovés *et al.*, 2001). Parbhoo *et al.* (1995) reported in rats that liver procyanidin extracts may induce cytochrome P-450, a protein that is capable of converting certain promutagens to mutagenic derivatives. Epidemiological data from Sachxi Province, China, and other parts of the world have shown that the incidence of oesophageal cancer was low in areas with sorghum consumption (Van Rensburg, 1981). In each country, the author studied 21 communities over a period of 6 years and found that consumption of sorghum showed lower mortality from oesophageal cancer than wheat and maize. These results suggest that anti-carcinogenic compounds present in sorghum lower the risk of oesophageal cancer.

3.3.4 Anti-hypertensive and cardiovascular disease prevention

Millet extracts prevent the oxidation of LDLs ('bad' choleresterols). Oxidized LDLs are believed to result in the formation of fatty streaks after being taken up by macrophages, leading to the development of atherosclerosis. The prevention of LDL oxidation is therefore important in the prevention of CVD and hypertension. The prevention of LDL oxidation by millets is attributed to their antioxidant properties, presumably resulting from phytochemicals present mainly in the seed coat. Low-tannin sorghum fed to guinea pigs was more beneficial in lowering cholesterol than were rolled oats, wheat or pearl millet (Awika *et al.*, 2004).

3.3.5 Coeliac disease

Coeliac disease is a common autoimmune disorder whereby predisposed people suffer an adverse reaction to gluten – proteins that are found in wheat, rye and other cereal grains. The abnormal response to gluten by the immune system leads to abdominal pain, diarrhoea and weight loss. Gluten-free grains, such as sorghum and other millets, are a healthy alternative to wheat etc. for those with coeliac disease. Millet products have been reported to be beneficial and will not modify the anti-transglutaminase antibody levels in coeliac patients even after prolonged consumption (Carolina *et al.*, 2007).

3.3.6 Obesity

Obesity is no longer a problem of only the wealthiest of countries; it is now a world health epidemic and many countries, including India, are suffering at an alarming rate of increase. Obesity is positively associated with several chronic diseases, including diabetes and CVD. Empirical evidence suggests that intake of foods containing high levels of dietary fibre decreases the incidence of obesity (Burkitt and Trowell, 1975; Alfieri *et al.*, 1995). Foods rich in dietary fibre improve the large bowel function and slow the digestion and absorption process, thereby reducing the risk of chronic diseases (Ali *et al.*, 1982; Schneeman and Tietyen, 1994). Sorghum is rich in dietary fibre and has unique chemical and physical characteristics (bulk to the diet, viscosity, water holding and absorption capacity), which determine the subsequent physiological behaviour. It increases satiety thereby reducing the risk of overeating.

3.3.7 Coronary heart disease

Empirical evidence (Slavin, 2003, Slavin and Slavin, 2004) suggests that regular consumption of whole millet grains reduces the risk of CVD (Anderson and Hanna, 1999). A long-term prospective study by Heidemann *et al.* (2008) showed that regular consumption of a balanced diet that includes whole grains, vegetables, fruits, fish and poultry reduces the risk of CVD and total mortality. In 2004, Jensen *et al.* (2004) analysed the diet records of 27,000 men aged 40–75 years for a period of 14 years and found that intake of whole grains at 40 g/day reduced the risk of coronary heart disease by 20%.

3.3.8 Other benefits

Treatment of stomach ulcers

Pearl millet is recommended for curing stomach ulcers. The most common cause for stomach ulcers is excess acidity in the stomach after food intake. Pearl millet is one of the very few foods that turns the stomach alkaline and prevents the formation of stomach ulcers or reduces the effect of ulcers. No incidents of duodenal ulcers are found among regular finger millet consumers (The Research Pedia, 2015).

Prevention of gallstones and allergic reactions

The high insoluble fibre content in pearl millet is known to reduce the risk of gallstone occurrence, reducing the production of excessive bile that often aggravates the condition. Pearl millet has a very low probability of causing allergic reactions; it can be safely included in the diets of infants, lactating mothers, the elderly and convalescents (The Research Pedia, 2015).

Finger millet for anaemia control

Finger millet is a very good source of iron. Its consumption helps in conditions of anaemia (Tatala *et al.*, 2007). If consumed regularly it also helps in keeping malnutrition, degenerative diseases and premature aging at bay. Green finger millet is recommended for blood pressure conditions, liver disorders, asthma and heart weakness. It is also recommended to lactating mothers with inadequate milk production (Arya, 2014).

3.4 Nutritional Evaluation of Food Products from Millets

Although the specific nutritional features of millet grains have been adequately documented, there is still much to be done to widen the scope of their use as food. With a view to promoting and initiating the commercialization of millet – specifically sorghum – as a health food, processing interventions have been made through the NAIP (discussed in Chapter 5). The pilot initiative has seen the development of novel sorghum ready-to-eat/-cook (RTE/RTC) products, such as multigrain *atta*, flakes, *rawa* and pasta, under the brand name *eatrite*. Nutritional evaluation and certification of these products has been carried out at NIN, under the Indian Council of Medical Research (ICMR), on the following areas:

1. Nutritional composition of sorghum processed foods.
2. Evaluation of organoleptic properties of sorghum-based Indian traditional breakfast and snack recipes.
3. GI and GL of sorghum foods.
4. Effect of sorghum diet on glycosylated haemoglobin and lipid profile in people suffering from diabetes.
5. Effect of sorghum diet on nutritional status of school children.
6. Amino acid profile of sorghum processed foods.
7. Determination of protein efficiency of sorghum.

3.4.1 Nutritional composition of sorghum products

Sorghum products such as flour, multigrain flour, coarse, medium and fine semolina, flakes, pasta, vermicelli and biscuits were evaluated for their nutritional composition. Nutrients such as energy, carbohydrates, protein, fat, fibre, ash, moisture, total carotenoids, beta carotene, B vitamins, vitamin C, calcium, iron, zinc, copper, manganese, magnesium, phosphorus and molybdenum were estimated with standard methods (Table 3.7). The carbohydrate content observations are the total available carbohydrates after deducting the total fibre present (Gopalan *et al.*, 2011). The methods used for the detection of various nutrients are more sensitive than those previously undertaken. Results show that all the sorghum-based foods are rich in energy (330–481 kcal/100 g). The available

Table 3.7. Proximate composition of sorghum processed foods (per 100 g). (Source: Dayakar *et al.*, 2014.)

Name of product	Moisture (g)	Protein (g)	Fat (g)	Total DF (g)	Insoluble DF (g)	Soluble DF (g)	Carbohydrates (g)	Energy (kcal)
Sorghum flour	13.8	6.2	2.8	9.69	8.1	1.59	76.15	355
Sorghum/ soya blend	7.89	11.92	2.62	12.71	9.77	2.94	63.22	330
Sorghum *rawa*	8.97	7.15	1.2	9.23	7.92	1.31	77.74	350
Sorghum pasta	11.47	8.39	1.38	5.56	4.82	0.74	76.21	355
Sorghum flakes	13.80	5.09	2.40	5.97	5.43	0.54	74.9	342
Sorghum biscuits	5.67	4.59	24.50	5.27	3.54	1.73	60.29	481

Note: DF, dietary fibre.

carbohydrate content was found to be highest in sorghum *rawa*, followed by pasta, flour, flakes, multigrain flour and biscuits. Protein, total dietary fibre, soluble fibre and insoluble fibre content were high in the sorghum soya blend, whereas fat content was found to be high in sorghum biscuits due to the incorporation of trans-free fat in the recipe formulation. The sorghum soya blend was found to be relatively low in availability of carbohydrates when compared with sorghum flour.

All products were further analysed for their mineral content (Table 3.8). Sorghum flakes contained the highest amount of calcium, iron, zinc and copper, multigrain flour contained the highest amount of manganese, while pure sorghum flour contained the highest amount of phosphorus and magnesium. Calcium, zinc and iron are deficient in the majority of the Indian population, a problem that could be curbed by consumption of mineral-rich millets. As the bioavailability of minerals in whole millets is poor, conversion into processed foods would be beneficial.

The vitamin composition of the products is given in Table 3.9. Thiamine content is highest in sorghum flour, followed by *rawa*, sorghum soya blend, biscuits and flakes. Niacin content is also highest in sorghum flour, followed by sorghum soya blend, biscuits, *rawa*, flakes and pasta. Riboflavin content is richest in sorghum biscuits, followed by pasta, *rawa*, flour, sorghum soya blend and flakes.

For all the sorghum-based products, the amino acid lysine is limited. People subsisting purely on sorghum foods carry the risk of lysine deficiency and a resulting reduced absorption of niacin, which can lead to the disease pellagra. Sorghum as a staple consumed at around 50% of whole cereal intake (McKevith, 2004) is safe and lysine deficiency can be avoided by combining sorghum consumption with lentils and legumes.

3.4.2 Glycaemic index and glycaemic load of sorghum products

Sorghum-based foods such as multigrain flour, coarse semolina, fine semolina, flakes, pasta and biscuits were evaluated for their GI and GL and were compared with wheat-/rice-based foods using standardized methodology. For this study, ten non-diabetic healthy volunteers, 20–40 years old, were recruited for evaluating each food item. The study was conducted according to the guidelines laid down in the declaration of Helsinki, and all procedures involving human subjects were also approved by the Institutional

Table 3.8. Mineral composition of sorghum processed products (mg/100 g). (Source: Dayakar *et al.*, 2014.)

Name of product	Calcium	Magnesium	Copper	Manganese	Iron	Zinc	Phosphorus
Sorghum flour	10.03	171	0.46	0.78	8.4	1.3	222
Sorghum/ soya blend	25.41	62.90	0.22	1.44	3.03	1.06	85.14
Sorghum *rawa*	5.75	86.02	21.11	0.91	5.1	1.3	150.0
Sorghum pasta	64.51	67.48	1.407	0.66	64.51	5.7	110.0
Sorghum flakes	93.15	80.51	27.7	0.53	87.78	8.78	110.0
Sorghum biscuits	68.80	56.10	0.25	0.57	2.2	BDL	107.7

Note: BDL, below detectable level.

Table 3.9. Vitamin composition of sorghum food products (mg/100 g). (Source: Dayakar *et al.*, 2014.)

Name of product	Thiamine	Niacin	Riboflavin	Total carotenoids	B-carotene	Vitamin C
Sorghum flour	2.31	3.1	0.38	ND	ND	ND
Sorghum/soya blend	0.45	1.93	0.15	ND	ND	ND
Sorghum *rawa*	0.615	1.58	1.093	ND	ND	ND
Sorghum pasta	BDL	1.15	1.28	ND	ND	ND
Sorghum flakes	0.07	1.28	0.02	ND	ND	ND
Sorghum biscuits	0.23	1.9	2.26	ND	ND	ND

Note: BDL, below detectable level; ND, not detectable.

Ethics Committee of the National Institute of Nutrition, Hyderabad. Raw materials used for the preparation of test foods were prepared at IIMR (formerly DSR), Hyderabad and recipes were prepared in the Metabolic Kitchen, NIN, Hyderabad. Fifty grams of available carbohydrate of a reference food and test food (sorghum-, wheat- or rice-based recipes) were given in a random order after an 8–10 h overnight fast (at least one week apart was considered between the tests), and postprandial blood samples were taken from finger pricks at 15, 30, 45, 60, 90 and 120 min after feeding to determine the GI and GL. Calculation of the GI of test food was measured using the positive incremental area under the curve (+iAUC).

$$\mathrm{GI}\,(\%) = \frac{\left(\begin{array}{c}\text{Area under 2 h glucose}\\ \text{response curve of test food}\end{array}\right)}{\left(\begin{array}{c}\text{Area under 2 h glucose}\\ \text{response curve of}\\ \text{reference food}\end{array}\right)} \times 100$$

The protocol used to measure GI was adapted as described by Wolever *et al.* (1991) and is in line with the procedure recommended by the FAO (1998). GL of a specific serving of each food was calculated by the formula

$$\mathrm{GL}\,(\%) = \frac{(\text{GI} \times \text{serving size of food} \times 50)}{\left(\begin{array}{c}100 \times \text{serving size of food}\\ \text{containing 50 g of available}\\ \text{carbhohydrate}\end{array}\right)}$$

Differences between the test foods was assessed using a paired *t*-test. Among the sorghum-based foods, the GI of coarse *rawa upma* ($P<0.05$), fine *rawa upma* ($P<0.05$), pasta ($P<0.01$) and *poha* ($P<0.01$) were significantly lower than their respective control food (wheat/rice) and no significant difference was found between the controls and multigrain *roti* or biscuits (Rajendra Prasad *et al.*, 2014). The GL of all sorghum-based foods was lower than that of wheat-/rice-based foods (Table 3.10). This information is useful in the labelling of sorghum-based products for commercialization, particularly to indicate their suitability as anti-diabetic foods.

3.4.3 Effect of sorghum diet on glycosylated haemoglobin and lipid profile in diabetics

Effect of sorghum diet on glycosylated haemoglobin and the lipid profile in diabetic patients was assessed with supplementation of sorghum diet for a period of 60 d. Volunteers ($n=150$) in the age range of 30–60 years suffering from type-2 diabetes were recruited for the study. The control group was given a regular rice diet and the experimental group was given a 50/50% sorghum/rice diet for a period of 2 months. The anthropometric indices such as height, weight and BMI were measured and the biochemical indices such as glycosylated haemoglobin, fasting blood glucose, haemoglobin, insulin, creatinine, triglycerides, and LDL and HDL cholesterol were assessed before and after supplementation of the sorghum diet, using standard methods. The biochemical indices showed that there was a significant decrease in glycosylated haemoglobin and fasting glucose levels (Table 3.11).

Table 3.10. GI (glycaemic index) and GL (glycaemic load) of test foods and + iAUC. (Source: Rajendra Prasad *et al.*, 2014.)

	+ iAUC (mg/dl)		GI		GL	
Foods	Sorghum	Wheat	Sorghum	Wheat	Sorghum	Wheat
Multigrain *roti*	146±8.16	138±6.28	68±8.63	64±9.24	35±6.2*	32±4.62*
Coarse semolina *upma*	114±1.93*	125±4.65*	53±2.84*	58±6.85*	23±1.24**	27±3.21**
Fine semolina *Upma*	119±7.04**	144±8.77**	56±9.83**	67±10.80**	26±4.87**	46±7.3**
Flakes *poha*	96±4.28**	158±8.17**	45±5.27**	74±4.87**	50±5.85**	75±3.5**
Pasta	100±3.58**	154±3.71**	46±6.47**	72±6.51	60±2.8**	108±6.2**
Biscuits	115± 4.39	122±4.42	54±6.30	57±11.40	23±8.4**	31±11.51**

Results are mean±SE; *, significant at 5% level; **, significant at 1% level.

Table 3.11. Mean glycosylated haemoglobin and lipid profile of diabetic patients before and after supplementation with sorghum. (Source: Rajendra Prasad *et al.*, 2014.)

Parameter	Before	After
glycosylated haemoglobin (g %) (normal, <7.0)	7.9±2.0*	7.3±1.5*
Fasting glucose (mg %)	161.3±50.35*	150.4±54.16*
Insulin (µ/ml)	24.9±16.54	24.6±18.2
Creatinine (mg %)	1.02±0.21	1.0±0.19
Cholesterol (mg %)	189±47.15	190±33.78
HDL cholesterol (mg %)	54.3±13.38	54.9±16.01
Triglycerides (mg %)	139.2±73.75	138.0±74.92

Results are mean±SE; *, significant at 5% level.

3.4.4 Amino acid profile of sorghum products

Sorghum multigrain flour has higher availability of total proteins and individual amino acids, especially methionine – an essential amino acid – which is found at twice the level of that in sorghum flour. The total amino acids present across all sorghum foods were almost similar, although they were found to be higher in sorghum flour than other foods; and the amino acid score was found to be high in multigrain flour (Table 3.13). This could be due to the inclusion of soybean in the flour formulation. In all sorghum-based products the amino acid lysine is the limiting factor – which is the case for most cereal-based products – and is thus the cause of poor quality protein.

3.4.5 Effect of sorghum diet on nutritional status of school children

To assess the effect of a sorghum diet on the nutritional status of school children (Rajendra Prasad *et al.*, 2015), girls (n =137) and boys (n =125) aged 9–12 years, residing at the social welfare hostel *Wanaparthy*, Mahabubnagar district, were recruited. The control group was given a regular rice diet and the experimental group was given a 50/50% sorghum/rice diet for 8 months. Anthropometric indices such as height, weight and BMI were measured, and biochemical indices were assessed before and after supplementation with the sorghum diet. There was a significant increase in height, and folic acid, retinol binding protein (RBP), iron, haemoglobin, ferritin and vitamin B12 levels with the sorghum diet. There was a significant decrease in total protein and albumin in both the experimental and control group of girls (Table 3.12). There was a significant decrease in height in the experimental group of boys compared to the control group. There was a significant increase in haemoglobin, ferritin, albumin, calcium, iron and folic acid in the sorghum diet compared to the rice-only diet. No significant difference was observed in total protein levels. Significant increase of RBP is observed, which is found to be higher in the control group than in the experimental group of boys. This study suggests that a sorghum-rich diet will improve growth as well as levels of serum ferritin, folic acid, haemoglobin, calcium, iron and retinol-binding protein among school children (Rajendra Prasad *et al.*, 2015).

3.4.6 Determination of protein efficiency of sorghum

A study was conducted by NIN under the NAIP pilot study in which six weanling male NIN-wistar rats of similar weight were taken in three groups to assess the protein efficiency of sorghum (for which Animal Ethics committee approval was obtained, P20/7-2011/GBR). One group was fed with control (casein starch diet), the second group received a sorghum diet and the third group a protein-free diet for 2 weeks. The total amount of dietary intake and faeces excreted were measured during the last 4 days of the experiment. The faeces were lyophilized, weighed and ground to measure the protein digestibility of sorghum. The following formula was used to calculate the true protein digestibility:

$$\text{True protein digestibility} = \frac{PI-(FP-MFP)}{PI} \times 100$$

Table 3.12. Nutritional status of school children before and after sorghum supplementation: (A) girls; (B) boys. (Source: Rajendra Prasad *et al.*, 2015.)

(A)

	With sorghum	Control	With sorghum	Control
	Before intervention		After intervention	
Height (cm)	129.8±7.43**	135.8±7.84**	135.3±9.00**	136.8±9.34 **
Weight (kg)	25.4±4.33**	28.4±4.45 **	27.4±4.26**	27.9±5.77 **
BMI (kg/m²)	15.2±1.63NS	15.5±1.64NS	15.4±1.48NS	15.8±1.64NS
Haemoglobin (g/dl)	10.9±1.88**	11.9±1.64**	12.4±1.57**	11.1±1.51**
Total protein (g/dl)	9.30±0.61**	11.0±1.25**	8.20±0.52**	9.0±0.81**
Ferritin (ng/ml)	26.9±13.19**	43.8±118.6**	39.6±26.17NS	34.5±12.23NS
Folic acid (mg/ml)	4.80±1.71**	4.50±1.57**	5.10±1.48NS	5.20±1.53NS
Iron (μg/dl)	54.8±26.91**	53.7±21.2**	68.9±41.09*	58.2±22.8*
Retinol-binding protein	89.6±26.28**	61.6±15.46**	98.9±26.0**	72.8±15.02**
Vitamin B12 (mg/ml)	295.0±44.93 NS	216.5±35.60*	375.0±176.70*	261.9±163.93 NS
Serum albumin (g/dl)	46.2±4.71**	44.4±4.21**	42.0±3.27NS	42.9±3.38NS

Results are mean±SE; NS, not significant; *, significant at 5% level; **, significant at 1% level.

(B)

	With sorghum	Control	With sorghum	Control
	Before intervention		After intervention	
Height (cm)	132.4±8.48**	134.2.3±8.67*	128.5±7.80**	133.3±7.90*
Weight (kg)	27.5±5.26**	28.6±5.81**	25.3±4.32**	27.9±4.71**
BMI (kg/m²)	14.9±1.330**	14.9±1.57**	14.7±2.02**	15.6±2.17**
Haemoglobin (g/dl)	11.8±1.650**	12.2±1.31**	12.6±1.15**	11.4±1.73**
Total protein (g/dl)	8.80±0.830**	9.30± 1.40**	8.80±0.61**	8.60±0.58**
Ferritin (ng/ml)	40.9±23.45**	65.3±15.43**	79.1±36.99**	44.3±14.02**
Folic acid (mg/ml)	4.30±1.450**	5.20±4.19**	5.20±2.52NS	5.20±4.19NS
Iron (μg/dl)	53.2±21.25**	74.2±36.79**	58.2±25.60NS	61.70±22.9NS
Retinol-binding protein	63.9±10.92**	33.6±17.10**	71.1±13.93NS	77.7±14.04NS
Vitamin B12 (mg/ml)	295.0±44.93 NS	216.5±35.60*	375.0±176.70*	261.9±163.93 NS
Serum albumin (g/dl)	47.3±4.68**	41.7±3.46**	44.4±3.58NS	44.9±3.19NS

Results are mean±SE; NS, not significant; *, significant at 5% level; **, significant at 1% level.

where PI = protein intake, FP = faecal protein and MFP = metabolic faecal protein.

True protein digestibility of casein starch diet was found to be 96.7 and true digestibility of sorghum was found to be 89.9. The amino acid score of casein was found to be 93 and the amino acid score of sorghum flour was found to be 42.74 (Table 3.13), suggesting that protein digestibility is higher in casein. However, the protein digestibility-corrected amino acid score (PDCAAS) of the casein starch diet and the sorghum diet was 89.9% and 38.4%, respectively, or 0.90 and 0.38 in ratio, respectively. Thus, the study concluded that the PDCAAS value was significantly greater in the rats on the casein diet compared with those on the sorghum diet (Table 3.14).

The amino acid score is given based on the ratio of the amount of the first-limiting dietary indispensable amino acid in the protein source to the amino acid requirement of the 1–2-year-old child corrected for protein digestibility based on true faecal nitrogen digestibility and using the growing rat as a model for the adult human. Amino acid scores thus found are corrected for true digestibility of individual amino acids. When the limiting amino acids are taken as lysine, threonine and tryptophan, the AAS are reported to be 0.38, 1.15 and 1.49, respectively (Table 3.14), for 1–2 years of age.

Table 3.13. Total amino acids, amino acid score, total essential amino acids and limiting amino acids of sorghum foods (as percentages). (Source: Dayakar *et al.*, 2014.)

Nutrient	Sorghum flour	Multigrain flour	Sorghum *rawa*	Sorghum pasta	Sorghum flakes	Sorghum biscuits
Total amino acid	99.65	97.4	99.81	99.42	99.6	98.44
Total essential amino acids	40.12	36.38	40.43	36.83	39.84	41.35
Amino acid score	42.74	61.71	32.64	35.09	34.97	41.75
Limiting amino acid	Lysine	Lysine	Lysine	Lysine	Lysine	Lysine

Table 3.14. Amino acid score (AAS) and protein digestibility-corrected amino acid score (PDCAAS) values for sorghum, calculated using 2002 WHO/FAO/UNU Expert Consultation recommendation. (Source: Dayakar *et al.*, 2014.)

Age group preference	Lysine scoring pattern	AAS	Threonine scoring pattern	AAS	Tryptophan scoring pattern	AAS	PDCAAS (based on lysine)
0–5 years	57	0.35	31	1	8.5	1.29	0.26
1–2 years	52	0.38	27	1.15	7.4	1.49	0.28
4–18 years	48	0.42	25	1.24	6.5	1.69	0.31
>18 years	45	0.44	23	1.35	6	1.83	0.33

3.5 Impact of Sorghum Grain Processing on Nutritional Composition

3.5.1 Effect of grain processing on nutrient composition of sorghum products

Sorghum grain used for human consumption is mainly subjected to primary and secondary processing. The grain consists of the pericarp, corneous endosperm, floury endosperm and the germ unit of the embryo. The pericarp consists of stone cells and contributes to dietary fibre, sorghum wax and some anti-nutritional factors like infectious agents, external contaminations and microscopic artefacts. During primary processing the grain will be subjected to cleaning, purifying and refining to remove the undesirable and anti-nutritional factors. The primary processing involves dehulling (pearling) and milling the grain into flour and semolina (fine and medium semolina). Secondary processing involves use of the primary processed raw material to process RTC and RTE products for processing such as flaking, extrusion, popping, baking and blending with other cereals.

Under NAIP, the effects of these processing techniques on the nutritional quality of the processed end products (Dayakar *et al.*, 2014) were examined. The nutrient composition of the sorghum whole and dehulled grain is given in Table 3.15. There was a decrease in the macronutrients (protein and fat) and micronutrients (iron, calcium, chromium, zinc and riboflavin) on dehulling. Similarly, the nutrient composition of the primary processed products is given in Table 3.16 and the effect of milling on the end product is tabulated. In general, a decrease of both macro- and micronutrients was observed, except in carbohydrate content, which showed a marginal increase. The decrease in nutrients is minimal during milling. There was also a decrease in fat composition, a quality that may be useful especially for those suffering from obesity.

The primary processed products are subjected to secondary processing for preparation of final products such as flakes, pasta, vermicelli, pops, biscuits, multigrain flour and multigrain *roti*/*dosa*. A decline in macronutrients (e.g. protein) and micronutrients (e.g. iron, zinc and riboflavin) during processing was observed, but this decline varied with product and was found to be variably distributed when compared to the rest of

Table 3.15. Nutrient composition of whole sorghum grain and dehulled sorghum grain. (Source: Dayakar *et al.*, 2014.)

Parameters	Whole grain	Dehulled grain
Moisture (%)	11.9	10.00
Ash (%)	1.6	1.7
Protein (%)	10.4	6.56
Fat (%)	1.9	1.1
Carbohydrates (%)	72.6	76.15
Iron (mg/100 g)	4.1	2.9
Calcium (mg/100 g)	25	12.09
Zinc (mg/100 g)	1.6	1.1
Riboflavin (mg/100 g)	0.13	0.8
Energy (kcal/100 g)	349.0	340

Table 3.16. Chemical, mineral and vitamin composition during the milling process of sorghum. (Source: Dayakar *et al.*, 2014.)

Parameters	Whole grain	Flour	Fine semolina (*idli rawa*)	Medium semolina (*upma rawa*)
Moisture (%)	11.9	13.8	10.17	08.97
Ash (%)	1.6	1.6	00.73	02.03
Protein (%)	10.4	06.2	06.65	07.15
Fat (%)	1.9	02.8	01.70	01.20
Carbohydrates (%)	72.6	76.15	77.75	77.74
Iron (mg/100 g)	4.1	8.4	10.57	5.102
Calcium (mg/100 g)	25	10.03	7.552	5.750
Chromium (mg/100 g)	0.008	0.008	01.27	1.476
Zinc (mg/100 g)	1.6	1.3	1.209	1.382
Riboflavin (mg/100 g)	0.13	0.38	0.11	1.09
Energy (kcal/100 g)	349	355	350	350

the nutrients. The change in the carbohydrate composition of processed products when compared to the raw material used is not very significant – this indicates that the calorific value was intact in the processed ready-to-eat foods as well. Products such as biscuits showed a large increase in fat content due to incorporation of fat in the recipe formulation (Table 3.17).

Multigrain flour was prepared by blending five different flours: sorghum, wheat, finger millet, black gram *dal* and fenugreek, and then subjected to a nutrient analysis. Significant decreases were observed in zinc, iron and riboflavin while most of the other nutrients remained unaltered. Much of the nutrient loss in sorghum flour is compensated for by the blending of different grains. Multigrain flour is one of the secondary processed products that can be recommended for all age groups. The nutrient composition of the multigrain flour by blending with three and five different grains is given in Table 3.18.

From the study it is observed that nutritional losses occur during processing, which is unavoidable. However, attention should be given to select product and processing of the semi-finished products to avoid further loss in nutrition and calorific values of the end product. Various genotypes of millets may differ in their original composition and these should be subjected to primary and secondary processing to identify those with minimal loss.

3.5.2 Assessment of sorghum semi-processed products based on nutrient superiority for marketability

The marketability of processed sorghum products as health and convenience foods primarily

Table 3.17. Macro- and micronutrient changes during sorghum processing. (Source: Dayakar *et al.*, 2014.)

Parameters	Grain	Flour	Fine *rawa*	Medium *rawa*	Flakes	Vermicelli	Pasta	Pops	Biscuits
Moisture (%)	11.9	13.8	10.17	08.97	13.80	8.430	11.47	5.870	5.670
Ash (%)	1.6	1.6	00.73	02.03	0.630	0.770	0.770	0.630	2.000
Protein (%)	10.4	06.2	06.65	07.15	05.09	8.39	8.39	5.040	4.590
Fat (%)	1.9	02.8	01.70	01.20	02.40	1.38	1.38	2.600	24.50
Carbohydrates (%)	72.6	76.15	77.75	77.74	74.99	76.21	76.21	83.06	60.29
Iron (mg/100 g)	4.1	8.4	10.57	5.102	87.78	64.51	64.51	2.402	2.254
Calcium (mg/100 g)	25	10.03	7.552	5.750	93.15	54.51	64.51	10.26	68.8
Chromium (mg/100 g)	0.008	0.008	01.27	1.476	00.90	0.200	0.215	1.400	0.510
Zinc (mg/100 g)	1.6	1.3	1.209	1.382	08.78	7.49	5.740	4.510	BDL
Magnesium (mg/100 g)	171	171	76.48	86.02	80.51	67.48	67.48	86.77	56.10
Riboflavin (mg/100 g)	0.13	0.38	0.11	01.09	0.020	1.28	1.28	0.150	2.260
Energy (kcal/100 g)	349	355	350	350	342	355	355	376	481

Note: BDL, below detectable level.

Table 3.18. Nutrient composition of sorghum flour, and after blending sorghum flour with other cereal grains. (Source: Dayakar *et al.*, 2014.)

Parameters	Sorghum flour	Multigrain flour (3-grain)	Multigrain flour (5-grain)
Moisture (%)	13.8	10.17	8.570
Ash (%)	1.600	1.530	1.500
Protein (%)	6.2	7.100	5.960
Fat (%)	2.8	2.400	2.600
Carbohydrates (%)	76.15	75.61	77.42
Iron (mg/100 g)	8.4	4.588	2.984
Calcium (mg/100 g)	10.03	10.61	15.94
Chromium (mg/100 g)	0.008	0.1460	0.300
Zinc (mg/100 g)	1.3	0.550	0.810
Magnesium (mg/100 g)	171	76.25	–
Riboflavin (mg/100 g)	0.38	0.140	BDL
Energy (kcal/100 g)	355	345	339

Note: BDL, below detectable level.

depends on their relative nutritional composition compared to existing on-shelf products (but also depends on shelf-life, price etc.). In order to assess processed sorghum's market worthiness we make an attempt to compare its nutrient composition with that of processed wheat and rice products.

Sorghum grain

Of the various nutritional parameters tested in sorghum grain as compared to wheat grain, macronutrients such as carbohydrates, fat and energy, folic acid and magnesium were nutritionally superior (Table 3.19). Moisture, thiamine, iron, copper, calcium, protein and zinc

Table 3.19. Nutrient analysis of sorghum grain as compared to wheat grain. (Source: Dayakar *et al.*, 2014.)

Parameters	Grain		Superiority/inferiority of sorghum compared to wheat
	Sorghum	Wheat	
Moisture (%)	11.9	12.8	Low
Ash (%)	1.6	1.53	Similar
Protein (%)	10.4	11.38	Low
Fat (%)	1.9	1.5	High
Carbohydrates (%)	72.6	71.2	High
Iron (mg/100 g)	4.1	5.3	Low
Calcium (mg/100 g)	25	41	Low
Chromium (mg/100 g)	0.008	0.012	Similar
Zinc (mg/100 g)	1.6	2.7	Low
Riboflavin (mg/100 g)	0.13	0.17	Similar
Thiamine (mg/100 g)	0.37	0.45	Low
Copper (mg/100 g)	0.46	0.68	Low
Magnesium (mg/100 g)	171	138	High
Folic acid (mg/100 g)	20.0	BDL	High
Energy (kcal/100 g)	349	346	High

Note: BDL, below detectable level.

were found to be relatively low in sorghum grain. The values of riboflavin, chromium and ash were found to be similar to those of wheat grains.

Sorghum flour

Sorghum flour had relatively high content of moisture, fat, iron, thiamine, riboflavin, folic acid, carbohydrate, ash, magnesium and energy compared to wheat flour (Table 3.20). The chromium content was found to be similar to that of wheat flour, while that of protein, zinc, calcium and copper were relatively low compared to wheat flour.

Sorghum semolina

Sorghum semolina contained relatively high values of ash, carbohydrates, chromium, iron, magnesium, thiamine, riboflavin, folic acid, moisture, fat, copper, zinc and energy in comparison to wheat semolina (Table 3.21), while the protein and calcium values were relatively low.

Sorghum flakes

Sorghum flakes had relatively high moisture, fat, calcium, folic acid, chromium, iron, copper and zinc values in comparison to rice flakes (Table 3.22), while the carbohydrate, ash, protein, magnesium, thiamine, riboflavin and energy values were relatively low.

Sorghum pasta

Sorghum pasta had relatively high fat, moisture, zinc, calcium, magnesium, copper, protein, iron and riboflavin values in comparison to refined wheat flour pasta (Table 3.23). The thiamine, folic acid and ash values were similar, while the carbohydrate content was relatively low in comparison to wheat pasta.

Sorghum vermicelli

Sorghum vermicelli contained relatively high values of iron, zinc, magnesium, fat, calcium, riboflavin, energy and thiamine in comparison to those of refined wheat flour vermicelli (Table 3.24). The ash and protein contents

Table 3.20. Nutrient analysis of sorghum flour as compared to wheat flour. (Source: Dayakar *et al.*, 2014.)

Parameters	Sorghum flour	Wheat flour	Superiority/inferiority of sorghum compared to wheat flour
Moisture (%)	13.8	11.67	High
Ash (%)	1.6	1.43	High
Protein (%)	6.2	9.51	Low
Fat (%)	2.8	1.5	High
Carbohydrates (%)	76.15	69.4	High
Iron (mg/100 g)	8.4	4.9	High
Calcium (mg/100 g)	10.03	48	Low
Chromium (mg/100 g)	0.008	0.006	Similar
Zinc (mg/100 g)	1.3	2.2	Low
Copper (mg/100 g)	0.46	0.51	Low
Magnesium (mg/100 g	171	132	High
Thiamine (mg/100 g)	2.31	0.49	High
Riboflavin (mg/100 g)	0.38	0.17	High
Folic acid (mg/100 g)	20	BDL	High
Energy (kcal/100 g)	355	341	High

Note: BDL, below detectable level.

Table 3.21. Nutrient analysis of sorghum semolina as compared to wheat semolina. (Source: Dayakar *et al.*, 2014.)

Parameters	Semolina		Superiority/inferiority of sorghum semolina compared to wheat semolina
	Sorghum	Wheat	
Moisture (%)	8.97	–	High
Ash (%)	2.03	0.57	High
Protein (%)	7.15	10.4	Low
Fat (%)	1.20	0.8	High
Carbohydrates (%)	77.74	74.8	High
Iron (mg/100 g)	5.1	1.6	High
Calcium (mg/100 g)	5.75	16	Low
Chromium (mg/100 g)	1.476	–	High
Zinc (mg/100 g)	1.3	–	High
Thiamine (mg/100 g)	0.615	0.12	High
Copper (mg/100 g)	21.11	–	High
Magnesium (mg/100 g)	86.02	–	High
Riboflavin (mg/100 g)	1.093	0.03	High
Folic acid (mg/100 g)	1.19	BDL	High
Energy (kcal/100 g)	350	348	High

Note: BDL, below detectable level.

were found to be similar, and moisture, fibre and carbohydrate values were relatively low.

From among the various food products analysed, sorghum flour was the nutritionally superior for the greatest number of traits, followed by sorghum semolina, sorghum pasta, sorghum vermicelli and finally sorghum flakes.

3.6 Conclusion

Millets contain many nutritional and health benefits. Nutritionally important components in millets such as complex carbohydrates, dietary fibre, minerals, vitamins and phytochemicals that include phenolic compounds are

Table 3.22. Nutrient analysis of sorghum flakes as compared to rice flakes. (Source: Dayakar *et al.*, 2014.)

Parameters	Flakes		Superiority/inferiority of sorghum flakes compared to rice flakes
	Sorghum	Rice	
Moisture (%)	13.8	12.2	High
Ash (%)	0.63	2.0	Low
Protein (%)	5.09	6.6	Low
Fat (%)	2.40	1.2	High
Carbohydrates (%)	74.99	77.3	Low
Iron (mg/100 g)	87.78	20	High
Calcium (mg/100 g)	93.15	20	High
Chromium (mg/100 g)	0.9	–	High
Zinc (mg/100 g)	8.78	–	High
Copper (mg/100 g)	27.789	0.37	High
Magnesium (mg/100 g)	80.51	101	Low
Folic acid (mg/100 g)	3.398	BDL	High
Thiamine (mg/100 g)	0.07	0.21	Low
Riboflavin (mg/100 g)	0.02	0.05	Low
Energy (kcal/100 g)	342	346	Low

Note: BDL, below detectable level.

Table 3.23. Nutrient analysis of sorghum pasta as compared to pasta made from refined wheat flour. (Source: Dayakar *et al.*, 2014.)

Parameters	Pasta		Superiority/inferiority of sorghum pasta compared to refined wheat flour pasta
	Sorghum	Refined wheat flour	
Moisture (%)	11.47	10.5	High
Ash (%)	0.77	0.7	On par
Protein (%)	8.39	7.21	High
Fat (%)	1.38	0.5	High
Carbohydrates (%)	76.21	77.69	Low
Iron (mg/100 g)	64.51	2.0	High
Calcium (mg/100 g)	64.51	22	High
Zinc (mg/100 g)	5.744	1.871	High
Copper (mg/100 g)	1.407	0.29	High
Magnesium (mg/100 g)	67.48	44.4	High
Riboflavin (mg/100 g)	1.28	BDL	High
Thiamine (mg/100 g)	BDL	BDL	On par
Folic acid (mg/100 g)	BDL	BDL	On par
Energy (kcal/100 g)	355	352	High

Note: BDL, below detectable level.

important positive contributors to human health. Furthermore, antioxidants and free radical scavenging active compounds in millets play a major role in preventing cancer and protecting the human system from cancer. The data generated thus far under the value chain can be used for labelling the products effectively for millet promotion. Further studies on the nutritional and health benefits of millets should focus on the changes in nutrients during processing.

Table 3.24. Nutrient analysis of sorghum vermicelli as compared to vermicelli made of refined wheat flour. (Source: Dayakar *et al.*, 2014.)

Parameters	Vermicelli		Superiority/inferiority of vermicelli based on sorghum compared to refined wheat flour
	Sorghum	*Maida*	
Moisture (%)	8.43	11.7	Low
Ash (%)	0.77	0.7	Similar
Protein (%)	8.39	8.7	Similar
Fat (%)	1.38	0.4	High
Carbohydrates (%)	76.21	78.3	Low
Fibre (%)	0.43	0.7	Low
Iron (mg/100 g)	64.51	2.0	High
Calcium (mg/100 g)	54.51	22	High
Zinc (mg/100 g)	7.49	BDL	High
Copper (mg/100 g)	BDL	BDL	Similar
Magnesium (mg/100 g)	67.48	42	High
Thiamine (mg/100 g)	0.73	BDL	High
Riboflavin (mg/100 g)	1.28	0.05	High
Folic acid (mg/100 g)	BDL	BDL	Similar
Energy (kcal/100 g)	355	352	High

Note: BDL, below detectable level.

References

Abdalla, A.A., El-Tinay, A.H., Mohamed, B.E. and Abdalla, A.H. (1998) Proximate composition, starch, phytate and mineral contents of 10 pearl millet genotypes. *Food Chemistry* 63(2), 243–246.

Abdelgadir, M., Abbas, M., Järvi, A., Elbagir, M., Eltom, M. and Berne, C. (2005) Glycemic and insulin responses of six traditional Sudanese carbohydrate-rich meals in subjects with type 2 diabetes mellitus. *Diabetic Medicine* 22, 213–217.

Alfieri, M.A.H., Pomerleau, J., Grace, D.M. and Anderson, L. (1995) Fiber intake of normal weight, moderately obese and severely obese subjects. *Obesity Research* 3(6), 541–547.

Ali, R., Staub, J., Leveille, G.A. and Boyle, P.C. (1982) Dietary fiber and obesity. In: Vahouny, G.V. and Kritchevsky, D. (eds) *Dietary Fiber in Health and Disease.* Plenum Press, New York, pp. 19–194.

Anderson, J.W. and Hanna, T.J. (1999) Whole grains and protection against coronary heart disease: what are the active components and mechanisms? *American Journal Clinical Nutrition* 70, 307–308.

Anju, T. and Sarita, S. (2010) Suitability of foxtail millet (*Setaria italica*) and barnyard millet (*Echinochloa frumentacea*) for development of low glycemic index biscuits. *Malaya Journal of Nutrition* 16, 361–368.

Annor, G.A. (2013) Millet starches: structural characteristics and glycemic attributes. Doctoral dissertation, The University of Guelph, Ontario.

Arya, J. (2014) *Food is Your Best Medicine*, 1st edn. Arya Publishers, Pune, India.

Asp, N.G. (1994) Nutritional classification and analysis of food carbohydrates. *American Journal of Clinical Nutrition* 59(1), 679–681.

Awika, J.M., Rooney, L.W. and Waniska, R.D. (2004) Anthoycanins from black sorghum and their antioxidant properties. *Food Chemistry* 90, 293–301.

Bhatt, A., Singh, V., Shrotria, P.K. and Baskheti, D.C. (2003) Coarse grains of Uttaranchal: ensuring sustainable food and nutritional security. *Indian Farmer's Digest* 34–38.

Burkitt, D.P. and Trowell, H.C. (1975) *Refined Carbohydrate Foods and Disease.* Academic Press, London, New York & San Francisco, pp. 25–31.

Carolina, C., Luigi, M., Nicola, C., Cristina, B., Luigi, D.G., Domenica, R.M., Paola, P., Natale, F., Scott, R.B., Brian, L. and Marco, L. (2007) Celiac disease: *in vitro* and *in vivo* safety and palatability of wheat-free sorghum food products. *Clinical Nutrition* 26, 799–805.

Carr, T.P., Curtis, L.W., Schlegel, V., Cuppett, S.L., Guderian, D.M. and Johnson, K.R. (2005) Grain sorghum lipid extract reduces cholesterol absorption and plasma non-HDL cholesterol concentration in hamsters. *Journal of Nutrition* 135(9), 2236–2240.

Chen, F., Cole, P., Mi, Z.B. and Xing, L.Y. (1993) Corn and wheat-flour consumption and mortality from esophageal cancer in Shanxi, China. *International Journal of Cancer* 53, 902–906.

Chhavi, A. and Sarita, S. (2012) Evaluation of composite millet breads for sensory and nutritional qualities and glycemic response. *Malaysian Journal of Nutrition* 18, 89–101.

Choi, Y.Y., Osada, K., Ito, Y., Nagasawa, T., Choi, M.R. and Nishizawa, N. (2005) Effect of dietary protein of Korean foxtail millet on plasma adiponectin, HDL-cholesterol, and insulin levels in genetically type 2 diabetic mice. *Bioscience, Biotechnology and Biochemistry* 69, 31–37.

Dayakar, R.B., Karthikeyan, K., Seetharama, N. and Hyma Jyothi, S. (2004) *Macro and Micro Level Changes in Consumption of Nutritious Cereals in India*. National Research Centre for Sorghum, Rajendranagar, India.

Dayakar, R.B., Patil, J.V., Hymavathi, T.V., Nirmal Reddy, K. and Rajendra Prasad, M.P. (2014) *Creation of Demand for Millet Foods through PCS Value Chain*. Final report of NAIP (ICAR). Directorate of Sorghum Research, Rajendranagar, India.

Deosthale, Y.G., Visweswara Rao, K., Nagarajan, V. and Pant, K.C. (1971) Varietal differences in protein and amino acids of grain bajra (*Pennisetum typhoides*). *Indian Journal of Nutrition and Dietetics* 8, 301–308.

Devi, P.B., Vijayabharathi, R., Sathyabama, S., Malleshi, N.G. and Priyadarisini, V.B. (2014) Health benefits of finger millet (*Eleusine coracana* L.) polyphenols and dietary fiber: a review. *Journal of Food Science and Technology* 51(6), 1021–1040.

Dicko, M.H., Gruppen, H., Traore, A.S., Van Berkel, W.J.H. and Voragen, A.G.J. (2005) Evaluation of the effect of germination on phenolic compounds and antioxidant activities in sorghum varieties. *Journal of Agricultural and Food Chemistry* 53, 2581–2588.

Doggett, H. (1989) Small millets – a selective overview. In: Seetharam, K., Riley, K.W. and Harinarayana, G. (eds) *Small Millets in Global Agriculture*. Oxford and IBH Publishing Company, New Delhi, pp. 3–18.

Duke, J.A. and Ayensu, E.S. (1985) *Medicinal Plants of China.* Reference publications, Inc. Algonac, Michigan.

Dykes, L. and Rooney, L.W. (2006) Sorghum and millet phenols and antioxidants. *Cereal Science* 44, 236–251.

Dykes, L. and Rooney, L.W. (2007) Phenolic compounds in cereal grains and their health benefits. *Cereal Food World* 52, 105–111.

FAO (1998) *Carbohydrates in Human Nutrition*. FAO Food and Nutrition Paper No. 66, Food and Agriculture Organization of the United Nations, Rome.

Gazzaz, S. S., Rasco, B.A., Dong, F.M. and Borham, M. (1989) Effect of processing on the thiamine, riboflavin, and vitamin B12 content of fermented whole grain cereal products. *Journal of Food Processing and Preservation* 13, 321–334.

Geetha, C. and Parvathi, E.P. (1990) Hypoglycemic effect of millet incorporated breakfast items on selected non-insulin dependent diabetic patients. *The Indian Journal of Nutrition and Dietetics* 27, 316–332.

Gómez-Cordovés, C., Bartolome, B., Vieira, W. and Virador, V.M. (2001) Effects of wine phenolics and sorghum tannins on tyrosinase activity and growth of melanoma cells. *Journal of Agricultural and Food Chemistry* 49, 1620–1624.

Gopalan, C., Rama Sastri, B.V. and Balasubramanian, S.C. (2011) *Nutritive Value of Indian Foods (NVIF)*. National Institute of Nutrition, Indian Council of Medical Research, Hyderabad, India.

Grimmer, H.R., Parbhoo, V. and McGrath, R.M. (1992) Antimutagenicity of polyphenol-rich fractions from *Sorghum bicolor* grain. *Journal of the Science of Food and Agriculture* 59, 251–256.

House, L.R. (1995) Sorghum and millets: history, taxonomy, and distribution. In: *Sorghum and Millets: Chemistry and Technology*. American Association of Cereal Chemists, St. Paul, Minnesota, pp. 1–9.

Hadimani, N.A., Ali, S.Z. and Malleshi, N.G. (1995) Physico-chemical composition and processing characteristics of pearl millet varieties. *Journal of Food Science and Technology* 32(3), 193–198.

Heidemann, C., Schulze, M.B., Franco, O.H., van Dam, R.M., Mantzoros, C.S. and Hu, F.B. (2008) Dietary patterns and risk of mortality from cardiovascular disease, cancer, and all-causes in a prospective cohort of women. *Circulation* 118, 230–237.

Ito, K., Ozasa, H., Noda, Y., Arii, S. and Horikawa, S. (2008) Effects of free radical scavenger on acute liver injury induced by D-galactosamine and lipopolysaccharide in rats. *Hepatology Research* 38, 194–201.

Jambunathan, R. and Subramanian, V. (1988) Grain quality and utilization of sorghum and pearl millet. In: *Biotechnology in Tropical Crop Improvement: Proceedings of the International Biotechnology Workshop*. ICRISAT, Patancheru, India, pp.133–139.

Jellum, M.D. and Powell, J.B. (1971) Fatty acid composition of oil from pearl millet seed. *Journal of Agronomy* 63, 29–33.

Jensen, M.K., Koh-Banerjee, P., Hu, F.B., Franz, M., Sampson, L., Grønbæk, M. and Rimm, E.B. (2004) Intakes of whole grains, bran, and germ and the risk of coronary heart disease in men. *The American Journal of Clinical Nutrition* 80(6), 1492–1499.

Kalinova, J. and Moudry, J. (2006) Content and quality of protein in proso millet plant foods for human nutrition. *Plant Food for Human Nutrition* 61, 43–47.

Kamath, M.V. and Belavady, B. (1980) Unavailable carbohydrates of commonly consumed Indian foods. *Journal of Science and Food Agriculture* 31, 194–202.

Kaur, K.D., Jha, A., Sabikhi, L. and Singh, A.K. (2014) Significance of coarse cereals in health and nutrition: a review. *Journal of Food Science and Technology* 51(8), 1429–1441.

Lakshmi, K.B. and Vimala, V. (1996) Hypoglycemic effect of selected sorghum recipes. *Nutrition Research* 16(10), 1651–1658.

Liang, S., Yang, G. and Ma, Y. (2010) Chemical characteristics and fatty acid profile of foxtail millet bran oil. *Journal of the American Oil Chemists' Society* 87, 63– 67.

Malleshi, N.G. and Klopfenstein, C.F. (1998) Nutrient composition and amino acid contents of malted sorghum, pearl millet and finger millet and their milling fractions. *Journal of Food Science and Technology* 35(3), 247–9.

Mbithi-Mwikya, S., Ooghe, W., Van Camp, J., Nagundi, D. and Huyghebaert, A. (2000) Amino acid profile after sprouting, autoclaving and lactic acid fermentation of finger millet (*Eleusine coracana)* and kidney beans (*Phaseolus vulgaris* L.) *Journal of Agriculture and Food Chemistry* 48(8), 3081–3085.

McDonough, C.M., Rooney, L.W. and Earp, C.F. (1986) Structural characteristics of *Eleusine coracana* (finger millet) using scanning electron and fluorescence microscopy. *Food Microstructure* 5, 247–256.

McKevith, B. (2004) Nutritional aspects of cereals. *British Nutrition Foundation Nutritional Bulletin* 29, 111–142.

Mittal, M. (2002) Development of finger millet and barnyard millet based convenience mixes for food products and their evaluation for nutritional quality, storage stability and acceptability. Ph.D thesis, Govind Ballabh Pant University of Agriculture and Technology, Pantnagar, India.

Mohan, B., Gopal, A., Malleshi, N.G. and Tharanathan, R. (2005) Characteristics of native and enzymatically hydrolyzed ragi (*Eleusine coracana*) and rice (*Oryza sativa*) starches. *Carbohydrate Polymers* 59, 43–50.

NABARD (National Bank for Agriculture and Rural Diversity) *Millet in your Meals*. Available at www.sahajasamrudha.org/images/Millet%20Booklet.pdf (accessed 24 February 2016).

Narasinga Rao, B.S. (2003) Bioactive phytochemicals in Indian foods and their potential in health promotion and disease prevention. *Asia Pacific Journal of Chemical Nutrition.* 12, 9–22.

NIN (2007) *Nutritive Value of Indian Foods*. National Institute of Nutrition, Hyderabad, India.

Osagie, A.U. and Kates, M. (1984) Lipid composition of millet (*Pennisetum americanum)* seeds. *Lipids* 19, 958–965.

Parbhoo, V., Grimmer, H.R., Cameron-Clarke, A. and McGrath, R. (1995) Induction of cytochrome P-450 in rat liver by polyphenol rich diet from a bird resistant sorghum rich diet. *Journal of the Science of Food and Agriculture* 69, 249–252.

Park, K., Ito, Y., Nagasawa, T., Choi, M.and Nishizawa, N. (2008) Effects of dietary Korean proso-millet protein on plasma adiponectin, HDL cholesterol, insulin levels, and gene expression in obese type 2 diabetic mice. *Biosciences Biotechnology and Biochemistry* 72(11), 2918–2925.

Prashant, S.H., Namakkal, S.R. and Chandra, T.S. (2005) Effect of the antioxidant properties of millet species on oxidative stress and glycemic status in alloxan-induced rats. *Nutrition Research* 25, 1109–1120.

Rai, K.N., Gowda, C.L.L., Reddy, B.V.S. and Sehgal, S. (2008) Adaptation and potential uses of sorghum and pearl millet in alternative and health foods. *Comprehensive Reviews in Food Science and Food Safety* 7(4), 320–396.

Rajasekaran, N.S., Nithya, M., Rose, C. and Chandra, T.S. (2004) The effect of finger millet feeding on the early responses during the process of wound healing in diabetic rats. *Biochimica et Biophysica Acta* 1689, 190–201.

Rajendra Prasad, M.P., Dayakar, R.B., Kalpana, K., Rao, M.V. and Patil, J.V. (2014) Glycaemic index and glycaemic load of sorghum products. *Journal of Science and Food Agriculture* 51, 12–16.

Rajendra Prasad, M.P., Dayakar, R.B., Kalpana, K., Madhari, R., Rao, M.V. and Patil, J.V. (2015) Impact of sorghum supplementation on growth and micronutrient status of school going children in southern India – a randomized trial. *Indian Journal of Pediatrics* 42, 51–52.

Ramanathan, M.K. and Gopalan, C. (1957) Effect of different cereals on blood sugar levels. *Indian Journal of Medical Research* 45, 255–262.

Ramulu, P. and Udaysekhara Rao, P. (1997) Effect of processing on dietary fibre content of cereals and pulses. *Plant Foods and Human Nutrition* 50, 249–257.

Ratnavathi, C.V., Biswas, P.K., Pallavi, M., Maheshwari, M., Vijaykumar, B.S. and Seetharama, N. (2003) Alternative uses of sorghum-methods and feasibility: Indian perspective in alternate uses of sorghum and pearl millet in Asia. In: *Proceedings of the Expert Meeting.* ICRISAT, Patancheru, India, pp. 19–24.

Ravindran, G. (1992) Seed proteins of millets: amino acid composition, proteinase inhibitors and in vitro digestibility. *Food Chemistry* 44 (1), 13–17.

Saldivar, S. (2003) Cereals: dietary importance. In: Caballero, B., Trugo, L., Finglas, P. (ed.) *Encyclopedia of Food Sciences and Nutrition*. ReinoUnido: Academic Press, Agosto, London, pp. 1027–1033.

Schneeman, B.O. and Tietyen, J. (1994) Dietary fiber. In: Shills, M. E., Olson, J.A. and Shike, M. (eds) *Modern Nutrition in Health and Disease.* Lea and Febiger, Philadelphia, Pennsylvania, pp. 89 –100.

Serna-Saldivar, S.O., Clegg, C. and Rooney, L.W. (1994) Effects of parboiling and decortication on the nutritional value of sorghum (*Sorghum bicolor* L. Moench) and pearl millet (*Pennisetum glaucum* L.). *Journal of Cereal Science* 19(1), 83–89.

Shinde, R.K.U. (2004) Utility of bran products in non-insulin dependent diabetes mellitus (NIDDM) patients. *Journal of Human Ecology* 16(3), 219–222.

Shobana, S., Sreerama, Y. and Malleshi, N.G. (2009) Composition and enzyme inhibitory properties of finger millet (*Eleusine coracana* L.) seed coat phenolics: mode of inhibition of α-glucosidase and pancreatic amylase. *Food Chemistry* 115, 1268–1273.

Shukla, K. and Srivastava, S. (2011) Evaluation of finger millet incorporated noodles for nutritive value and glycemic index. *Journal of Food Science and Technology* 51(3), 527–34.

Singh, R. and Popli, S. (1973) Amylose content and amylolytic studies on high-yielding varieties of bajra (*Pennisetum typhoides*). *Journal of Food Science and Technology* 10, 31–33.

Slavin, J. (2003) Why whole grains are protective: biological mechanisms. *Proceedings of the Nutrition Society* 62(01), 129–134.

Slavin, J. and Slavin, J. (2004) Whole grains and human health. *Nutrition Research Reviews* 17(1), 99–110.

Tatala, S., Ndossi, G., Ash, D. and Mamiro, P. (2007) Effect of germination of finger millet on nutritional value of foods and effect of food supplement on nutrition and anemia status in Tanzanian children. *Tanzania Health Research Bulletin* 9(2), 77–86.

Taylor J.R.N. and Emmambux, M.N. (2008) Products containing other specialty grains: sorghum, the millets and pseudocereals. In: Hamaker, B.R. (ed.) *Technology of Functional Cereal Products*. Woodhead Publishing, Abington, UK, pp. 281–335.

Taylor, J.R.N. and Schussler, L. (1986) The protein compositions of the different anatomical parts of sorghum grain. *Journal of Cereal Science* 4, 361–369.

Taylor, J.R.N., Belton, P.S., Beta, T. and Duodu, K.G. (2014) Increasing the utilisation of sorghum, millets and pseudocereals: developments in the science of their phenolic phytochemicals, bio-fortification and protein functionality. *Journal of Cereal Science* 59(3), 257–275.

Thompson, L.U. (1993) Potential health benefits and problems associated with anti-nutrients in foods. *International Journal of Food Resources* 26, 131–149.

The Research Pedia (2015) Available at: www.theresearchpedia.com/health/superfoods/health-benefits-of-pearl-millet (accessed 16 May 2015).

Van Rensburg, S.J. (1981) Epidemiological and dietary evidence for a specific nutritional predisposition to esophageal cancer. *Journal of the National Cancer Institute* 67, 243–251.

Vasan, P., Dutta, N., Mandal, A.B., Sharma, K. and Kadam, M.M. (2008) Comparative digestibility of amino acids of maize, sorghum, finger millet and pearl millet in cockerels and Japanese quails. *British Poultry Science* 49(2), 176–180.

Vithal, D.S. and Machewad, G.M. (2006) Processing of foxtail millets for improved nutrient availability. *Journal of Food Processing and Preservation* 30, 269–279.

Warsi, A.S. and Wright, B.C. (1973) Sorghum growth and its composition in relation to rates and methods of nitrogen application II – pattern of N accumulation. *Indian Journal of Agronomy* 18(3), 277–281.

Wolever, T.M., Jenkins, D.J.A., Jenkins A.L. and Josse, R.G. (1991) The glycemic index: methodology and clinical implications. *American Journal of Clinical Nutrition* 54, 846–854.

Wursch, P. (1997) The role of viscous soluble fiber in the metabolic control of diabetes. *Diabetes Care* 20(11), 1774–80.

Yeung, H.C. (1985) *Handbook of Chinese Herbs and Formulas*. Institute of Chinese Medicine, Los Angeles, California.

4

Value Chain: On-Farm Production Interventions

The National Agricultural Innovation Project (NAIP), orchestrated by the Indian Institute of Millet Research (IIMR) (the then Directorate of Sorghum Research, DSR) with a public–private partnership consortium, implemented interventions across the value chain for pre- and post-harvest sectors of millet. These interventions included developing specific-trait and high-yielding varieties and popularizing them among farmers, adoption of specific improved technologies, assured buy-back and linkages with processors, development of novel ready-to-eat (RTE) convenience foods and various millet recipes. The components of the value chain at the on-farm production stage are detailed in this chapter.

4.1 Identification of Genotypes for Specific End Products

Sorghum for food in India is grown in the *rabi* (dry season), which produces white grain, free of blemishes. This grain is highly prized for processing. Sorghum grown in the *kharif* (rainy season) has staining and moulding of its grains that reduces the quality of food. In this context, IIMR screened a total of 420 genotypes (160 germplasm lines, 200 elite SPV lines and 60 released hybrids/varieties) for various physical, biochemical and nutritional grain characters to assess their suitability for various semi-processed end products. The number of genotypes analysed for each grain character, its range and distribution of values are given in Table 4.1.

4.1.1 Analysis of physical attributes of sorghum genotypes

The physical grain attributes studied were grain size, grain weight, endosperm texture and grain hardness, important traits that affect the end-use quality of sorghum. The suitability of these genotypes for each of the semi-processed end products is given in Appendices 2–4. Of the various genotypes identified for their suitability based on their grain physical traits, the released varieties may be directly tested for their milling suitability and proportion of recovery, since it has yield advantage and other desirable attributes as compared to the existing germplasm lines. Subsequently, germplasm and other improved lines that showed greatest superiority for grain physical traits may be incorporated into breeding programmes for further improvement.

Out of the 420 genotypes tested for four different physical grain traits (grain size, grain weight, endosperm texture and grain hardness):

20 genotypes with bold (bulging) grains were most suitable for *poha* and pops; 106 genotypes with corneous endosperm were most suitable for coarse semolina preparation; 18 genotypes had the most suitable grain size, weight and floury endosperm for *roti* and other bakery products such as cakes and biscuits; and 25 genotypes were identified that had the most suitable grain weight, grain size and intermediary endosperm for fine semolina and cold extruded products (pasta and vermicelli). The released varieties were tested for their suitability for various end products; those most suited for specific end products are tabulated in Tables 4.2 and 4.3.

The grains were tested for their suitability for popping. Unlike maize, sorghum grain has the ability to pop after preconditioning of the grain. In the study, this process of preconditioning was standardized. Of seven cultivars, four had <70% pops and three had >70% popping ability – CSV 18 (84.5%), PKV Kranti (79%) and CSV 216R (76.2%), and were thus identified as the best for popping.

Table 4.1. Summarized information on genotypes performance with regard to physical and biochemical characters. (Source: Dayakar *et al.*, 2014.)

	No. of genotypes analysed	Range in values	Value distribution (no. of genotypes)
Physical characters			
Grain size	430	2–4 mm	<3.35 mm (138); 3.35–4 mm (272); >4 mm (20)
Grain weight	430	1.22–6 g	<3 g (153); 3–3.9 g (207); >3.9 g (7)
Endosperm texture[a]	430	1–9	1 (2); 3 (108); 5 (230); 7 (81); 9 (9)
Grain hardness	180	1–18	–
Biochemical characters			
Protein	156	8.13–13.80%	>9% (117)
Fat	174	1.42–3.18%	>2.5% (64)
Starch	77	51.70–80.75%	>68% (10)
Protein digestibility	244	22.44–72.57%	50–60% (63); 60–70% (68); >70% (6)
Starch digestibility	24	1.2–2.41 mg glucose/h	–
Phytic acid	127	2.4–6.7%	0 (0); <5% (67); ≥5% (60)
Amylose	153	8.35–25.05%	<10% (16); 10–20% (111); >20% (21)

Note: [a], proportion of corneous (vitreous or hard) fraction of endosperm with respect to the floury (soft) endosperm fraction.

Table 4.2. Sorghum genotypes/cultivars with high semolina recovery. (Source: Dayakar *et al.*, 2014.)

Genotype	100-seed weight (g)	Grain hardness	Semolina recovery (%)
RS 627	2.438	61.2	41.5
RS 673	2.476	81.7	44.2
CSV 13	2.966	84.0	43.2
CSV 15	3.414	65.0	44.3
SPV 462	3.506	68.7	38.0

Table 4.3. Sorghum cultivars suitable for flour making. (Source: Dayakar *et al.*, 2014.)

Cultivar	Grain lustre	Size of mark of germ	Texture of endosperm
CSV 18R	Lustrous	Large	¾ farinaceous
CSV 22R	Lustrous	Medium	¾ farinaceous
CSV 216R	Lustrous	Medium	¾ farinaceous
PKV Kranti	Lustrous	Medium	¾ farinaceous
Phule Revathi	Lustrous	Medium	¾ farinaceous

4.1.2 Analysis of biochemical attributes of sorghum genotypes

Proximate analysis involves the partitioning of compounds in a food item into ash, moisture, fat, protein, carbohydrate and mineral content, based on the chemical properties of the compounds. Out of these six categories, we have reported fat, protein and starch. With regard to proximate analysis, we recorded 64 genotypes with >2.5% fat, 117 genotypes with >9% protein, and ten genotypes with >68% starch. The details of proximate and biochemical analysis of parental lines, released varieties/hybrids and germplasm lines are given in Appendices 5–7. However, genotypes with high fat content should be tested for the presence of essential fatty acids for therapeutic importance. With regard to biochemical analysis, out of 125 genotypes tested, three genotypes – CSH 16, CSV 13 and RS 29 – had low percentages of phytic acid. Phytic acid is considered to be an anti-nutritional factor that limits the absorption of minerals, especially divalent ions into the human system when sorghum is consumed. Selection of genotypes with low phytic acid is desirable. However, although phytic acid at higher levels is believed to inhibit mineral absorption in humans, recent findings state that phytic acid in lower levels is known to act as an antioxidant (Osman, 2004).

Protein digestibility ranged from 22 to 73% in 244 genotypes. In general, released varieties/hybrids showed low protein digestibility as compared to germplasm lines, indicating scope for further improvement. The *in vitro* protein digestibility is a very important biochemical parameter to assess nutritional quality and had not been explored in Indian sorghum genotypes prior to the present study. Amylose content across 153 genotypes ranged from 8.35% (ICSV 745) to 25.05% (SPV 1687). Starch digestibility varied from 1.21 (SPV 107) to 2.41 mg glucose/h (SPV 436). Among the released cultivars, GJ35, GJ 40, JJ 1022, PSB3, PSB 9, PSR 23, P. Dagadi and CO 26 had high amylose content (>20%). From among the various released genotypes studied, none of them showed superiority for all traits. Some genotypes, i.e. AKR 354, PVK 809, P. Dagadi and JJ 741, showed superiority for three or more biochemical traits. Among the germplasm lines, EP 102 recorded the highest protein content and protein digestibility. These genotypes could be given priority for selection for specific end products and also for breeding programmes for further improvement.

4.1.3 Identification of high protein digestibility and amino acid profiles in sorghum genotypes

Protein quality is an important aspect for determining the nutritional quality of sorghum. Among protein quality, protein digestibility is an important nutritional trait that enhances the value of sorghum for food. The protein digestibility in most of the sorghum genotypes was around 40–50%. The objective of the present study was to identify stable lines for protein digestibility. Forty-nine elite sorghum genotypes were evaluated for protein digestibility and related grain quality traits. Turbidity assays for rapid identification of high protein digestible lines were used to screen the test genotypes. A wide range of variability was observed for this grain protein digestibility, from 25.15% (27B) to 69.7% (SPV 1775), but the hybrid parental lines of the released hybrids showed low variability for grain protein digestibility, indicating significant scope for its improvement. The genotypes SPV 1775 (69.7%), SPV 1758 (65.2%), SPV 462 (65.39%) and Sakkari Mukri (64.45%) were found to have high grain protein digestibility. The lines identified with high protein digestibility (SPV 1775, SPV 462) are crossed with the existing parental lines (296B, 27 B, NR 486, RS 627, AKR 150, C 43) of hybrids for development of new hybrid parental lines with superior grain protein digestibility. The newly developed lines are in the advanced (F4) stage of generation advancement. At stage F6, the lines will be evaluated for protein digestibility and grain quality and will also be tested for its suitability for use in hybrid development.

In addition to protein digestibility, an attempt has been made to study the amino acid profile of sorghum cultivars. Quality proteins are readily digestible and contain essential amino acids in quantities required by human beings. Grains of most cereals, including sorghum, are said to contain inadequate levels of essential amino acids, especially lysine, followed by threonine, tryptophan and methionine. Breeding for improved amino acid composition has been attempted in cereals, and commercially exploitable high lysine varieties are now available in maize, but there are no commercially available high lysine cultivars in sorghum. To breed sorghum lines with superior protein quality through improvement of the essential amino acid composition, it is necessary to understand the amino acid profile in our released cultivars.

The analysis was done using the near infrared spectroscopy (NIRS) method as per the calibrations developed by Evano feed industries. Among the various amino acids, high levels were observed for leucine, phenylalanine, valine and arginine, while tryptophan, cystine, methionine and lysine showed low levels. Wide ranges were observed among the cultivars for leucine, isoleucine, valine, phenylalanine, threonine and arginine, while low variability was observed for lysine, methionine and tryptophan (limiting amino acids). Among the nine cultivars tested, M35-1 and Phule Vasudha were superior for most of the amino acids (Fig. 4.1). Lysine (the most limiting amino acid) levels in Phule Vasudha were 0.26%, which compares well to the generally reported levels in sorghum of 0.18–0.21%. Purdue University scientists have reported sorghum lines with high lysine content (Edmondson and Busche, 2015). A high lysine sorghum mutant P721Q was reported to have been obtained by chemical mutagenesis of the normal line P721N. Subsequently, P851171 and P850029 were developed from P721Q with superior protein quality. These results indicate scope for utilization of high lysine sorghum lines for improving protein quality. Further, large-scale indigenous germplasm will be screened for identification of high lysine lines for its use in breeding programmes.

4.1.4 Variability and correlation analysis for biochemical grain quality traits in sorghum genotypes

Forty-two elite genotypes and cultivars were tested for levels of fat, protein, starch, phytic acid, amylose and protein digestibility. Table 4.4 shows the variability in the parameters measured; high variability observed for the traits indicates its usefulness for selection of genotypes for specific end uses. Among the grain quality traits studied, high variance was observed for protein digestibility, followed by amylose and starch content. Further, to establish associations among different grain quality traits, correlation coefficients were calculated (Table 4.5). There were no significant correlations among the grain quality traits except for amylose with phytic acid (a negative correlation).

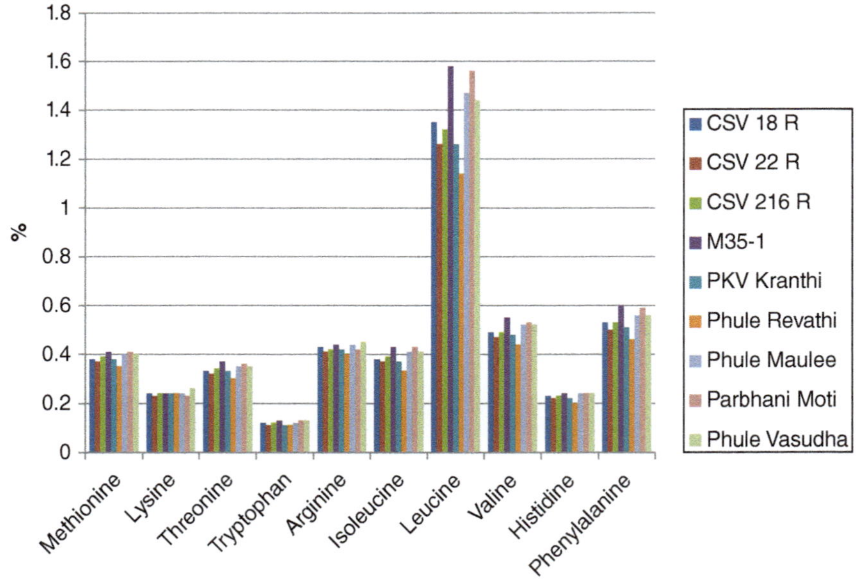

Fig. 4.1. Amino acid profile of the *rabi* sorghum cultivars.

Table 4.4. Variability analysis for grain quality traits in sorghum genotypes. (Source: Dayakar *et al.*, 2014.)

Parameters	Fat (%)	Protein (%)	Starch (%)	Phytic acid (mg/100 mg)	Amylose (%)	Protein digestibility (%)
Min.	1.94	7.46	54.70	2.40	8.35	40.00
Max.	3.50	13.80	72.50	6.70	20.96	67.10
Mean	2.68286	9.73024	64.1405	4.4119	14.08	47.7076
Variance	0.113826	1.39201	13.2748	1.44888	12.58	43.0887
SD	0.337381	1.17984	3.64346	1.20369	3.54	6.56419

Table 4.5. Correlation coefficients for grain quality traits. (Source: Dayakar *et al.*, 2014.)

Parameters	Fat (%)	Protein (%)	Starch (%)	Phytic acid (%)	Amylose (%)
Protein (%)	−0.13	–	–	–	–
Starch (%)	−0.18	−0.06	–	–	–
Phytic acid (mg/100mg)	0.04	0.02	0.09	–	–
Amylose (%)	0.05	0.01	0.08	−0.26*	–
Protein digestibility (%)	−0.07	0.09	−0.04	0.17	−0.03

Note: *, significantly correlate.

4.1.5 Shelf-life studies of *rabi* sorghum grain cultivars

The objective of this study was to identify the optimum storage period and conditions for sorghum grain without loss in quality, and to study the biochemical changes under artificial and natural insect infestation. For this, five *rabi* cultivars – M35-1, CSV 216 R, CSV 18, DSV 5 and CSV 22 – were taken for study. The genotypes were tested for various proximate compositions like rancidity, moisture, water activity, and ash content in both control (without insect) and insect-treated samples. The results indicated that the five genotypes gave no indication of rancidity and have a shelf life of 165 days without loss in quality, except for CSV 18, which has a shelf life of 135 days. No significant change was observed in ash content during the storage. In insect-treated samples, percentage starch decreased and was found to be at a maximum in DSV-5, while the protein content increased in all the cultivars stored up to 165 days.

With regard to shelf-life studies in sorghum flour, there was no significant change in fat content while an increase in protein content was observed in insect-treated samples but not in untreated samples. Starch content showed a decreasing trend during the first 60 days of storage. No significant change was observed with regard to moisture levels, and alcoholic activity significantly increased over storage, reaching 0.35% (mean) at 30 days of storage, while the prescribed safety limit is 0.18%. From among the various genotypes studied, the shelf life of CSV 216 R was greatest with regard to alcoholic activity.

From the above results it is clear that sorghum grain can be stored for up to 165 days without much loss in quality, while sorghum flour cannot be stored beyond 90 days while retaining its quality. The above results need to be considered while storing the sorghum samples. However, as only five cultivars are included, the results are not very conclusive; further studies of other genotypes would be helpful in providing a more meaningful conclusion.

4.2 End Product-Specific Sorghum Cultivation through Backward Integration: a Pilot Project

A four-year pilot project on the value chain in sorghum foods was conducted: (i) during *kharif*, of which 2 years (2008 and 2009) were in Adilabad, Andhra Pradesh, and the following 2 years (2010 and 2011) were in Nanded, Maharashtra; and (ii) during *rabi*, in Parbhani,

Maharashtra (Fig. 4.2) from 2008–09 to 2011–12. A baseline survey was conducted in these districts before starting the project with 100 farmers (25 farmers in each village), so as to establish benchmarks for the impact analysis of interventions.

From each selected district, 500 participating farmers were identified under different landholding categories, covering 20 *e-Choupal* stations; *e-Choupal* being an initiative of ITC Agri-Business Division (ABD) Ltd that provides internet access to farmers. Each farmer was provided with 3 kg of seed of improved cultivars (CSH-14, CSH-16, CSH-23, CSV-20 and SPH-1148 in *kharif* (Fig. 4.3) and M35-1 CSV 216 R (Phule Yashoda) and SPV 1411 (Parbhani Moti), Phule Vasudha and Phule Revati in *rabi*). Thus, a total of 1000 farmers were brought under 400 ha in a year. The end product-specific on-farm production was facilitated along the lines of successful models of ITC's *e-Choupal*. In this model, farmers were given market buy-back assurance for Fairly Average Quality (FAQ) grain; if they did not find a market or appropriate price, ITC would procure the grain. In Nanded, Maharashtra, through computers placed in the *sanchalak*'s (village representatives of *e-Choupal*) house, accessible to all participating farmers in the village, ITC has facilitated integrated farm extension services on procurement, extension, dissemination of knowledge and information, including daily market prices.

4.2.1 Sorghum yield in *kharif* season

Under the pilot project, introduction of improved sorghum varieties produced mixed results in yield. *Kharif* sorghum production experienced a

Fig. 4.2. Pilot project introducing improved cultivars during *rabi*. (A) The farmers' orientation and (B) a farmer's field in the project area (Parbhani, Maharashtra).

Fig. 4.3. Pilot project introducing improved cultivars during *kharif*. (A) The farmers' orientation and (B) a farmer's field in the project area (Nanded, Maharashtra).

decrease in grain and stalk yield by 22% and 41%, respectively, in 2008 (Table 4.6). In 2009, the grain and stalk yield decreased again by 2% and 35%, respectively. In the third year, the *kharif* sorghum on-farm production was shifted to Nanded in Maharashtra, where the ITC also have a well-established *e-Choupal* network. Here, in 2010, *kharif* sorghum production increased in grain and stalk yield by 26% and 59%, respectively, and again in 2011 it increased by 53% and 62%, respectively. Over these 2 years, the seeds (CSH 16, CSH 14 and SPH 1148) provided in Nanded were found very suitable. The farmers followed the recommended package of practices given by the IIMR during the orientation programmes, which resulted in substantial increases in grain and stalk yield.

4.2.2 Sorghum yield in *rabi* season

The use of improved cultivars increased grain and stalk yield to around 13% and 16%, reaching 1597 kg/ha and 6478 kg/ha, respectively, during 2008–09 (Table 4.6). During 2009–10, grain yield increased again and reached 1612 kg/ha whereas the stalk yield was lower (5610 kg/ha) than the previous year, though was still more than the baseline. So grain yields increased by 13% and 14%, respectively, over the baseline in 2008–09 and 2009–10. In 2010–11, yields of grain and stalk declined relative to the baseline by 47% and 6%, respectively, due to heavy fog and extremely cold weather during the crop growth, affecting almost all the crops in the entire region. In 2011–12, grain and stalk yields showed upward trends and reached 1952 kg/ha and 5960 kg/ha, an increase of 38% and 7%, respectively, over the baseline.

4.2.3 Economics of improved sorghum cultivars in *kharif* season

The cost of cultivation and gross returns were slightly higher in 2008 (36% and 24% higher,

Table 4.6. Economics of the pilot project introducing improved cultivars to participating sorghum farmers, 2008–2012. (Source: Dayakar *et al.*, 2014.)

Year	Cost of cultivation (INR)	Grain yield (kg/ha)	Stalk yield (kg/ha)	Gross returns (INR)	Net returns (INR)	Output–input ratio
Kharif						
Baseline (Adilabad)	8,830	1,418	7,188	15,309	6,479	1.73
2008	11,975	1,107	4,264	18,939	6,964	1.59
± over baseline	36%	–22%	–41%	24%	7%	8%
2009	11,272	1,391	4,640	24,783	13,511	2.19
± over baseline	28%	–2%	–35%	62%	109%	27%
Baseline (Nanded)	12,956	1,796	6,604	19,008	6052	1.47
2010	11,387	2,270	473	30,085	18,698	2.64
± over baseline	–12%	26%	59%	58%	209%	80%
2011	12,687	2,743	10,697	54,257	41,570	4.28
± over baseline	–2%	53%	62%	185%	587%	191%
Average	11,830	1,878	5,019	32,016	20,186	2.71
Rabi						
Baseline	7,972	1,420	5,588	23,234	15,262	2.91
2008–09	10,302	1,597	6,478	33,174	22,872	3.22
± over baseline	29%	13%	16%	43%	50%	11%
2009–10	12,624	1,612	5,610	33,796	21,172	2.68
± over baseline	58%	14%	0.39%	46%	39%	–8%
2010–11	11,241	759	5,234	38,535	27,294	3.43
± over baseline	41%	–47%	–6%	66%	79%	18%
2011–12	19,217	1,952	5,960	51,146	31,929	2.66
± over baseline	141%	38%	7%	120%	109%	–9%
Average	13,346	1,480	5,821	39,163	25,817	2.93

respectively) as compared to the baseline, whereas in 2009 cultivation costs and gross returns were up 28% and 62%, respectively, over the baseline (Table 4.6). Similarly, net returns were slightly higher than baseline in 2008 (INR 6964), whereas in 2009 net returns were double the baseline income.

Cost of cultivation decreased in 2010 and 2011, gross returns, net returns and output–input ratio increased over the baseline at an exceptional rate: by 58%, 209% and 80%, respectively in 2010, and by 185%, 587% and 191%, respectively, in 2011. The net returns in 2011 doubled those of 2010. One important aspect of intensive cultivation was the increase in income despite not very encouraging grain yields due to assured and higher market output price, even for the *kharif* produce.

4.2.4 Economics of improved sorghum cultivars in *rabi* season

With the use of improved cultivars, net returns increased over the baseline during *rabi* to the tune of 50%, 39%, 79% and 109% during 2008–09, 2009–10, 2010–11 and 2011–12, respectively (Table 4.6). During 2008–09, the sorghum economy showed a positive impact on net returns and reached an output–input ratio of 3.22. During 2009–10 the economy continued to gain positive results, though the net returns were less than the previous year (2008–09) owing to a lower stalk yield, despite the marginal increase in grain yield from the first year, and the output–input ratio was 2.68. Whereas net returns were double that of the baseline during 2010–11 and 2011–12, output–input ratios were 3.43 and 2.66, respectively.

Thus, it could be inferred that there was an increase in the net returns of the *rabi* participating farmers over baseline incomes. In addition, the assurance of buy-back gave motivation to the participating farmers to employ gainful crop management practices which led to higher income. In fact, the income of those participating farmers whose produce was procured increased by 44% over the baseline income, while for those whose produce was not procured, income increased by 36%. From the study, we could infer that sorghum yield and income levels were impacted by cultivation on an intensive scale with backstop technology of improved product-specific cultivars and use of a public-private partnership (PPP) model of farm extension services. The end product-specific production also had another advantage of linking up with entrepreneurs who actually bought the identity-preserved produce, which was procured and aggregated for engaging in small-scale production of specific sorghum processed and value-added products.

4.2.5 Comparative yield and economics of *kharif* and *rabi* sorghum

From Table 4.6, it can be inferred that average grain yields of *kharif* sorghum are more than that of *rabi* sorghum, while the inverse is true for stalk yields. But the net returns during *kharif* are less when compared to *rabi* because the price of the grain during *kharif* was lower, due to its inferior quality as a result of moulding from heavy rainfall at the time of harvesting.

The net returns increased year-on-year in both seasons. *Kharif* net returns were directly related to yields and increased linearly from year to year as the yields increased.The price of the commodity showed a positive increase from 2008 to 2009 but a slight decrease in price was observed during 2010. Even though the price decreased, net returns increased because of an increase in yields. In 2011, the highest yields were recorded with higher prices that fetched maximum net returns. In the case of *rabi*, the price was inversely related to yield as it showed a positive trend from 2008–09 to 2011–12 and then it declined. Net returns showed a slight decrease from 2008–09 to 2009–10, whereas they increased from 2009–10 and were recorded at their highest during 2011–12. Even though price decreased, yield was highest during 2011–12, with maximum net returns being achieved.

4.2.6 Factorization of income from on-farm sorghum production

The average increase in net income of the farmers (over two seasons) amounted to 114% over the baseline income (Table 4.7), with the average income of participating farmers being INR 12,237/ha in comparison to INR 10,764 /ha of the benchmark data. The results reveal that the average net income of *rabi* farmers exceeds that of *kharif* farmers in absolute terms, although the converse is true when considering their incremental net incomes in percentage over the baseline, a consequence of the very low baseline net income of *kharif* farmers compared to that of the *rabi* farmers.

With buy-back assurance given to the farmers (Fig. 4.4), almost all of them followed

Table 4.7. Factorization of income from on-farm end product-specific sorghum production. (Source: Dayakar *et al.*, 2014.)

Serial no.	Parameters	Net income (INR/ha)
I	**Benchmark data**	
A	*Kharif* 2007	6,480
B	*Kharif* 2009	6,052
C	Average of *kharif* (2007 and 2009)	6,266
D	*Rabi* 2007	15,262
E	Average (C and D)	10,764
II	**Actual on-farm production (*kharif* and *rabi*)**	
F	Average of *kharif* 2008 and 2009	10,237
G	Average of *kharif* 2010 and 2011	30,134
H	Average of *kharif* 2008, 2009, 2010 and 2011	20,186
I	Average of *rabi* 2008–09, 2009–10, 2010–11 and 2011–12	25,817
J	Average (of H and I)	23,001
K	Incremental net income in *kharif* over baseline (H–C)	13,920 (222%)
L	Incremental net income in *rabi* over baseline (I–D)	10,555 (69.16%)
M	Average incremental net income over baseline (J–E)	12,237 (114%)

Fig. 4.4. Buy-back procurement of the farmers' produce.

the recommended package of practices, though obviously there were some constraints in the procurement of inputs required. The price of sorghum increased year by year and the market price could be assured to farmers through buy-back assurance. This was the first time the farmers had focused on sorghum and had spent time adopting all the practices. Up until the pilot project, they were unfamiliar with the many alternative uses of sorghum; armed with knowledge of all food products prepared from sorghum and the impact of adopting the recommended practices, farmers are now enthusiastic about sorghum as a crop and are likely to continue the same kind of practices in future.

Before, farmers used to grow sorghum in marginal lands but now they are growing sorghum in fertile lands. It is also important to note that, under the subproject, ITC only opted for consultancy charges and focused on achieving higher returns for the farmers by negotiating on their behalf to receive maximum price for their grain.

4.3 Conclusion

The NAIP pilot project interventions for on-farm specific end-production improvement shows a positive result, with improved grain yield and increased net income of farmers both in *rabi* and *kharif* seasons due to the buy-back procedure, which ensures a market for the dryland farmers and the required quality grain for the processor. The identified specific end-product cultivars thus developed can be readily processed for development of value-added end products (such as sorghum flour, semolina, biscuits and pops) of high quality. The improved technologies and diversified processing interventions for the development of various novel value-added millet products are discussed elaborately in the subsequent chapter.

References

Dayakar, R.B., Patil, J.V., Hymavathi, T.V., Nirmal Reddy, K. and Rajendra Prasad, M.P. (2014) *Creation of Demand for Millet Foods through PCS Value Chain*. Final report of NAIP (ICAR). Directorate of Sorghum Research, Rajendranagar.

Edmondson, H. and Busche, M. (2015) The future of food: identifying genetic mechanisms affecting sorghum protein digestibility. *Journal of Purdue Undergraduate Research.* 5: 31–37.

Osman, M.A. (2004) Changes in sorghum enzyme inhibitors, phytic acid, tannins and in vitro protein digestibility occurring during Khamir (local bread) fermentation. *Food Chemistry* 88(1), 129–134.

5

Value Chain: Processing Interventions

5.1 Background and Scope

Although millets are nutritionally rich like other fine cereals, their consumption, especially of sorghum, has significantly declined over the last three decades. This is mainly due to the rather laborious and time-consuming process involved in the preparation of millet, as well as government policy to supply fine cereals at subsidized prices. It has therefore become necessary to reorient the efforts on millets to generate demand through value addition of processed foods. Value addition in food processing has a high degree of interdependence with forward and backward linkages and so can play an important role in accelerating economic development. Through value addition, shelf life and storage quality of food can be improved. Value addition preserves many of the health-promoting compounds.

Sorghum offers much potential for its use in different food products. The functional advantages of sorghum include its white/light colour, bland flavour and its requirement for processing preparations similar to those of rice, making it excellent for use in snacks, breakfast cereals, meals and porridges. Sorghum does not contain gluten and its slower hydrolysis makes it attractive to diabetics, coeliacs and ethnic groups. Sorghum is eaten in a variety of forms that varies from region to region. In general, it is consumed as whole grain or processed into flour from which traditional meals are prepared. The use of sorghum-based composite flours in bread-making has been widely studied (Rao and Shurpalekar (1976); Hugo *et al.* (2003); Ratnavathi *et al.* (2003); Chavan *et al.* (2015), for *papad*). Sorghum malt can be used to make biscuits and weaning foods; the addition of 40% sorghum malt to biscuits has been shown to cause a reduction in stack height and an increase in spread due to increased water absorption, helping to improve biscuit texture and taste (Kulkarni *et al.*, 1987). The use of sorghum in common traditional foods such as *idli* (a steamed product) and *dosa* (a leavened product) has also had successful results (see Subramaniam *et al.*, 1983).

Millets are milled to remove the inedible husk or outer bran prior to cooking. The dehulled seeds are either used as such, or further milled to grits and flours for specific end-users. Gopaldas *et al.* (1982) developed ready-to-eat (RTE) mixes for young children using malted and roasted finger millet. Pathak *et al.* (2000) studied the effect of foxtail millet in combination with fenugreek seeds and pulses in the development of *dhokla* and *laddu* mixes in polyethylene pouches with a shelf-life of 1 month. Devaraju *et al.* (2003) developed pasta with finger millet flour using the extrusion process and reported that the product could be stored for up to 3 months.

The health benefits of millets have also been widely explored. Radha and Vijayalakshmi

(2007) studied the effect of little millet on cardiovascular patients for a period of 60 days and reported a significant reduction in total cholesterol. Itagi (2003) developed composite mixes suitable for diabetics using foxtail and little millet along with wheat, black gram and fenugreek seeds. Barnyard millet was reported to be most effective in reducing blood glucose and lipid levels compared to other millets (foxtail, kodo and little millet) as revealed by Krishna Kumari and Thayumanavan (1997).

Thus, millets, with both functional and health benefits, offer many opportunities for use in a diverse range of products. However, as yet there has been no systematic study on their development as nutritious, energy-dense traditional indigenous foods that have a long shelf-life as per the palate of Indian consumers. This chapter deals in detail with the envisaged interventions across value chain for post-harvest sectors of millets and the development of novel RTE convenience foods and various millet recipes under the NAIP pilot project. In this, special focus has been given to sorghum, the most important crop in terms of area, production and productivity when compared to other millets.

5.2 Recent Millet Food Processing Initiatives by ICAR

In 2008, the Indian Institute of Millets Research (IIMR) (formerly the Directorate of Sorghum Research) initiated interventions to develop and standardize millet product technologies, to improve the nutritional quality and consumer acceptability of millet and to improve shelf life with the objective of promoting and commercializing millet products in the country. Under the National Agricultural Innovation Project (NAIP) of the Indian Council of Agricultural Research (ICAR), special focus was given to sorghum, with some interventions on pearl millet by IIMR (Fig. 5.1) and Acharya NG Agricultural University (ANGRAU).

At ANGRAU pearl millet products were developed, such as pearl millet flour, multigrain flour, semolina and RTE extruded snacks. The shelf life of pearl millet and flour was evaluated using different packaging materials and irradiation. Sorghum products were developed, such as multigrain *roti* (two types), multigrain flour, biscuits (14 varieties: salty, sweet, groundnut and coconut; gluten-free salty and sweet biscuits; transfat-free salty, sweet, groundnut and coconut; and low calorie salty, sweet, groundnut and coconut), RTE extruded snacks, gluten-free crunches, gluten-free pasta (two varieties: soy and flax seed incorporated) and designer semolina (rich in resistant starch). The shelf life of multigrain flour was extended using hurdle technology, where nitrogen packing technology and food additives (as discussed below) are used for ensuring that pathogens are eliminated or controlled.

At IIMR, suitable sorghum genotypes for food processing and the shelf life of the grains

Fig. 5.1. The inauguration of the Centre of Excellence on Sorghum at IIMR by Shri. P. K. Basu, Secretary, Ministry of Agriculture.

were evaluated. Sorghum products were developed using different technologies, e.g. soaking, malting, dehulling, parboiling, milling, baking, flaking, fermentation, and extrusion cooking. Dehulled sorghum, sorghum flour, multigrain flour, coarse, medium and fine semolina, parboiled semolina, flakes, vermicelli and pasta, biscuits (gluten free, transfat-free and cocoa), protein rich snacks, chocolate fur cake, pure sorghum cake and *lassi* (sweet and salty) were developed and standardized. Studies were conducted on the extension of shelf life of sorghum-rich multigrain flour using different packaging materials, and also to assess suitable packaging material for commercialization. Technology was also developed for pasta flavourings (containing a range of spices in order to develop taste for the Indian palate; this is discussed in detail later in the chapter) and for instant foods such as *idli, dhokla, upma, dosa, pongal*, vermicelli *kheer* mix, flakes *kheer* mix, *peda* and *masala* sorghum flakes and pops (two types: *aam chat masala* and *pudina masala*). Products such as sorghum *peda*, bran soup and bran *peda* were developed using byproducts of sorghum. More than 30 recipes were developed and standardized at IIMR. The acceptability of some products was either equivalent or better than on-shelf products made from other cereals. Also, at the Central Food Technological Research Institute (CFTRI), the shelf life of sorghum flour was assessed using infrared heating and sorghum pasta was developed.

5.3 Why is Processing Intervention Needed in Sorghum?

Processing of grains enhances the consumer acceptability and adds to its convenience. Although sorghum is known for its nutritional quality (Table 3.1), its consumption has declined due to the unavailability of RTE and ready-to-cook (RTC) sorghum products on the market. It is regarded in many countries as an inferior food. A decrease in sorghum consumption is found to be proportional to an increase in expenditure. Also, increased income is accompanied by increased consumption of wheat and rice, as products made from these fine cereals are easy to prepare and have better keeping quality. At the same time people now tend to eat a greater variety of foods. Technological change could perhaps change the scenario for sorghum, improving its production and utilization.

Processing also improves the food value in terms of increasing the variety of products, and improving carbohydrate and protein digestibility. In India, other Asian countries and also in most of the African countries, sorghum is primarily used as food. The food products differ between countries and between different regions of the same country. Most of the products fall into the category of traditional food products. In India, the major proportion of sorghum produced is pulverized and cooked in the form of *roti* (unleavened dry pan cake or flat bread). Thick porridge, dumping, or *mudde*, as well as thin porridge, or *ambali*, are the other popular sorghum foods. Sorghum semolina is used to prepare *kuskus* in Africa and *nucchu, idli* and *dosa* in India. Decorticated sorghum is cooked like rice and the product is known as *guggari* in India but it is not a common sorghum product. Fermented sorghum products are also popular in Africa. In China and Vietnam, vermicelli and noodles are prepared from sorghum. Malting and brewing of sorghum is extensively practised in South Africa and Nigeria. It is also processed to prepare flaked, popped and extrusion-cooked convenience foods (Fig. 5.2). The well-filled milky grains of sorghum can be harvested along with the ear head, after which the product may be carefully exposed to fire to partially cook the grains and then brushed to remove them from the ear head. These grains are garnished with spice or sugar syrup and eaten as fresh sorghum snacks (FAO, 1995). Sorghum grains are naked caryopses and are free from non-edible husk tissue, unlike paddy, barley, oats and small millets. However, a few traditional varieties and hybrid sorghums contain non-edible glumes covering mostly the germ portion. Like other cereals, sorghum is not an RTE cereal and needs processing to make it suitable for cooking and product development (Desikachar and Malleshi, 1987).

The harvesting practices followed for sorghum in India and other developing countries are still unhygienic, as the grains get mixed with foreign matter that includes soil, mud, stones,

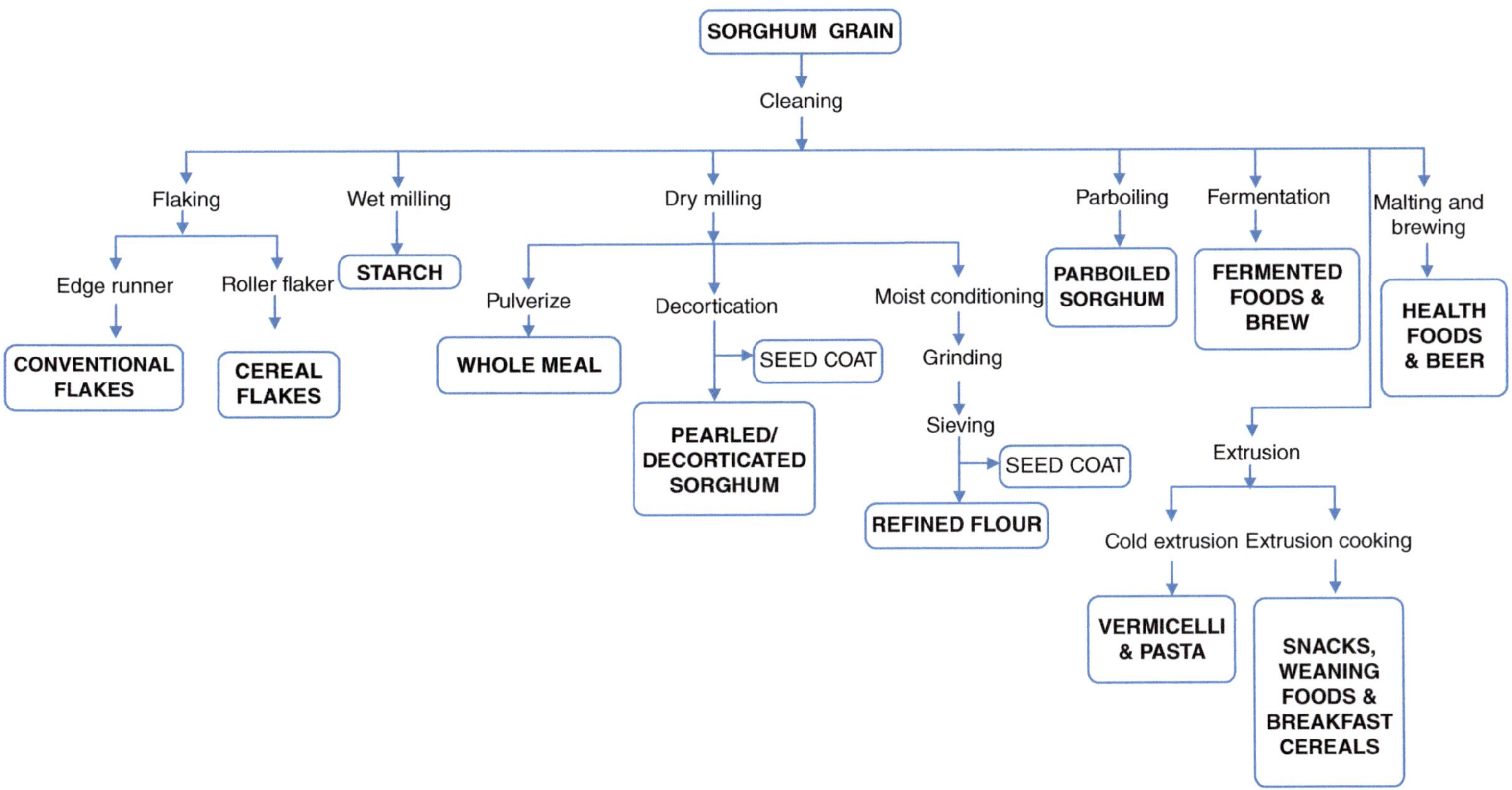

Fig. 5.2. Flow chart of sorghum processing.

and other kinds of seeds, which also makes the grains prone to microbial contamination. In view of this, the grains need to be cleaned, and processing of sorghum invariably includes pre-cleaning. This also helps to minimize the risk of infestation during storage. Some of the cultivars of sorghum contain a high proportion of tannins, polyphenols, phytate, etc. which, when present in large proportions, act as anti-nutritional factors, i.e. they reduce nutrient availability. Most of these are localized in the pericarp or the peripheral endosperm and hence judicious decortication or pearling (that is, the removal of the bran) reduces their concentration without affecting the overall nutrition. In view of this, the processing of sorghum mainly includes milling, popping, flaking, malting and fermentation as well as cold and hot extrusion, all of which reduce the concentration of anti-nutrients. The flow chart in Fig. 5.2 gives a broad picture of sorghum processing for food and allied purposes.

One of the constraints identified in developing a variety of sorghum products is the drudgery of processing sorghum. Traditional processing methods of hand pounding are not efficient, as part of the husk remains in the grain and also, because dehusking is not done uniformly, it cannot be ground to fine quality flour. Therefore, the use of millet has been limited to traditional preparations like *roti* and rice (Geervani and Vimala, 1993). The absence of processing technologies that yield shelf-stable products is the major limitation in the development of value added sorghum products (Ali *et al.*, 2003).

5.4 Structure of the Sorghum Grain

The sorghum kernel is a naked caryopsis and consists of three main anatomical parts, namely the pericarp (outer layer), endosperm (storage tissue) and germ (embryo), with their respective proportion of seed mass at about 6%, 84% and 10% (Fig. 5.3). However, their relative proportions depend on the cultivar and environmental conditions. The pericarp originates from the ovary wall and is divided into three histochemical tissues: the epicarp, mesocarp and endocarp. Sorghum is the only cereal grain known to have starch in the mesocarp layer. The endosperm is composed of the aleuronic layer and the peripheral, corneous (glassy) and floury areas. The corneous and floury endosperm cells are composed of starch granules, a protein matrix, protein bodies and cell walls rich in cellulose, β-glucans, and hemicelluloses. Endosperm happens to be the main storage tissue. The starch granules are polygonal and often contain dents from the protein bodies. The size of the starch granule varies from 4 μm to 25 μm. The ratio of corneous to floury endosperm determines the kernel texture in terms of grain hardness; the higher the corneous portion, the harder the kernel will be and vice versa. The corneous portion will be translucent, whereas the floury layer will be opaque. The aleuronic tissue is made up of thick cell walls and happens to be a rich source of protein, oil and minerals. The enzyme proteins are also mostly located in this tissue. The germ consists of the embryonic axis and scutellum and it contains reserve nutrients and serves as the bridge or connecting tissue between the endosperm and germ. The embryo is a very good source of protein, minerals, oil and vitamins (Waniska and Rooney, 2000).

5.5 Sorghum Processing

Processing involves the partial separation and/or modification of the three major constituents of the sorghum grain – the germ, the starch-containing endosperm and the protective pericarp. Various traditional methods of processing are still widely used, particularly in those parts of the semi-arid tropics where sorghum is grown primarily for human consumption. Most of the traditional processing techniques are laborious, monotonous and manual. They are almost entirely left for women to do. To some extent, the methods that are used have been developed to make traditional foods to suit local tastes and are appropriate for these purposes. Traditional techniques that are commonly used include decorticating (usually by pounding followed by winnowing or sometimes sifting), malting, fermentation, roasting, flaking and grinding. These methods are mostly labour intensive and give

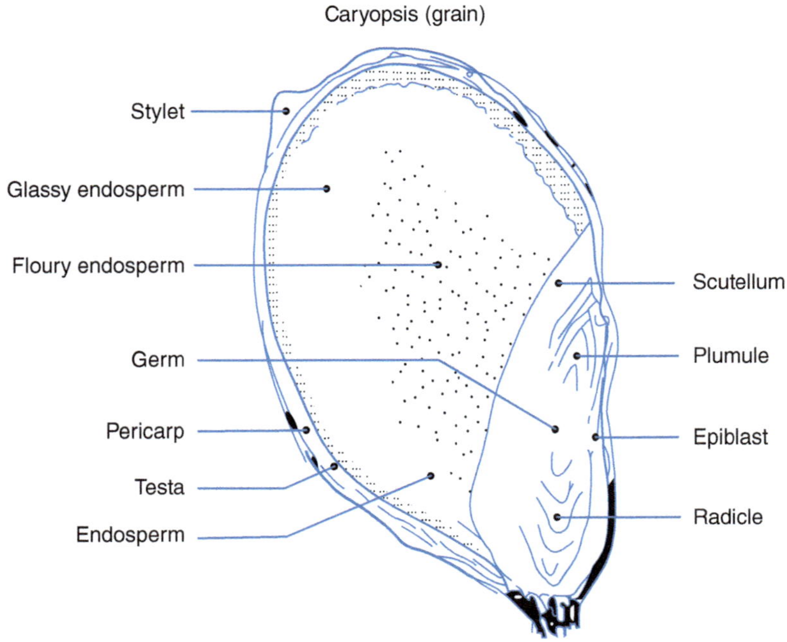

Fig. 5.3. Structure of the sorghum grain. (From Sautier and O'Deye, 1989.)

a poor-quality product. Sorghum would probably be more widely used if processing were improved and if sufficient good-quality flour were made available to meet the demand (Eastman, 1980).

Sorghum processing methods have been either fine-tuned or developed by IIMR using modern equipment to prepare good quality processed products like dehulled sorghum, semolina or *suji*, flakes, extruded products (vermicelli and pasta), biscuits, sorghum-rich multigrain *roti* and sorghum-rich multigrain flour, to improve the nutritional quality as well as the consumer acceptability of sorghum. There are three types of sorghum grain processing:

1. **Primary processing:** cleaning, grading (depends on size of the grain), destoning (removal of stones) and dehulling (polishing) of the grain. The bulk of these operations can be done mechanically. The removal of waste, stones and glumes from the grain is necessary as it improves the storage of the grain and consumer acceptability for usage. It also includes grinding or milling to prepare flour and semolina.

2. **Secondary processing:** the use of the primary processed raw material to make different RTE (biscuits, flakes, pops and multigrain *roti*) and RTC sorghum products such as flour, sorghum-based multigrain flour, semolina (fine, medium) and parboiled sorghum.

3. **Tertiary processing:** the use of the secondary processed raw material to produce commercialized products. Tertiary processing leads to high value-added RTC and RTE products like those that are extruded (vermicelli and pasta), *roti* and all recipes prepared.

5.5.1 Sorghum milling technology

Milling is the single most important aspects of sorghum processing. Sorghum grains are rarely cooked in the form of grain, and even then only after surface screening. The cooking time needed to obtain a soft-textured edible sorghum product will also be considerably higher than that needed for rice. Hence, sorghum is invariably pulverized to prepare flour or fine grits and so the process of sorghum milling is more or less comparable to that of wheat milling for the preparation of wheat *atta* and semolina. Milling of sorghum involves the preparation of whole meal, size grading to semolina or flour and also pearling or decortications followed by size grading (Fig. 5.4).

Sorghum used to be cleaned manually and then milled, but nowadays it is cleaned mechanically even at the household level, using a destoner machine (Fig. 5.5). The machine is integrated with an aspirator and grader which help in cleaning and grading the material. Stones, metal pieces, glass, mud particles and other high-density impurities as well as straw, chaffs and such other low-density impurities are separated from sorghum in one operation. Even small and lightweight pebbles which are of the size of a grain can be separated, thus ensuring optimal cleaning.

Historically, sorghum flour was prepared using a grinding stone (a flat slab and stone roller) with or without wetting (Fig. 5.6), a pestle and mortar, or the hand pounding method (Fig. 5.7). Normally, these practices were followed to prepare the meal for the day. Subsequently, grinding sorghum manually in stone discs became popular and each house would have a hand-operated stone grinder to prepare the flour (Fig. 5.8). These methods of sorghum milling were highly time consuming and caused a lot of drudgery – mostly to women who normally prepared food for the family.

But with the advent of milling technology and easily available small-scale cereal milling machinery, milling or pulverization of sorghum began to be carried out at mechanized mills

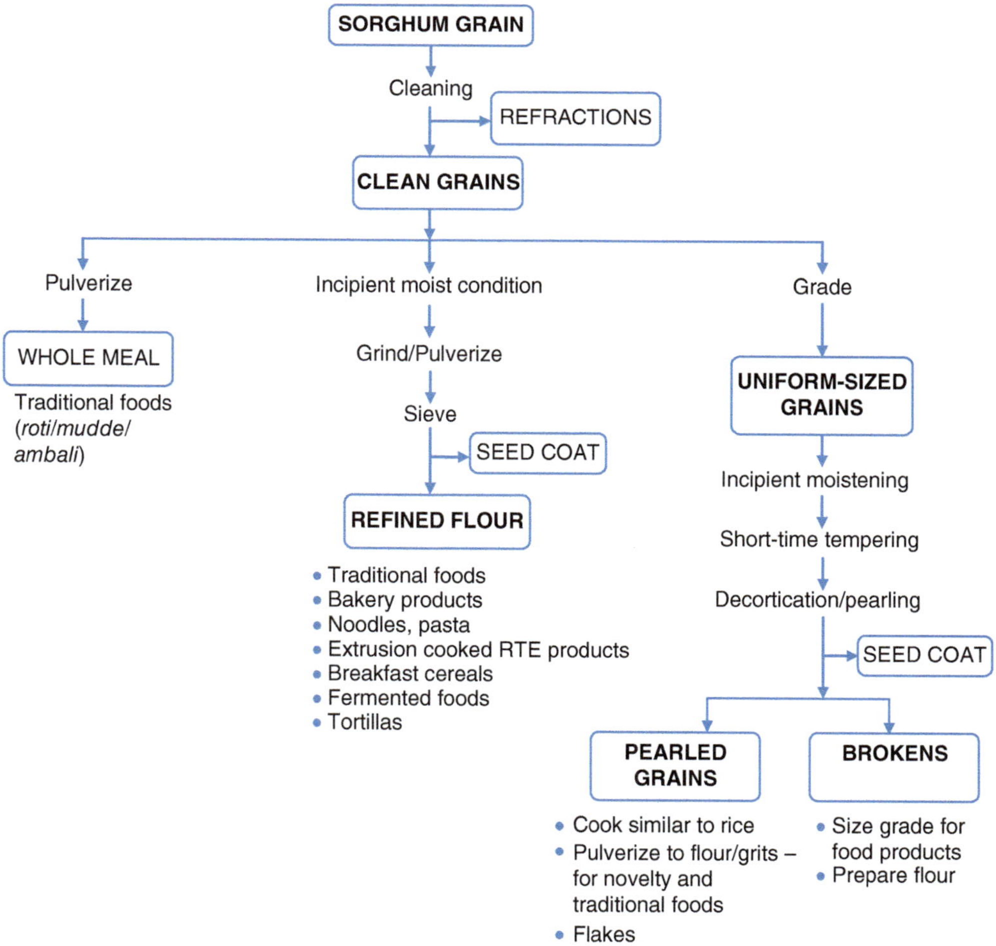

Fig. 5.4. Flow chart of the milling process (pulverizing and decortications) for sorghum.

Fig. 5.5. Destoner machine for cleaning millet grains.

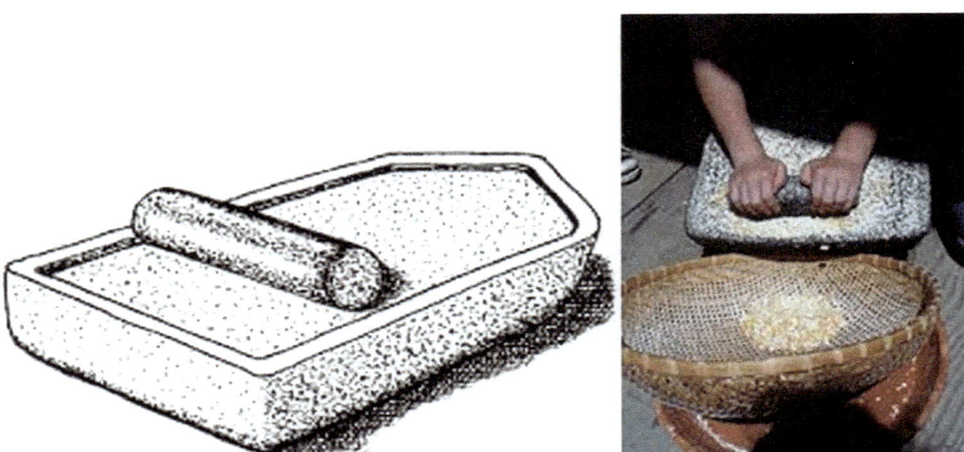

Fig. 5.6. Grinding stone for hand pounding.

(Eastman, 1980). For this, the pre-cleaned sorghum is pulverized or ground in stone discs, iron discs (Fig. 5.9) or in a carborundum disc mill (Fig. 5.10) to prepare wholemeal (flour). In the carborundum disc mill, one of the plates will be stationary and the other rotates at about 1440 rpm. The plates are of 12–18″ (30–45 cm) diameter in the case of custom mills and about 36″ (91 cm) in the case of industrial mills. The grains are fed through the feed hopper and get crushed and pulverized between serrated discs by shear and attrition action. The particle size of the products could be

Fig. 5.7. Traditional hand-operated mortar and pestle.

tailored by adjusting the gap between the discs. Normally, the flour prepared from these milling systems would be of finer particle size, as more than 95% of the flour is less than 300 μm.

The millet grains should be dry with less than 12% moisture content otherwise clogging of the flour occurs between the plates. The mill output in the form of flour is used as it is, or sometimes it is sieved to separate out coarse bran and the glumes, in which the yield of the flour would be about 95% of the original weight of the material. During grinding, considerable heat is generated and a substantial proportion of starch damage occurs. The meal thus prepared is mostly used for traditional foods such as *roti, mudde* and *ambali*. In the case of finger millet, it has been reported that the flour resulting from a disc mill contains a higher proportion of damaged starch compared to a hammer or roller flour mill (Katti *et al.*, 2008) and the same may

Fig. 5.8. Traditional hand-operated stone grinder. (From Martin, 2007.)

also hold good for sorghum. Since whole meal contains bran and other seed coat constituents such as sugars and amino acids, the *rotis* prepared from whole meal develop a highly desirable characteristic aroma during baking and will have a mild sweet taste.

The whole meal sorghum may be prepared irrespective of size, shape and texture of the grains. Varietal variations also have no direct bearing on the process but some of the physical properties of the flour will be influenced by the colour and hardness of the grains. Normally, the softer the cultivars, the finer the flour will be. Varieties with coloured seed coats yield slightly darker, grey or mild reddish flour. Sometimes sorghum gets wet due to rains at pre-harvest and shows incipient sprouting. The flour from such grains will have poor dough qualities because of the presence of active amylases (Murthy *et al.*, 1984).

Hammer mills or comminuting-type mills are also used for grinding of sorghum at a commercial level (Fig. 5.11). One such unit exists in IIMR, Hyderabad, wherein the pulverized sorghum gets sifted using a vibro-sifter to obtain low seed-coat containing flour. However, the whole meal flour prepared either in a disc or hammer mill will be slightly darker in colour with specks and will exhibit poor shelf life. The sorghum seed coat matter is mostly cellulosic and will be of a coarse fibrous nature. It affects the culinary and sensory qualities of the products. This is more so in the case of sorghum flour intended for use in non-traditional food products such as bakery goods and vermicelli. Hence, it is desirable to separate out most of the seed coat matter to prepare refined flour for traditional and non-traditional food products. There is no doubt that pearling or decortication is the most ideal method of preparation of seed-coat free sorghum but it leads to loss of considerable endosperm due to breakage, especially in the case of sorghum with a softer endosperm. Moreover, it needs special machinery for the separation of seed coat and also size grading machinery (the process is described in detail elsewhere).

The easiest process for preparing refined sorghum flour or grits involves incipient moist conditioning of the grains and a short period of tempering, grinding and sieving; the grains are mixed with about 3% additional water, tempered for about 10 min and then pulverized and immediately sieved to separate out the coarse seed coat matter from the flour (pulverized endosperm). Water mixing and a few minutes of tempering hydrates the seed coat matter only, without affecting the endosperm texture. This imparts a leathery texture to the cellulosic and hemi-cellulosic matter of the seed coat, leading to its reduced fragmentation during grinding. As a result, the seed coat comes out in the form of flaky material, which can be separated by sieving or aspiration. The process enables preparation of good quality semolina and white flour depending on the desired degree of fineness. However, the flour and semolina may contain a small proportion of finely ground seed coat. In the process, the major portion of the germ also gets pulverized and mixes with the flour. Since the flour contains most of the pulverized germ as well as the bran portion, it will be nutritionally superior to fully refined flour. But because of the high oil content which is in the form of ruptured globules derived from germ and the aleuronic tissue, the keeping quality of the flour will be poor. The refined flour prepared by this method is more suitable for traditional food products as well as bakery and extruded foods. However, the semolina can be graded into fine and medium grades and is useful for *upma*, vermicelli and similar products (a brief account of sorghum semolina is given elsewhere).

CFTRI, Mysore, has developed a simple integrated machine having provision for mixing water with the grains, tempering, grinding and sieving of the pulverized material to separate it into seed coat, coarse semolina, fine semolina

Fig. 5.9. Iron disc mill.

and flour. This machine is known as a 'mini grain mill'. The mill comprises the simple secreted iron disc mill or commonly used plate grinder, to which a water mixer and a triple deck sifter are attached (Fig. 5.12). The grains are poured into the hopper, from which they move through a screw conveyer to the feed hopper of the mill. At the entry of the grains in the screw conveyer, a predetermined quantity of water is added through a regulator. The water moistens the seed coat of the grains uniformly while they traverse to the hopper and get tempered in the process. From the hopper, the material is fed to the discs. The gap between the discs is suitably adjusted and the pulverized material falls over the deck and gets screened by the sieves of different mesh sizes. From the top, the screens will be of 16, 24, 44 and 60 mesh, respectively, fixed to a vibratory deck. The pulverized material gets fractionated; the seed coat comes out as tailings at the top

Fig. 5.10. Carborundum disc mill.

sieve whereas coarse semolina, fine semolina and the flour fractions are collected separately from the second, third and fourth sieves, respectively. The gap between the discs could be adjusted to prepare flour or semolina fractions of desired yield; the smaller the gap, the higher the yield of finer material and vice versa. Thus the mill enables preparation of seed coat-free semolina of different particle size, and sorghum flour almost free from seed-coat contamination in one unit operation (Shankara *et al.*, 1985). It is a very useful small-scale mill for sorghum semolina and flour preparation, which is easy to operate and could be used even at the farm level and in rural areas.

Roller flour milling is universally used for grinding wheat into *maida* (refined wheat flour), *atta*, semolina and bran (Munck, 1995). Wheat is equilibrated to about 16% moisture to toughen the bran, to soften the endosperm and to loosen the germ. During milling the bran and germ separate out and the endosperm is size graded to desired products such as flour and semolina. Milling of sorghum in a roller mill is similar to wheat milling. The moistened and tempered grains are crushed between the rolls consisting of break (corrugated) and reduction (smooth) rolls. In the break rolls, the grains break open and the softer portion of the endosperm comes out as flour almost free from bran fraction. Subsequently, the tailings are passed through reduction rolls to extract the endosperm in the form of semolina or flour. Corrugations and also the clearance between the rolls are adjusted to different degrees so that the endosperm is extracted as semolina and flour, whereas the seed coat is obtained as flaky bran and the aleuronic layer comes out as relatively fine flaky material known as 'shorts'. Mills of this type are in use from the lab to industrial level. The Quadrumat Jr. Mill (Fig. 5.13), a precision lab roller mill at IIMR, Hyderabad, was successfully used to prepare refined flour suitable for cookies and also for vermicelli and pasta from sorghum.

The flour from a roller mill contains lower levels of starch damage and is more suitable for bakery purposes. Normally, it makes bland *roti* and is not preferred for conventional sorghum foods. It is difficult to obtain a clean separation of purely white endosperm of sorghum from the bran fraction and the final flour will have a few specks of shattered bran which may be difficult to remove by sieving (Hulse *et al.*, 1980). A milling process involving the friction-decortication of sorghum followed by roller milling has been developed in Denmark (Hallgren *et al.*, 1992). If the raw material is properly conditioned, the process produces more or less intact sorghum endosperm (grits) and embryo removal (Dlamini *et al.*, 2007). Industrial-scale sorghum processing by Swiss milling technology company Bühler advocates hydrothermal treatment followed by roller-milling, or removal of the outer layers in a vertical whitener followed by hammer-milling.

5.5.2 Wet milling for starch extraction

Starch can be extracted from sorghum by wet milling. The process is similar to that for starch extraction from maize (Watson, 1970). The pigments of sorghum may discolour the final starch and to prevent this, low-tannin sorghums without testa are preferred (Yang and Seib, 1996).

Fig. 5.11. Integrated hammer mill or comminuting-type mill for millet milling at IIMR.

5.5.3 Sorghum dehulling/decortications

Many of the sorghum cultivars possess slightly to intensely coloured seed coats. Moreover, the harvesting practices in India are still largely on muddy ground, often causing surface contamination. In view of these, pearling or decortication of sorghum is highly desirable before milling, to obtain safe and healthy products from it. The demand for sorghum flour and other products will increase if a supply of these with acceptable texture, taste and colour can be guaranteed. Decorticated sorghum will have better consumer acceptability and could also be cooked similarly to rice (Fig. 5.14). Thus decorticated sorghum will have multipurpose usage and can offer scope for value addition.

A dry mechanical system is one way in which good quality flour and semolina could be supplied in sufficient quantities to meet commercial requirements. Efforts have been made to decorticate sorghum grain in the dry state using different kinds of dehullers or decorticators. However, sorghum pearling is carried out by dry milling. Normally, the mills are equipped with emery or carborundum discs or cones. The cones could be vertical or horizontal and are covered by screens of suitable openings which facilitate screening out of the seed coat and bran portion and also screen out small broken endosperm particles to get pearled grains. The feed and abrasive rate as well as the frictional pressure applied is adjusted to prepare pearled grains of the desired degree of decortication (Kebakile *et al.*, 2008). The degree of polish or decortication could be adjusted to the desired level by applying weights on the lever, which helps to develop pressure inside the dehuller and also helps in the retention of the material inside the mill.

Generally, the degree of polish or decortication varies from 8 to 15%. It means that 8–15% of the grain matter is separated out during milling. Based on the seed coat content and hardness of the sorghum, the degree of decortications or the polish and endosperm breakage varies. Normally, sorghum with a higher proportion of floury

Fig. 5.12. The mini grain mill.

endosperm provides higher levels of brokens (a grade in milling consisting of grains that are broken during the milling process) or flour. The pearled grains generally retain the germ.

Sorghum needs gentle abrasion so that the peripheral seed coat detaches and separates out during dehulling or decortication without causing much breakage and loss of the endosperm matter. Several prototype dehullers have been developed for the dehulling of sorghum. Many of the machines use some form of carborundum stone to abrade or wear off the seed coat. In most cases the stone is rotated inside a metal housing. As the kernels move through the unit between the stones and the housing, the seed coat is abraded. The carborundum stones may be in the shape of a cylinder, a truncated cone or a series of discs mounted on a central shaft. The metal housing, which conforms to the shape of the carborundum stones, may be perforated to allow the fine abraded material (seed coat) to be removed either with or without the aid of a fan. Amongst such mills the IDRC dehuller (Fig. 5.15) is most popular. It consists of a series of carborundum discs mounted on a central shaft at about 1.5–3.0 cm intervals housed in a metal case with provision for the material to be fed through a hopper and an opening at the bottom to discharge the milled grains. The discs rotate at variable speed and the kernels are pressed against the discs and continuously debranned until all the bran is removed to the desired degree.

The dehulled kernels are separated from the seed coat by other means, such as external aspiration or hand sieving. In the case of the PRI dehuller a series of carborundum discs is mounted on a horizontal shaft. The rotor is mounted in a metal case with a clearance of 1.5–3.0 cm at the sides and the bottom of the rotor. A screened air inlet is provided along one side of the top and there is an air outlet at the opposite side. By connecting the air outlet to an aspiration system, the fine particles formed during abrasion are automatically removed. The unit may be operated without either air inlets or outlets, in which case the finely ground material leaves the metal case.

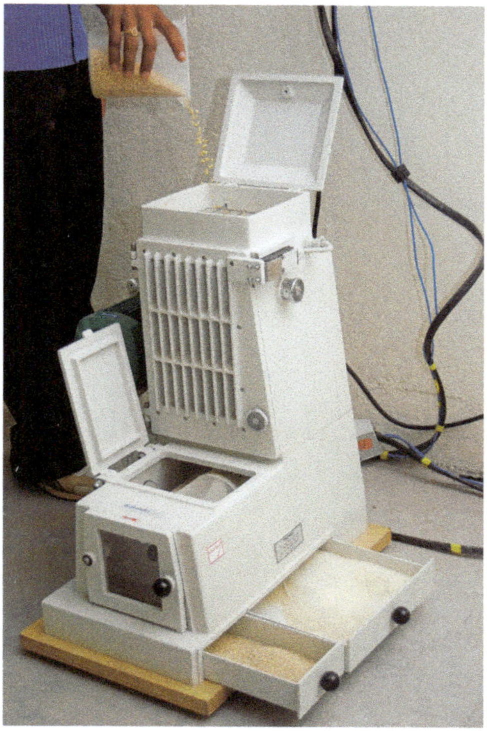

Fig. 5.13. The Quadrumat Jr. mill.

The grains are fed into one end of the dehulling unit at a uniform rate and dehulled grains exit at the other end. The dry dehuller or debranner for pearling sorghum (developed in Canada by Palyi-Hansen – a horizontal debranner) has a multiple combination of hollow discs covered on both sides with abrasive corrugated blades. The discs rotating at variable speed are arranged along with the horizontal shaft. The kernels are pressed against the discs and continuously debranned until all the bran is removed as desired. The bran particles are removed by the air pressure through the screen. The discharge of the stock from the debranner is controlled by opening the clamped door, which is regulated by the counter weight (Rasper, 1979).

An improvement in decortication efficiency has been shown by the process of moist-conditioning and pearling carried out at CFTRI, Mysore. This involves uniform application of about 3% water, tempering for about 10 min and then subjecting to decortication. The moisture wets the seed coat, rendering it leathery and allowing the seed coat to peel off easily in a flaky form. This process also reduces kernel breakage. Furthermore, the milled grains do not require drying because the water addition is just sufficient to moisten the seed coat without penetration into the endosperm. The decorticated grains are suitable for cooking, similar to rice but with a longer cooking time.

Fig 5.14. Decorticated and cooked sorghum.

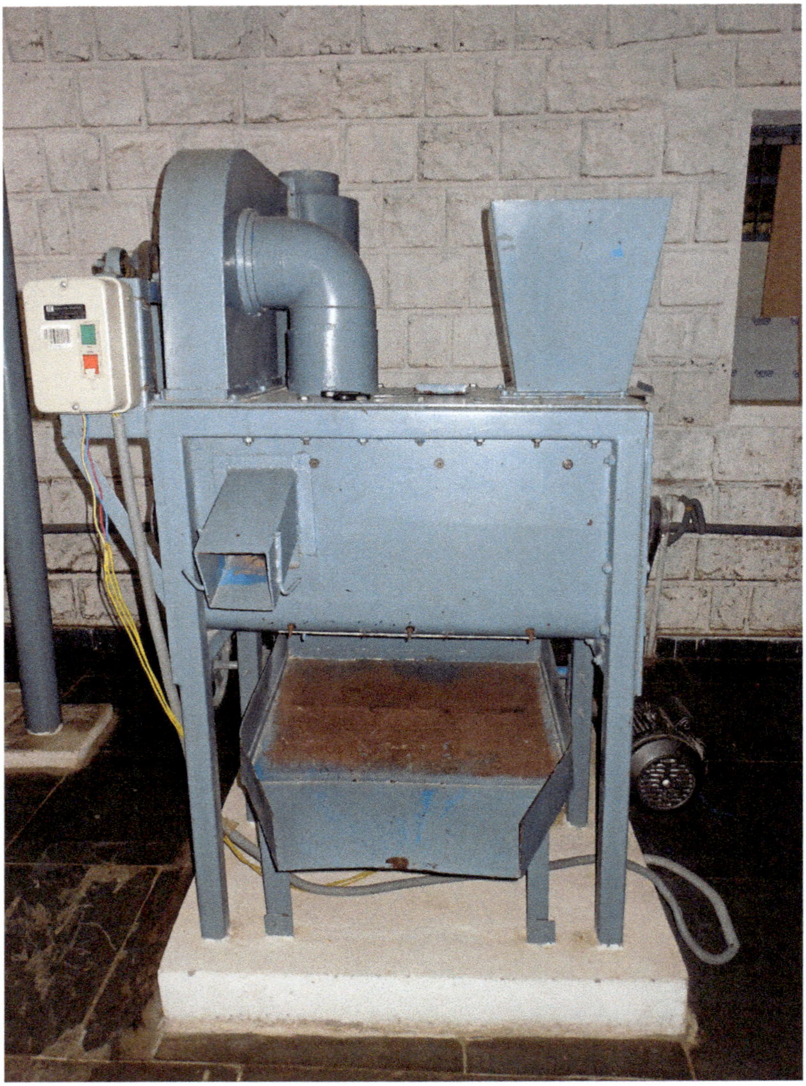

Fig. 5.15. IDRC dehuller.

The commonly used rice milling machinery for custom milling, the rice huller (the Engelberg huller, a horizontal ribbed rotor, rotating in a cylinder), or the commercial cone polisher could also be used for pearling of sorghum. Viraktamath *et al.* (1971) moistened sorghum with 3–5% extra water to soften the outer bran in a horizontal paddle-type screw-conveyer/mixer, conditioned it for 5–15 min, and then pearled the grains by passing them repeatedly through the huller mill or by passing them through more than one cone in the commercial rice mill. The milled material was screened and aspirated for separation of bran, brokens and hulls/glumes. The polished sorghum was also milled in a roller flour mill normally used for wheat. *Suji* (semolina), *maida, atta* and bran fractions were collected. The culinary properties of the milled fractions were studied by preparing *roti, dosa, upma, idli* and other traditional products using conventional methods, and the organoleptic acceptability of the prepared products was assessed by a taste panel. It was reported that, as degree of polish increased, the protein content of the material decreased. The pearled grains were suitable for cooking similarly to rice, and the semolina and flour were suitable for traditional food products such as *upma* and *roti*. The pearled grains could be ground into semolina or flour for various food preparations and value added products. Since the demand exists for high quality stable sorghum flour with low fibre and fat content, there is a need for the development of a simple machine for decortication without the loss of endosperm.

5.5.4 Effect of parboiling on milling and cooking quality

Parboiling of rice is practised on an industrial scale in the Indian subcontinent whereas bulgurization, the process of parboiling soft wheat, is carried out in the USA and many of the Gulf countries. Parboiling of rice is gaining popularity worldwide, including some of the African countries, because of its technological and nutritional advantages. Parboiling of cereals heals cracks and hardens the endosperm, thereby drastically reducing breakage during milling and ultimately increasing yield of head grains (grains that are not broken during milling). Further, parboiling retains micronutrients such as minerals and B-group vitamins, as they bind with other macromolecules and enhance the resistant starch content, with leaching of soluble sugars. In addition to these technological and nutritional benefits, the cooking characteristics of parboiled millets improve and the cooked grains are less sticky. The process also improves the storage life of the grains, minimizing the risk of insect and microbial attack during storage, and facilitates secondary processing for preparation of expanded products.

Parboiling of sorghum is not commonly practised but the exploratory studies carried out in different laboratories have been promising (Serna-Saldivar *et al.*, 1994), indicating that its pearling or decortication characteristics improve considerably on parboiling. Sorghum cultivars with soft endosperm, floury textured sorghum and high tannin *milo*-type sorghum can be parboiled to enhance their food product making qualities. However, highly pigmented/high tannin varieties turn slightly dark on parboiling due to polymerization of polyphenols, although the endosperm will be fairly white.

Parboiling of sorghum (Fig. 5.16) is mainly a hydrothermal treatment and involves hydrating the grains – normally to their equilibrium moisture content – steaming to gelatinize the starch, and then finally drying the grains. The clean grains are soaked in cold water for 12–18 h or for about 6 h in hot water to hydrate the grains to their capacity. The water is drained off and the soaked material is steamed for about 30 min at atmospheric pressure or for about 10 min at higher pressure. The seed coat prevents the grains from bursting open during steaming, although the starch gets gelatinized and the grain swells to some extent. Thereafter, the steamed material is sun-dried or hot air-dried to a safe storage moisture level (12–14%) at less than 80°C (Fig. 5.17).

Dry heat parboiling can also be applied to sorghum, which involves subjecting hydrated grains to 'high temperature short time' (HTST) treatment in hot sand. The parboiled grain will normally have a slightly opaque appearance. The process can be applied to all kinds of sorghum, irrespective of shape, size, variety, agro-climatic conditions of cultivation and harvesting method. Steaming destroys insects and hence mildly infested grains can also be parboiled. A study conducted as part of the NAIP pilot project examined the milling characteristics of parboiled sorghum grain. It found that good quality decorticated grains with retained germ can be prepared using parboiled grain. Sorghum grain was soaked until saturated, drained and then steamed or heated to gelatinize the starch. It was then dried, dehulled and 'milled' (decorticated). Whole sorghum grain was suspended in tap water for 16 h at room temperature, steamed under 20 kg/cm^2 pressure for 20 min, tempered for 30 min to cool and dried at 60°C for 24 h in a tray dryer (Sandeep Instruments, Delhi). The milling yield of parboiled grain was 82.5% whereas the yield of control was only 65% (Table 5.1). Thus, hardening the endosperm during parboiling enables preparation of semolina with the desired particle size (Dayakar *et al.*, 2014). Furthermore, parboiling allowed sorghum to preserve all the nutritive substances and vitamins.

However, parboiled sorghum is not an RTC or RTE product and requires milling, mainly size grading to semolina, pearling to obtain decorticated grains or further processing to prepare flakes or expanded cereals. In the same study, grains were milled into medium and fine semolina and assessed for their physicochemical characteristics, shelf life and sensory values.

Fig. 5.16. Cottage-scale parboiling unit.

SORGHUM GRAIN

↓

Soak in cold water for about 12 h or in hot water for about 5 h

↓

HYDRATED GRAINS

↓

Drain off the water

↓

Steam for about 30 min at atm. pressure or about 10 min at 3–5 bar pressure

↓

Equilibrate for about 1 h

↓

Dry (sun drying or mechanical drying)

↓

PARBOILED SORGHUM

Fig. 5.17. Flow chart of the parboiling process in sorghum grain.

Parboiling treatment significantly increased the firmness of the cooked kernels (Fig. 5.18) and decreased the cooking losses when compared with the control (Table 5.2).

Moisture content, ash and alcohol acidity were increased in all samples with a storage period of 90 days. However, with parboiling the milled semolina shelf life increased to 120 days. The organoleptic acceptability of parboiled fine semolina and medium semolina scores also remained the same, up to 3 months, whereas the control fine and medium semolina showed changes. Hence it can be concluded that parboiling increases the shelf life of the sorghum semolina without any deterioration, having up to 3 months' storage time.

5.5.5 Standardization of particle size for different products

Particle size of the milled cereals depends upon the nature of cereal, the milling or grinding methods and the machinery used. Sorghum cultivars can have hard (vitreous) or soft (floury) endosperm and this endosperm texture influences

Table 5.1. Total material loss due to parboiling during milling (%).

Grain type	Boiling loss	Ruptured kernels	Bran	Total material loss[a]	Milling yield
Control	0.0	0	35	35.0	65
Parboiled	1.3	0.2	16	17.5	82.5

Note: [a], total material loss = boiling loss + ruptured loss + bran loss.

Fig. 5.18. (A) Dehulled sorghum and (B) parboiled sorghum.

Table 5.2. Cooking quality of the parboiled grain. (Source: Dayakar *et al.*, 2014.)

Grain type	Weight of semolina (%)	Water absorption (ml)	Cooking time (min)	End product weight (g)	Observations
Native sorghum grain	50	400	25	233	Sticky texture
Parboiled sorghum grain	50	400	35	216	Discrete grains, non-sticky

particle size after milling or grinding. The floury type yields a majority of finer particles whereas the vitreous type yields slightly coarse flour under comparable conditions of pulverization. Vitreous sorghum also facilitates the preparation of bigger size grits and a higher proportion of semolina. The hard portion of the endosperm contains a rigid protein matrix in the form of cell walls and hence influences its food preparation characteristics. Because of these features, the particle size also influences the product-making quality of the flour. The finer the particles, the higher will be the hydration capacity at ambient conditions. Normally, softer grains contain relatively lower protein. The hydration capacity of the dough also depends on the damaged starch content of the flour. Pulverizing the hard endosperm to prepare finer flour generally causes greater damaged starch (Hulse *et al.*, 1980). The damaged starch behaves similar to partially gelatinized starch and, as a result, the flour absorbs higher levels of water during dough preparation and the dough will become stickier. The grinding machinery also influences the particle size as well as the damaged starch content of the flour. Normally, the flour prepared in a disc or plate mill produces higher proportions of damaged starch. Sorghum flour of finer particle size (60 mesh) with fairly high levels of damaged starch will be more suitable for *roti* because the dough can be rolled easily.

On the other hand, the flour prepared in a hammer or comminuting mill will be more suitable for *mudde*. Generally, grits from the harder endosperm will be better raw material for pasta and vermicelli. Medium-size grits are more suitable for products such as *nucchu* and cous cous. But for extrusion cooking both flour and fine grits could be used. Medium to fine sorghum semolina could be used conveniently as a substitute for rice/wheat semolina in traditional products such as *idli*, *dosa* and *upma*. The semolina from mild hydrothermally treated sorghum will be less sticky compared to normal semolina and hence will be more amenable for *upma*-type food products. To date, there is no mill exclusively used for the preparation of sorghum semolina and it is mostly prepared using the available rice and wheat milling machinery. Because of this, the yield of semolina will be at most 30% of the dry matter of sorghum kernels. Hence, there is a need to develop appropriate milling machinery that is suitable for semolina production.

IIMR has made an attempt to prepare and standardize coarse, medium and fine semolina and flour from sorghum for various recipe preparations using particle size analysis. For this, the particle size distribution in the different products was measured by sifting them through standard BSS test sieves (Jayant). Two or three mesh sizes for each sorghum product were used: coarse semolina (12, 14 and 16), medium semolina (22 and 30), fine semolina (44 and 60) and flour (60 and 85). Results showed that sorghum semolina possessed good particle size distribution compared to the control (wheat semolina) samples (Table 5.3), indicating the former's suitability for commercialization.

5.5.6 Formulation of designer semolina for enhanced resistant starch

There is considerable interest in the nutritional implications of resistant starch (RS) in foods. The term 'resistant starch' was used to designate mainly enzyme-resistant retrograded amylose but was expanded to all forms of starch that escape digestion and absorption in the small intestines. Its positive effect on nutrition is based on the RS fermentation process, during which short-chain fatty acids – primarily acetates, propionates and butyrates – are produced. These fatty acids directly affect the large intestine by decreasing the pH value, which prevents the growth of pathogenic micro-organisms, and increases the potential for nutrient absorption through stimulation of colonic blood flow.

Besides physiological benefits to humans, RS has been reported to have potential as a unique ingredient that can yield high quality foods. For example, application tests of RS showed improved crispness and expansion in certain products with better mouth feel, colour

Table 5.3. Standardization of sorghum semolina particle sizes compared with a control (wheat) sample. (Source: Dayakar *et al.*, 2015.)

Product	Mesh	Micron size (μm)	Passing (%)	Held (%)
Control coarse semolina	12	1405	96.56	3.44
	14	1204	71.73	28.26
	16	1003	53.61	46.38
Control medium semolina	22	710	73.49	26.51
	30	500	13.063	86.93
Control fine semolina	44	355	35.86	64.13
	60	250	40.45	59.54
Sorghum coarse semolina	12	1405	99.11	0.88
	14	1204	82.24	17.75
	16	1003	37.94	62.05
Sorghum medium semolina	22	710	39.97	60.02
	30	500	30.36	69.63
Sorghum fine semolina	44	355	38.92	61.07
	60	250	57.72	42.27

and flavour as compared to products produced with traditional or insoluble fibres. Application tests of RS_3 (the type of resistant starch that is retrograded) after hydrothermal treatment was found to be due to increased interaction between starch components. It has been shown that after debranching of starch, the linear chains can contribute to a high RS content (debranching is when the complex starch carbohydrate is broken down using enzymes or heat treatment where carbon chains align and aggregate to form perfectly crystalline structures, thereby leading to the formation of resistant starch). In addition, partial acid hydrolysis and debranching of amylopectin are very effective in generating RS from various starches. The advantages of debranching over mild acid hydrolysis include a shorter processing time, better processing control and higher RS yields.

In the study carried out by ANGRAU, consortium partner of IIMR in the NAIP subproject, dehulled sorghum was soaked and autoclaved and then incubated with debranching enzymes; cooling and autoclaving cycles were then followed. The control used autoclaved dehulled sorghum without the debranching enzyme; the debranching enzyme (2%) is used to enhance the percentage of resistant starch content (Table 5.4). The study demonstrated an improved yield of RS from 12 g/100 g in the control method to 16.4 g/100 g in the present method with the usage of enzymes. This is a total increase of 27% in the RS content (Table 5.4). Further, the method with enzymes reduced the number of freeze–thaw cycles, thus reducing the overall cost and time (three freeze–thaw cycles and 3 days in the control versus two freeze–thaw cycles and 2 days in the experiment using debranching enzymes).

Table 5.4. Effect of enzyme on the yield of resistant starch (RS) content. (Source: Dayakar *et al.*, 2014.)

Treatment	Process time (min)	% increase	RS content (g/100 g)
Control (no treatment)	–	–	12
Debranching enzyme (2%)	70	27	16.4

5.5.7 Development and standardization of sorghum-rich multigrain flour

Sorghum-rich multigrain flour was developed and standardized with sorghum, wheat, finger millet, black gram *dal* and fenugreek (Dayakar *et al.*, 2014) (Table 5.5). Among the different blends studied, T4 was found to have satisfactory functional characteristics, such as nutritional composition, bulk density, water absorption and solubility index, and overall organoleptic acceptability was found to be greater than other sorghum-rich formulations. T4 was then analysed for shelf life using tertiary butylhydroquinine (TBHQ; 0.02% per 100 g sample) on moisture (%), water activity (a_w), alcoholic acidity and acceptability studies at 0, 30, 60, 90, 105 and 120 days' storage period by a panel of judges using a five-point hedonic scale. No significant difference was found between control and TBHQ-added samples for moisture content, a_w and alcoholic acidity. Organoleptic properties (colour, flavour, texture, taste and overall acceptability) of T4-blend *roti* were better than those of the control samples during a storage period of 120 days with the addition of TBHQ.

Table 5.5. Formulations used for the development of sorghum-rich multigrain flour. (Source: Dayakar *et al.*, 2014.)

	Flour (%)				
Formulation	Sorghum	Wheat	Finger millet	Black gram	Fenugreek
T1	20	75	1	2	1
T2	30	65	1	2	1
T3	40	55	1	2	1
T4	50	45	1	2	1
T5	60	35	1	2	1
T6	100				

5.5.8 Intervention on sorghum flours

Sorghum flour (whole and dehulled) and multigrain flour were treated with mould inhibitor and acid inhibitor at different percentages (Table 5.6) and placed in high density polyethylene (HDPE) pouches and kept at room temperature. All the samples were tested for moisture, microbial content, free fatty acids, pH and alcoholic acidity periodically for up to 6 months.

Both bacterial and fungal growth were monitored for 6 months during storage and it was found that up to 4 months the counts were in permissible limits. Beyond 4 months the growth crossed the limit (50,000 colony-forming units, cfu) and more counts were seen in the flours without additives. However, the least microbial growth was seen in the T7 formulation, indicating that sorghum-incorporated flour with additives is more resistant than those of the other formulations. The moisture content of the flour was within the permissible limits (14%) and ranged from 9.6 to 11.5%. Flour formulations without the additives exhibited lower moisture changes during storage. For up to 4 months, the free fatty acid (FFA) and alcohol acidity of flour was within the permissible limit. FFA was lowest in the T7 formulation and highest in T10 and a similar trend was seen in the case of alcoholic acidity. Overall, the study demonstrates that incorporation of mould inhibitor and acidity regulator could arrest significant changes in FFA, alcoholic acidity and microbial growth, which are detrimental to the quality of the flour. Among the studied flour combinations, the sorghum flour-incorporated formulation (T7) with 2% acidity regulator and mould inhibitor best maintained the quality characteristics during storage, followed by T5, T1 and T3. However, with improved packaging having higher moisture and oxygen barrier properties it might further enhance the shelf life.

5.5.9 Storage studies on sorghum rich multigrain flour in different packages

The effect of storage of sorghum rich multigrain flour (SRMF) packed in low density polyethylene (LDPE), high density polyethylene (HDPE) and polypropylene (PP) under accelerated environment (38°C and 90% RH) on moisture (%), total ash (%), water activity (a_w), alcoholic acidity (%) and bacterial count was studied. All the parameters values increased on storage. Among the plastic packages, LDPE and HDPE gave the least protection, while PP gave the maximum protection against spoilage of SRMF during storage. The bacterial growth was observed on SRMF after 30 days of storage and was highest in the LDPE pack and lowest in the PP pack. Data on the organoleptic qualities showed that the quality of SRMF in HDPE and PP packaging remained unchanged for 30 days. SRMF packed in LDPE started deteriorating after 15 days of storage, as determined by the development of a musty flavour.

Table 5.6. Sorghum flour and multigrain flour formulations with and without flour additives. (Source: Dayakar *et al.*, 2014.)

Formulation	Whole sorghum flour (%)	Dehulled flour (%)	Wheat flour (%)	Defatted soy flour (%)	Finger millet flour (%)	Barley flour (%)	Mould inhibitor (%)	Acidity regulator (%)
T1	99.3	–	–	–	–	–	0.5	0.2
T2	100	–	–	–	–	–	–	–
T3	–	99.3	–	–	–	–	0.5	0.2
T4	–	100	–	–	–	–	–	–
T5	59.3	–	30	10	–	–	0.5	0.2
T6	60	–	30	10	–	–	–	–
T7	29.3	–	40	20	5	5	0.5	0.2
T8	29.3	–	40	20	5	5	–	–
T9	–	–	99.3	–	–	–	0.5	0.2
T10	–	–	100	–	–	–	–	–

5.5.10 Enhancing shelf life of sorghum flour and multigrain flour

Shelf life of sorghum flour was assessed using infrared radiation (at ANGRAU). A central composite design was used to optimize infrared heating parameters, set at 80–120°C for 5–15 min. Free fatty acids, lipase enzyme activity and peroxidase enzyme activity were estimated using the neutral alkaline, titrimetric and spectro photometric methods, respectively. Pasting profiles of native and treated samples were studied using the Brabender Micro Visco-Amylo-Graph (Duisburg, Germany) for 10% slurry concentration (10% flour, 90% water). Colour was measured using the Hunter colour measuring system. It was found that optimum conditions for lowest free fatty acid content, lipase activity and peak viscosity value were achieved at a processing time of 8.5 min and a temperature of 120°C (Table 5.7).

Sorghum flour processed at 120°C for 8.5 min exhibited a 70.98% and 15.1% decrease in moisture content and peroxidase enzyme activity, respectively, compared to control flour. Colour of control and treated flours, i.e. whiteness (L*) and average colour value (ΔE) ranged from 97.42–87.65 and 13.06–14.13, respectively. Pasting indices, i.e. hot peak viscosity and cold paste viscosity ranged from 272–248 BU (Brabender unit) to 582–511 BU, respectively, indicating the product has favourable pasta texture characteristics similar to traditional wheat pasta. The results convey that sorghum flour treated with infrared radiation at 120°C for 8.5 min can have a desirable storage stability.

5.6 Development of Sorghum Products

Flour, coarse, medium and fine semolina were prepared from the grain in a hammer mill (Centrifugal Compact Pulverizer, Centrifugal Products, Gujarat) through rotating blades that grind the grains in a grinding chamber and pass it through a screen which separates the flour from the larger and ungrounded particles. The recovery of sorghum flour was 85%. Sorghum grain was pulverized in a *chakki* mill (Sowbhagya Industries, Guntur) to obtain coarse semolina (1003 μm), medium semolina (500 μm) and fine semolina (250 μm) and separated using a vibration shifter (Sowbhagya Industries, Guntur). The recovery of coarse semolina was 20–25%, medium semolina 20–25%, fine semolina 25%, and the remaining 25% was flour. However, flour and semolina recovery depends on the cultivar and milling process.

5.6.1 *Roti, mudde* and *peda*

Roti

Roti, an unleavened pancake, is the most popular sorghum food product in India. *Rotis* are eaten along with cooked grain legumes seasoned with spice, fresh vegetables, spiced oilseed powders or

Table 5.7. Shelf life of sorghum flour with infrared treatment. (Source: Dayakar *et al.*, 2014.)

Infrared treatment						Colour	
Temperature (°C)	Time (min)	FFA%	Peroxidase value (OD at 470 nm)	Lipase activity (μeq.mg/h)	Moisture (%)	L*	ΔE
Control		0.2±0.14	0.64±0.12	90±0.02	15.81±0.02	88.73	13.06
80	10	0.15±0.07	0.63±0.30	85±0.02	8.18±0.02	88.57	13.27
86	6.5	0.14±0.01	0.66±0.14	85±0.03	9.58±0.14	88.76	13.11
	13.5	0.12±0.02	0.67±0.14	70±0.01	7.32±0.12	88.23	13.56
100	5	0.15±0.01	0.62±0.20	80±0.01	8.26±0.05	88.54	13.09
	10	0.11±0.02	0.63±0.21	70±0.20	6.28±0.02	88.58	12.97
	15	0.08±0.02	0.62±0.22	65±0.01	4.68±0.41	88.52	13.24
114	6.5	0.07±0.14	0.60±0.20	45±0.05	6.09±0.00	88.44	13.47
	13.5	0.04±0.14	0.52±0.12	20±0.05	5.03±0.04	88.54	13.15
120	8.5	0.03±0.00	0.53±0.16	10±0.21	4.79±0.04	87.65	14.12

Notes: OD, optical density; L*, whiteness; ΔE, average colour value.

chutneys. Thus the *roti* meal is generally nutritionally balanced. Normally, whole meal sorghum is used to prepare *roti* (Fig. 5.19). The preparation of *roti* differs from *chapathi*, a similar product made from wheat; because the sorghum protein does not contain gluten, the cohesive dough cannot easily be rolled.

In view of this, the starch needs to be partially or fully gelatinized to prepare the dough for rolling or flattening. Accordingly, sorghum flour is mixed with a suitable quantity of boiling water, heaped for a few minutes and then kneaded. During this process the starch content of the sorghum flour gets partially gelatinized and enables the preparation of dough. While kneading, a small quantity of cold water is added to prepare homogenous and cohesive dough that can be rolled into a flat sheet.

Normally, about 50 g of dough is flattened into 2–3 mm thick and 20–25 cm circular sheets either traditionally, by beating with the palm of the hand, or by rolling using a rolling pin. It is then baked over a hot plate by contact heat. The special feature of sorghum *roti* baking is that, soon after putting it on the hot plate, water is smeared on its surface using a wet cloth, allowed to dry and then the sheet is turned over and baked. Baking continues until the *roti* swells forming two distinct layers and develops a characteristic aroma (Murty and Kumar, 1995). Sometimes, half-baked *roti* is also exposed to direct heat, which facilitates swelling to their capacity. Nowadays, *roti* making has been mechanized, with machinery developed for rolling and baking; it is especially used in catering establishments. Often the flour/boiling water mixture is cooked for a few minutes before being kneaded, flattened and baked. *Rotis* prepared by this method are slightly thicker but will be of a highly soft texture. Sorghum *rotis* are normally consumed fresh while they are warm, otherwise they become brittle-hard on cooling and drying.

IIMR, Hyderabad, in active collaboration with a private entrepreneur, has developed a simple *roti*-making machine useful for households and small catering units (Fig. 5.20). This unique machine is able to flatten the dough into circular sheets, which can then be taken away

Fig. 5.19. Traditional sorghum *roti* preparation.

Fig. 5.20. Manual sorghum *roti*-making machine: hand operated (left) and foot operated (right).

for baking. The flattening process is almost comparable to hand beating and hence the texture of *roti* is similar to those made by the traditional method. The institute also has *roti*-making machines which can deliver about 1000 pieces per h (Fig. 5.21). Such machines are readily used in several industrial and educational catering establishments. In the automatic machine the dough is sheeted and moves onto a PTFE-coated belt. The sheet is then cut into a circular shape of 6–9″ diameter and baked soon after in the same machine, thus producing them in a hygienic manner. *Rotis* thus produced are packed and marketed. The freshly prepared *rotis* will have about 30% moisture and will have poor keeping quality, but when the same *rotis* are dried to about 6% moisture they can be stored and eaten along with *sambar* as and when required. Such *rotis* are very handy for the travelling population.

SHELF LIFE EXTENSION OF SORGHUM-BASED *ROTI*. The shelf life of different formulations for the development of sorghum *roti* was evaluated for a period of 7 days (Table 5.8). *Roti* quality was also evaluated in comparison with the whole sorghum flour control. The thickness of the *roti* was greater in the formulations T2, T3 and T4, where hydrocolloid was incorporated, and the spreading was more extensive in T4 because of the addition of hydrocolloid and oil. Except in the control, all *rotis* had a shelf life of 7 days without any mould growth and not much change in texture. However, in T1 where maltodextrin was used, the *roti* texture was good for up to 4 days, after which the *roti* became brittle and very hard to chew. Hence, the combination was modified with the addition of hydrocolloid and oil. Sensory scores showed that, among the formulations for improving shelf life, T4 scored high in terms of texture, taste and overall acceptability among the sorghum formulations; however, acceptance for control was highest.

Mudde

Mudde is an important traditional food product made from sorghum flour or fine grits. To prepare *mudde*, normally thin cold water slurry (about

Fig. 5.21. Automatic sorghum *roti*-making machine.

Table 5.8. Formulations used for sorghum-based *roti*. (Source: Dayakar *et al.*, 2014.)

Formulation/ ingredients (%)	Control	T1	T2	T3	T4	T5
Whole sorghum flour	60	60	60	60	60	60
Wheat flour	30	29.2	29.2	28.4	28.4	25.4
Defatted soy flour	10	10	10	10	10	10
Mould inhibitor	–	0.3	0.3	0.3	0.3	0.3
Acidity regulator	–	0.2	0.2	0.2	0.2	0.2
Maltodextrin	–	0.3	–	–	–	–
Hydrocolloid	–		0.3	1	1	1
Oil	–	–	–	–	1	1
Yeast	–	–	–	–	–	1
GMS	–	–	–	–	–	1

Note: GMS, glycerol monostearate.

1.5%) is heated and as soon as it starts boiling, a predetermined quantity of fresh sorghum flour is added to it and is left undisturbed for about 2 min. This facilitates steaming of the flour and improves its cooking quality. Then it is mixed while heating and this is followed by mixing-cum-kneading using a wooden rod until a homogeneous mass is formed. The mass is made into balls of about 200 g each and served with *sambar* or thin *curry* (a spiced product). Generally, a small portion (about 5–10 g) of the *mudde* is made into small balls with the fingers and is dipped in *sambar* and swallowed without chewing. *Mudde* is a fully cooked food product and interestingly it is the only cereal food that is simply swallowed without chewing; there is no initiation by the salivary enzymes in the mouth, there is slow digestion in the stomach and subsequently in the intestine. Hence, consumers experience satiety for a longer time when compared with other food products. Preparation of *mudde* without any lumps needs high culinary skill. A similar product from finger millet is also very popular in south India. It is prepared not only in households but also in big industrial catering services and institutional messes.

Sorghum bran peda

Peda is a popular Indian sweet traditionally presented as semi-soft balls. Sorghum bran *peda* is made from sorghum bran powder (45%), sugar (25%), ghee (14.5%), milk powder (15%) and cardamom (0.5%). The *peda* can be stored for 7 days at room temperature. The sorghum bran is roasted and ground to fine powder, to which sugar, milk powder and cardamom are added. Ghee is then added slowly and the resulting mixture is made into small balls, which are decorated with almond or cashew nuts. The organoleptic properties of the bran *peda* have shown an overall acceptability of 4 and rice *peda* have an overall acceptability of 3.5. Thus it is concluded that the sorghum bran *peda* is preferred to rice *peda*.

5.6.2 Sorghum bakery products

Bakery products such as bread, biscuits or cookies are extremely popular convenience foods in almost all countries. Internationally, refined wheat (*maida*) forms the main cereal product and raw material for these goods. *Maida* is a fully refined wheat product and does not contain any of the nutrients from its seed coat or wheat bran and the germ. On the other hand, sorghum flour even from dehulled grains provides a considerable proportion of dietary fibre and complex carbohydrates. Hence, sorghum flour as a raw material for bakery products will be nutritionally superior to refined wheat flour (Ratnavathi and Patil, 2013).

Accordingly, development of bread and bakery products from non-gluten cereals, including sorghum, have been developed mainly to provide consumers with their nutritional advantages. But wheat has an advantage due to its special protein (gluten) that forms cohesive dough even with cold water and exhibits very good rheological properties. Sorghum proteins lack this characteristic feature and hence its flour has limitations for the preparation of bakery products. In addition, the gelatinization temperature of sorghum starch is slightly higher than that of wheat starch and because of that the baking temperature would be higher compared to wheat. Also, the hydration characteristics of sorghum flour differ from wheat flour. These factors combine to cause a harder texture, a cracked surface and a slightly grainy mouth feel to sorghum biscuits (Fig. 5.22). Furthermore, sorghum dough will not hold the gas formed during yeast fermentation and because of that, the loaf volume of sorghum bread will be considerably lower when compared to wheat bread.

Fig. 5.22. Sorghum bakery products: (A) biscuits and (B) bread.

Considering these adversities, composite flours have been developed comprising wheat blended with sorghum flour, and their suitability for baking bread and cakes and also other types of products have been standardized (Fig. 5.22). Varietal variations of sorghum also exist with respect to the baking characteristics and hence identifying suitable varieties for bakery and other speciality foods have been carried out (Ratnavathi *et al.*, 2003). It has been observed that the composite flour comprising 80% wheat and 20% sorghum makes bread comparable to normal wheat-based bread in terms of loaf volume and other characteristics (Rao and Shurpalekar, 1976; Hugo *et al.*, 2003). On the other hand, cookies could be prepared with composite flour comprising wheat and sorghum at any ratio; cookies could even be prepared with 100% sorghum flour (Dayakar *et al.*, 2014). Steam-treated and also malted sorghum flour that contains higher levels of damaged starch makes better cookies (Rooney and Taylor, 2000). It is also observed that the flour of soft sorghum makes better quality biscuits. Sorghum flour prepared from a plate mill is more suitable for bakery products probably because of the high proportion of damaged starch.

Sorghum biscuits

The biscuits prepared at IIMR using composite flour were readily acceptable and exhibited desirable physicochemical characteristics. Consumption of bakery products is increasing globally and, at the same time, demand for non-wheat bakery products is also on the rise. In view of this, the use of sorghum flour in such products holds great promise. But in order to render it suitable for bread, biscuits and other bakery products, it is necessary to identify suitable cultivars and also to develop technological innovations to modify sorghum flour's functional properties.

In order to optimize the proportions used in biscuit production, different proportions were used to maximize the acceptability for commercialization at the large scale. At IIMR, pure sorghum, transfat-free and chocolate biscuits were standardized. Pure sorghum cookies were prepared with fat (16.5%), sugar powder (9%), water (24%), skimmed milk powder (3%), sorghum flour (45.0%), salt (0.4%), baking powder (0.6%) and essence (1.5%), while chocolate biscuits were prepared with sorghum flour (31%), wheat (24%), sugar (15%), fat (12%), skimmed milk powder (1.6%), cocoa powder (1.6%), chocolate chips (10%), flavouring (3.3%), salt (0.3%) and baking powder (1.2%). Transfat-free biscuits were prepared with trans-free fat (16.5%), sugar powder (9%), water (24%), skimmed milk powder (3%), sorghum flour (45.0%), salt (0.4%), baking powder (0.6%) and essence (1.5%).

EFFECT OF MILLING AND PARTICLE SIZE OF FLOUR ON SORGHUM BISCUITS. Sorghum (cultivar M35-1) was milled using either a hammer or a *chakki* mill to obtain flour of different particle size, i.e. 75 (200 BSS), 106 (150 BSS), 150 (100 BSS), 180 (85 BSS) and 250 (60 BSS) μm, and were used to prepare biscuits. Biscuits were subjected to physicochemical analysis using standard methods and their organoleptic properties were evaluated using a nine-point hedonic rating. Colour, texture and organoleptic properties showed that biscuits prepared from 250 μm and 180 μm particle-sized flours were more highly acceptable than the biscuits prepared from other particle-sized flours. Also, biscuits prepared from a *chakki* mill were more acceptable than hammer-milled flour biscuits (Table 5.9). It can be concluded that highly acceptable sorghum biscuits can be prepared from 180–250 μm particle-size flour from a *chakki* mill (Fig. 5.23).

OPTIMIZATION OF TECHNOLOGY FOR COMMERCIALIZATION OF SORGHUM-BASED BISCUITS. Sorghum biscuits such as salty, sweet, groundnut, coconut, gluten-free salty and sweet biscuits, and transfat-free salty, sweet, groundnut and coconut biscuits were prepared at ANGRAU. Proportions of the different ingredients were prepared and standardized for highest acceptability (Table 5.10). Gluten-free biscuits and transfat-free biscuits (Table 5.11) and low calorie biscuits were also prepared and standardized for highest acceptability (Table 5.12). Since there is growing demand for low calorie biscuits, an attempt was made to develop and standardize four kinds of sorghum biscuits by replacing the cane sugar with low calorie sweetener that gives 50% of the energy of regular sugar. The standardized formulations had the highest acceptability. Nutritional composition analysis was then carried out for the four formulations in Table 5.10, the results of which are given in Table 5.13.

Table 5.9. Physical characteristics of sorghum biscuits using either the *chakki* traditional mill or the hammer mill. (Source: Dayakar *et al.*, 2014.)

Particle size (μm)	Thickness (mm)	Weight (gm)	Texture (kg/s)**	Colour		
				L*	a*	b*
Chakki traditional mill						
60	1.087	1.182	5.35±1.02	68.89±0.01	6.41±0.04	29.7±0.03
85	1.128	1.182	5.46±0.39	70.54±0.006	5.15±0.01	27.25±0.02
100	1.392	1.198	4.76±1.31	57.79±0.006	10.83±0.03	30.33±0.02
150	0.985	1.251	3.09±1.29	64.46±0.01	9.35±0.03	31.49±0.03
200	1.269	1.22	2.95±0.09	69.83 ±0	7±0	29.99±0.02
Hammer mill						
60	1.186	1.214	5.41±0.93	57.79±0.02	10.56±0.005	30.66±0.07
85	1.06	1.218	3.81±0.19	52.69±0.005	13.77±0.01	32.85±0.11
100	1.06	1.257	3.8±1.07	58.13±0.01	10.79±0.02	10.79±0.09
150	1.035	1.31	2.66±0.25	54.96±0.02	12.2±0.04	32.58±0.03
200	1.053	1.269	4.08±1.07	57.45±0.02	10.9±0.02	25.4±0.1

Note: L*, whiteness; a*, redness; b*, yellowness; **, the first peak of the force-distance plot per time of the texture analyzer was interpreted as hardness of the product, expressed in kg/s.

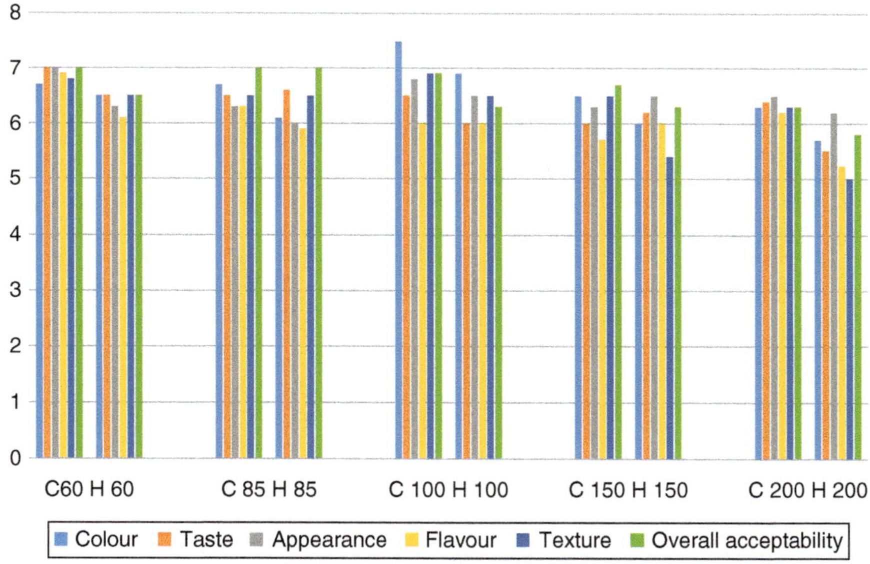

Fig. 5.23. Sensory evaluation of sorghum biscuits prepared from different particle sizes of flour. C, traditional (*chakki*) mill flour; H, hammer mill flour. Numbers on the *x*-axis indicate mesh size.

5.6.3 Fermented sorghum products

Fermentation is one of the traditional food processing technologies in India. Here, foods based on fermented cereals and legumes are popular breakfast dishes, e.g. *idli*, *dosa*, porridge and *dokla*. Brew from cereals is also a fermented product, but this aspect is not discussed here. A good number of fermented foods are also prepared for consumption as regular food in African countries (Osman, 2004; Abd Elmoneim *et al.*, 2005). In India, fermentation of rice is extensively carried out for making *idli* and *dosa* but these products can also be prepared from sorghum. For *idli* and *dosa* preparation, sorghum grits and black gram dal are used in the ratio of 75:25. Both are

Table 5.10. Formulations of sorghum-based biscuits.

Ingredients/formulation (%)	Sweet	Salty	Coconut	Groundnut
Dehulled sorghum flour	33	41	29	25
Refined wheat flour	13	16	11	10
Skimmed milk powder	5	6	4	4
Sugar (powdered)	24	12	27	23
Fat	25	22	18	23
Desiccated coconut	–	–	11	–
Salt	–	3	–	–
Groundnut powder	–	–	–	15

Table 5.11. Formulations of gluten-free and transfat-free sorghum-based biscuits.

Type of biscuits	Gluten-free		Transfat-free			
Ingredients/formulation (%)	Sweet	Salty	Sweet	Salty	Coconut	Groundnut
Dehulled sorghum flour	43	26	33	45.2	29	25
Pearl millet flour	–	26	–	–	–	–
Refined wheat flour	–	–	12.5	–	12.1	12
Skimmed milk powder	5	–	5	6.4	5.7	5
Sugar (powdered)	24	13.3	25.2	24.1	26.5	23
Trans-free fat	25	27	24.3	21.3	15.5	20
Desiccated coconut	–	–	–	–	11.2	–
Salt	–	2.7	–	3	–	–
Groundnut	–	–	–	–	–	15
Calcium caseinates	3	5	–	–	–	–

Table 5.12. Formulations of low calorie sorghum-based biscuits.

Type of biscuit	Low calorie			
Ingredients/formulation (%)	Sweet	Salty	Coconut	Groundnut
Dehulled sorghum flour	33	41.2	28.8	25
Refined wheat flour	12.5	16	11.1	10
Skimmed milk powder	5	4.2	4.4	4
Low calorie sugar	25.2	12.1	26.6	23
Trans-free fat	24.3	24	17.7	23
Desiccated coconut	–	–	11.4	–
Salt	–	2.5	–	–
Groundnut	–	–	–	15

soaked in excess water separately for about 8 h and then wet-ground in a special stone grinder to a soft textured mass (batter). Both batters are then mixed and left to ferment, preferably overnight. During fermentation, considerable amounts of gas evolve and are trapped, due to mucilaginous properties of the galactomannan matrix of black gram.

The batter is transferred in small portions to special *idli* moulds with cup-like indentations, and steamed at atmospheric pressure for about 10 min. During this process, the starch content of the batter gets cooked and solidifies and the mass takes the shape of the mould. Then it is scooped out to get the *idlis* (Fig. 5.24). The product is of a spongy and soft texture with mild sourness. It is consumed with spicy adjuncts such as *chatney* or *sambar*, mainly as a breakfast item. Alternatively, the batter may be evenly spread on a hot plate in a thin circular shape and baked. Normally, oil is smeared onto the plate

Table 5.13. Nutrient composition of sorghum-based biscuits per 100 g. (Source: Dayakar *et al.*, 2014.)

Nutrient	Sweet	Salty	Coconut	Groundnut
Moisture (%)	6.64	16.8	9.23	4.7
Protein (g)	14	9.1	8.73	9.13
Fat (g)	27.7	27	27	29
Energy (kcal)	468	499	484	524
Dietary fibre (g)	5	6	4.5	4.9
Calcium (mg)	92.1	162.8	121.4	85.4
Iron (mg)	0.68	1.8	0.8	0.8
Zinc (mg)	0.41	0.91	0.82	0.6
CHO (g)	40.7	55.1	51.6	56.6
Magnesium (mg)	62	78	55.3	50.3
Ash (mg)	2.238	1.608	2.441	2.373
FFA	0.225	0.282	0.282	0.282

Note: CHO, carbohydrate; FFA, free fatty acid.

Fig. 5.24. Sorghum *idli.*

and also sprinkled on the batter. After about 2 min of baking the batter forms a rigid mass and is turned over and again baked (Fig. 5.25). Since free sugars are formed during fermentation (sometimes a small quantity of sugar is also added to the batter) mild caramelization occurs in addition to the Maillard reaction; as a result a highly desirable and characteristic aroma develops. *Dosa* is normally taken along with chutney or vegetable curry.

Fig. 5.25. Sorghum *dosa.*

Fermentation is a kind of *in vivo* biotransformation and improves the overall nutritional quality of sorghum (Kazanas and Fields, 1981; Mahgoub and Elhag, 1998): the bioavailability of minerals improves due to reduction in the polyphenols and phytate contents, and the carbohydrate and protein digestibility enhance and some of the B-group vitamins synthesize (Chavan *et al.*, 1988). Since both *idli* and *dosa* are products of the blend of sorghum and black gram, they are nutritionally balanced with respect to protein quality and dietary fibre. The products are comparable to rice-based *idli* and *dosa* in terms of texture and sensory acceptability.

5.6.4 Sorghum malt products

Malting is one of the traditional processing methods of cereals. Although barley has pride of place in the malting and brewing industry, sorghum and finger millet malting for food and brewing is practised both in India and many African countries. Sorghum malt is used for preparation of a local brew as well as clarified beer in South and West Africa. Sorghum malt has fairly good proportions of both alpha and beta-amylases (Malleshi, 1987). In South Africa and Nigeria, sorghum cultivars suitable for malting and brewing have been identified and are used as an alternative cereal to barley for malting. Large quantities of sorghum malt is processed to prepare traditional *Bantu* beer in Africa.

Grains with high viability (more than 95% germinability) are a prerequisite for malting. The malting process involves three main unit operations, namely steeping the grains in water, germination and kilning. The quality of the malt depends on each of these steps. Steeping hydrates the grains and facilitates germination, germination enriches the seed with hydrolytic enzymes, and kilning cures the sprouts. The grains are washed and then soaked (steeping) in excess water for 12–24 h. During soaking, it is desirable to change the steep water at least once. This minimizes the microbial growth and reduces the development of an 'off' flavour. Replenishing the steep water provides oxygen to the grains and also facilitates a better degree of germination. The addition of calcium hydroxide to steep water is recommended to minimize mould growth. At the end of the steeping period, the material is washed again and the water is drained off. Following that, the grains are spread for germination. The germination bed should be about 2–3″ thick for uniform germination. During germination the *chitted* grains or the sprouts are mixed gently to dissipate heat and to expose the material to fresh air. Germination may be continued for 2–7 days, depending on the temperature. The duration of germination also depends on the end use of the malt. At the ambient conditions of tropical countries (about 30°C) the germination process is rapid and 48–72 h is optimum. Longer periods of germination lead to excessive metabolic loss. Subsequently the sprouts are sun-dried or dried in a mechanical dryer. This is to terminate the process of germination by dehydration. Generally, the sprouts are dried to 12–14% moisture levels after which the rootlets are removed by gentle brushing. However, in the case of industrial-scale malting, a brushing machine may be used. The product thus prepared is termed 'green malt'. The green malt is further subjected to mild heat treatment, preferably by exposing to hot air or in a rotary grain roaster by contact heat at about 65–70°C until the desirable characteristic malt aroma develops. This cures the malt and improves its sensory acceptability as well as the keeping quality. The product obtained after this treatment is the malt (Fig. 5.26).

The malt is not an edible product and needs further processing, mainly to prepare flour. Whenever the malt is intended for preparation of malt extract or for brewing purposes then it is first crushed. But for weaning food or malt-based beverages it is milled to separate out the seed coat matter. To prepare malt flour free from seed coat, decortication of malt is not feasible because malting softens the grain and as a result it breaks during decortications. Thus, the malt is sprayed with about 5% additional water, mixed well so that the seed coat of the material gets moistened, pulverized and sieved using a 44–60 mesh. This process yields malt flour rich in amylase – termed 'amylase-rich flour' – which is almost free from cellulosic- and polyphenol-rich seed coat matter.

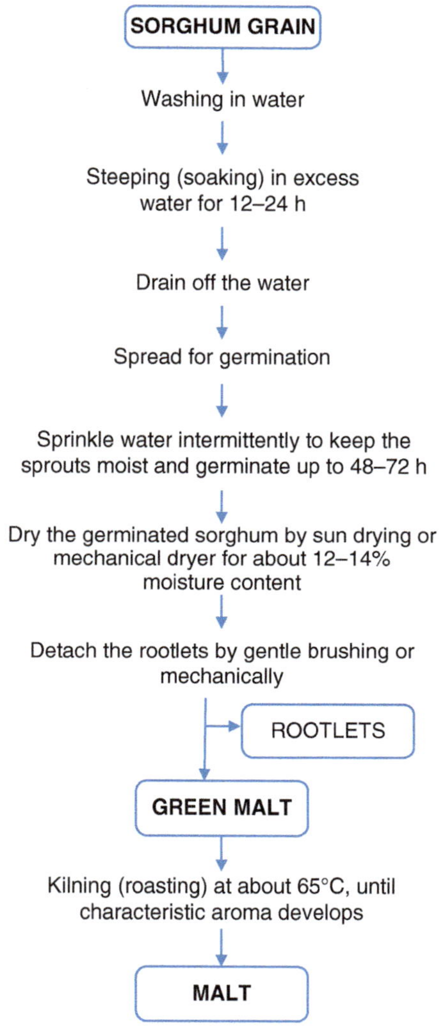

Fig. 5.26. Flow chart for malting of sorghum.

The malt flour is also used to prepare milk-based beverages or for preparation of low-bulk dietary, nutrient-dense weaning and supplementary foods. Malt flour may also be used as a source of amylases to lower the dietary bulk of the weaning foods prepared by dry roasting, extrusion cooking or roller drying methods. Malting of sorghum improves its overall nutritional quality and also the bioavailability of nutrients, including vitamins and minerals (Mahgoub and Elhag, 1998; Elmaki *et al.*, 1999). The carbohydrate and protein digestibility also improves due to lowering of polyphenols concentration and phytate content. In addition, germination enhances some of the vitamin content.

During germination, the α-amylase degenerates whereas β-amylase gets activated. The fairly good ratio of α- to β-amylases in sorghum malt helps to hydrolyse starch during cooking of malted sorghum flour in water (Murthy *et al.*, 1984), i.e. they enhance pre-digestion of the food during cooking itself. However, it may be appropriate to mention a word of caution about the use of sprouted sorghum for food because it is reported that sorghum rootlets contain 'cyanogenic' compounds and hence the rootlets should be fully detached from the dry sprouts (Panasiuk and Donald, 1984).

Sorghum malt is a rich source of easily digestible carbohydrate and blended with legume flour, preferably in a 2:1 ratio, it forms a nutritionally balanced weaning or supplementary food (Malleshi and Klopfenstein, 1998). Part of the legume flour could alternatively be replaced with milk solids or defatted soy flour. It needs to be cooked in water (about 25% w/v) to render it of a consistency suitable for feeding to an infant. During cooking the amylases hydrolyse the starch to low molecular weight carbohydrates and as a result the water-holding capacity of the food will be low, i.e. the food will be not only easily digestible but also nutrient dense.

5.6.5 Sorghum *lassi*

Sorghum *lassi* was prepared by IIMR in collaboration with the National Diary Research Institute (NDRI), Karnal, and for which three sorghum cultivars (M35-1, VJH and PV) were selected. Nutritional profiles of raw sorghum flour vs germinated and pearled sorghum flours are given in Table 5.14. The procedure followed for the preparation of salty and sweet sorghum *lassi* is given in Fig. 5.27.

The study revealed that germination helped in reducing the content of phytic acid (an antinutritional factor) and significantly enhanced the protein content of sorghum grain. Sorghum *lassi* (sweet) prepared without stabilizer had superior organoleptic properties compared to that prepared with stabilizer. *Lassi* prepared with germinated sorghum flour of M35-1 was more acceptable than others (VJH and PV) in terms of physicochemical and organoleptic properties. Acceptable sorghum *lassi* can be prepared from germinated sorghum flour of M35-1 and can be explored for commercialization. The idea of implementing technology to sorghum *lassi* was borrowed from NDRI, Karnal, who had, prior to this study, successfully produced *lassi* from pearl millet.

5.6.6 Sorghum flakes

Flaking is one of the traditional methods of cereal processing; indeed, rice flakes were mentioned in *Mahabharata* when Sudhama offered them to Lord Krishna. Flakes are a convenient food product often used as snacks. At present there are three kinds of cereal flakes: (i) conventional flakes, (ii) flakes prepared using a roller flaker and (iii) RTE breakfast cereals (Fig. 5.28) (Lu and Walker, 1988). The first category of flakes is largely confined to rice, the second category includes flakes from almost all cereals, including oats and barley, whereas the third category of flakes is produced mainly from maize. Currently, sorghum flake products are limited, but it is

Table 5.14. Nutritional profiles of raw sorghum flour vs germinated and pearled sorghum flour. (Source: Dayakar *et al.*, 2014.)

	Cultivar					
	M35-1		VJH		PV	
Parameter	Raw	Germinated and pearled	Raw	Germinated and pearled	Raw	Germinated and pearled
Moisture (%)	7.76	8.06	8.67	7.08	9.04	5.65
Protein (%)	9.39	9.42	11.24	11.37	12.1	12.56
Fat (%)	2.74	0.59	3.12	1.34	1.77	0.94
Ash (%)	1.18	0.29	1.31	0.69	1.85	0.46

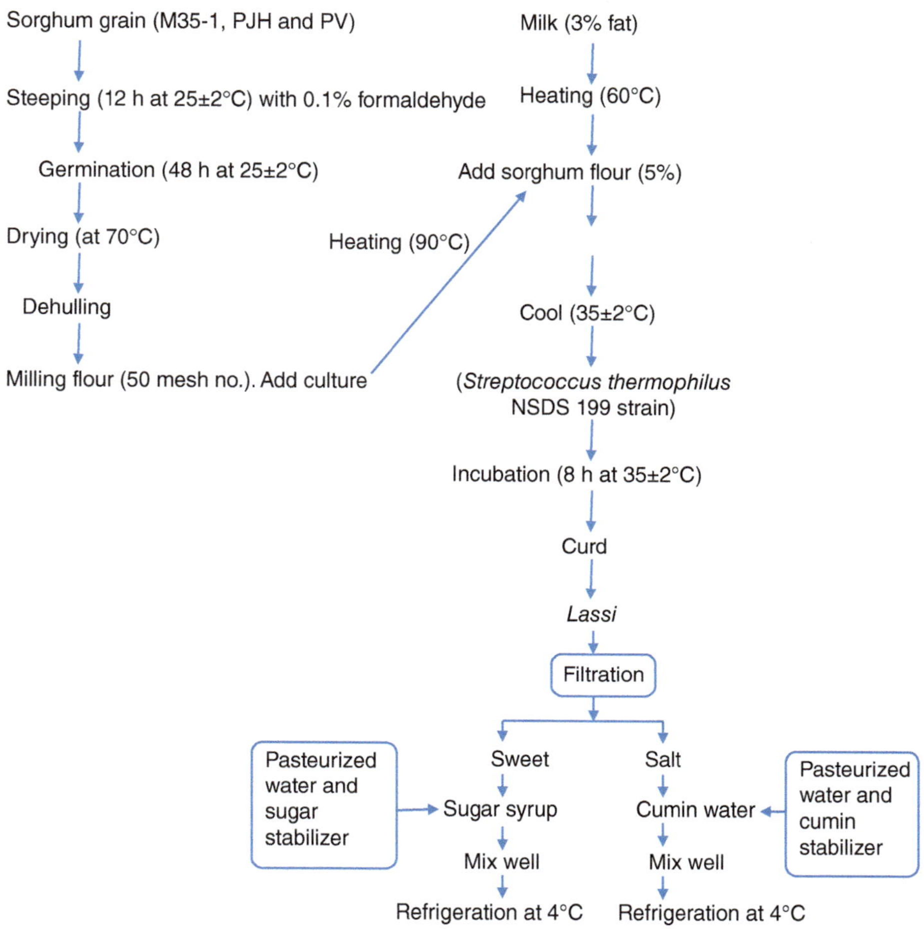

Fig. 5.27. Flow chart for preparation of sorghum *lassi*.

possible to prepare all three types of flakes from sorghum. Sorghum flakes, similar to those of rice, have been successfully produced on a semi-industrial scale from a roller flaker (Fast *et al.*, 1990). A brief account of the methods of preparation, the quality characteristics and the uses of each of these flake types follows.

Conventional flakes

Conventional sorghum flakes are marketed in some of the sorghum-growing areas in India. The product resembles rice flakes in its physical properties and is termed 'beaten' sorghum. The process involves soaking the grains in excess water for 10–12 h to hydrate to their equilibrium moisture content. In the case of using hot water for soaking, the soaking time could be reduced to about 5 h (the addition of a small amount of lime to the soak water is advised to minimize the microbial growth and to retain freshness). After soaking, the water is drained off and the soaked grains are left for about 1 h for surface drying and equilibration. Subsequently, about 2 kg of soaked material is dropped in a dry heat roasting machine maintained at about 180–200°C and constantly agitated by the rotating paddle (Fig. 5.29). In this process, the material comes into contact with the hot surface of the roaster and heat is transferred to the core of the grain. The process is continued until the grains get cooked or the starch content is gelatinized. During this time, the moisture content drops to about 20%. Care is taken to prevent grains from popping or bursting open and also to prevent over-drying of grains below 18% moisture content.

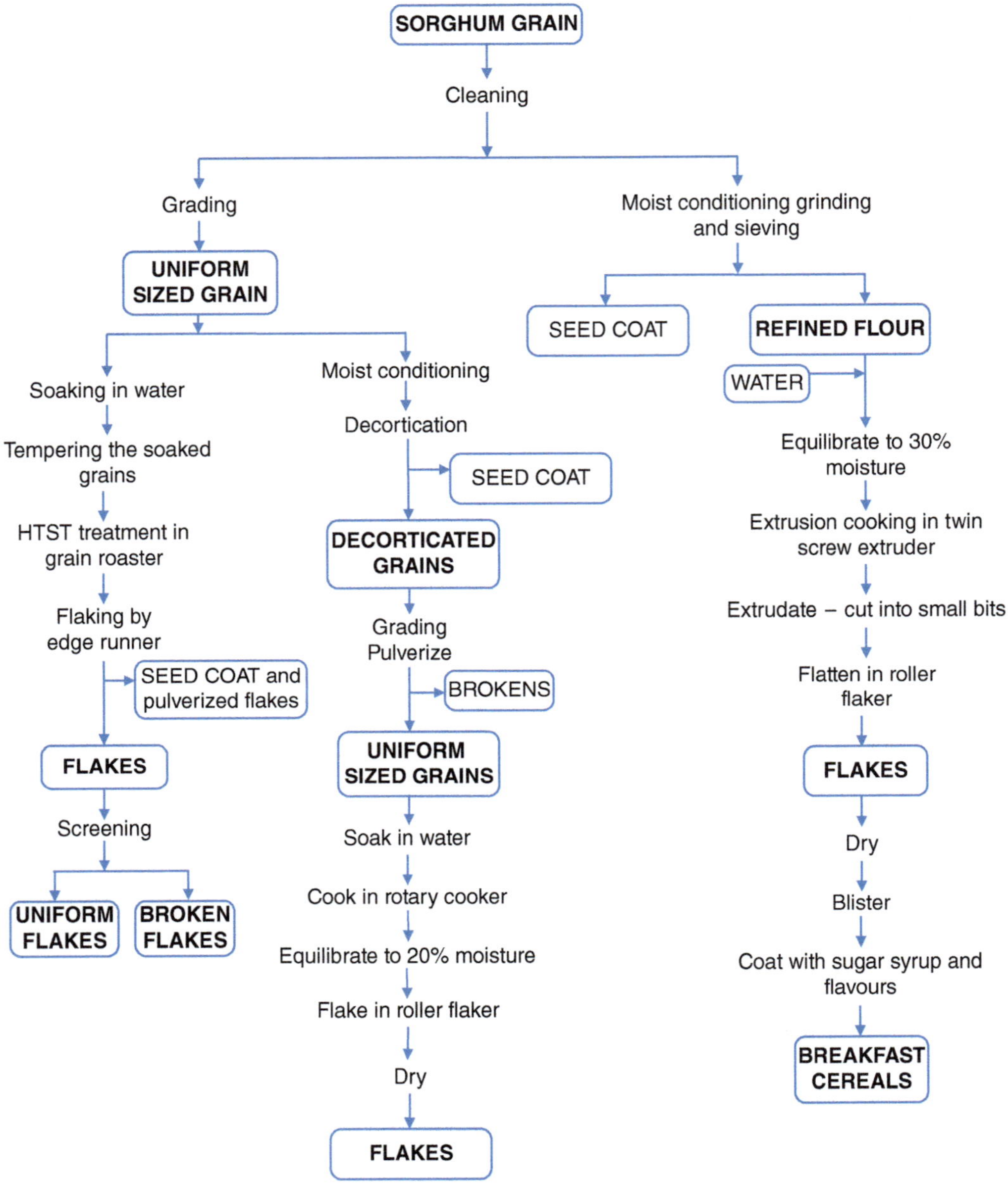

Fig. 5.28. Flow chart for preparation of flakes and breakfast cereals from sorghum.

The grain is then taken out and transferred to the edge runner machine (Fig. 5.30) where the roasted sorghum is subjected to repeated pressing between the idle roller and flat surface of the wall of the machine. This process gradually flattens the grain and at the same time the seed coat gets dried and pulverized and drops through the perforated screen. The flattened sorghum or the flakes of desired thickness are collected manually. The duration of flattening depends on the kind of flakes to be produced, i.e. thin or thick flakes.

The flaked material will be fairly circular in shape but generally fissured at the edges, and it is screened over perforated decks to separate out finer brokens. The thickness of the flakes will be in the order of <1 mm (Fig. 5.31A). The flakes are further dried by spreading in an open place or using a mechanical dryer, and then packed.

Fig. 5.29. Dry grain roaster for sorghum flaking.

Sometimes, the flakes are graded to obtain high value, uniform-sized flakes. The seed coat along with the pulverized endosperm material forms the byproduct. Normally, the yield of high quality flakes will be about 60% of the mass of sorghum used for flaking (Shankara *et al.*, 1984).

Flakes of this kind were successfully produced at IIMR, Hyderabad, and also by a progressive farmer-entrepreneur on a semi-industrial level. The marketing trials indicated their ready acceptance. The physicochemical and product development properties of the flakes were studied. Sorghum flakes are soft textured and hydrate easily just after sprinkling with cold or hot water and could be seasoned with spice and garnishing to prepare very tasty snacks or breakfast items (Fig. 5.32). They can be consumed after mixing with milk and sugar and also with buttermilk or curd. Thinner flakes can just be seasoned with a little oil and spice and may be used as delicious dry snacks. Sorghum flakes can partially or fully substitute rice flakes in traditional flake-based products such as *bisibele bath*, *imli poha*, *idli* and *dosa*. Some recipes using sorghum flakes are presented in Appendix 8.

Flakes made from roller flaker technology

Cereal flakes used for muesli formulations and preparation of deep-fried snacks generally are prepared using the roller flaker machine. Both native and decorticated grains could be used for flaking in this method, and maize grits are commonly processed in this way. To prepare sorghum roller flakes (Fig. 5.31C), dehulled sorghum is graded and soaked in water (cold or hot) until it attains equilibrium moisture content. Then the water is drained off and grains are steamed to gelatinize the starch for about 30 min at atmospheric pressure or for about 10 min at 3–4 bar pressure. The steamed grains are dried to a moisture content of about 20% and are then flattened by passing them between the rolls of the heavy-duty roller flaker machine (Fig. 5.33). The grains get flattened with a single impact of the rollers. The flattened material is dried and graded. Flakes prepared by this method will be of uniform shape, without any kind of fissuring. But some flakes form cones or buck after drying, although the degree of buckling depends on the thickness of the flakes. The gap between the rollers can be adjusted to prepare flakes of desired thickness. It is always desirable to grade the material prior to flaking to prepare uniform sized flakes.

Fig. 5.30. Edge runner machine for sorghum flaking.

Roller flakes are generally translucent, hard and lustrous and are more suitable for consumption as snacks after deep frying or garnishing with spice and condiments. The flakes can be subjected to HTST treatment to impart blistering and make them RTE products. The process can

Fig. 5.31. (A) Conventional flakes, (B) extruder flakes and (C) roller flakes.

Fig. 5.32. *Masala* flakes (*chudwa*/*poha*).

Fig. 5.33. Roller flaker machine.

be applied to any type of sorghum grain irrespective of variety, size and shape. However, the colour of the pericarp slightly affects the colour of the flakes. The yield of flakes can be 80–90% (pearled grains).

Optimal conditions for manufacture and nutritional content of sorghum flakes

The optimized conditions for flaking under an edge runner machine are: soaking time of 16 h, roasting temperature of 310°C, roasting time of 90 s, incubation time of 20 s, flaking time of 210 s and sample weight for replication of 500 g. The nutritional composition of sorghum flakes is presented in Table 5.15, suggesting that they are rich in protein, dietary fibre, calcium, iron and magnesium.

5.6.7 Pop sorghum and sorghum puffs

Popcorn, the product from maize, is a well-known snack worldwide. Sorghum can similarly be popped to prepare popped sorghum (pop sorghum), another RTE product. Popcorn preparation uses a special cultivar of maize and contact or microwave heating. In contrast, popping of sorghum is still a primitive method using hot sand as the heat transfer media. Preparation of pop sorghum is a HTST treatment and is a simple and dry process. It is one of the traditional processes applied extensively to sorghum and is still a cottage industry in India, with the entire process carried out manually.

The process involves adding extra water to sorghum to raise its moisture content to 16–18%, tempering for 4–6 h, followed by dropping the grains in sand heated to about 240°C in a wide-mouth concave pan and agitating continuously using an iron ladle until all the grains pop (about 1 min) (Fig. 5.34). During this treatment the grains pop instantaneously and are separated from the sand using a perforated pan (Fig. 5.35). The popping process occurs due to rapid transfer of heat from the sand to the grain and transforming its moisture content into steam. The steam gelatinizes the starch and exerts very high pressure inside the grain because the seed coat acts as a barrier for its escape. This results in the grain bursting open, leading to expansion of the endosperm and formation of popped sorghum (Hoseney *et al.*, 1983). Equilibrating the grains to about 18% moisture levels facilitates formation of steam sufficient to gelatinize the starch and also to develop enough pressure inside the grain to burst open the endosperm (Khedker *et al.*, 2008). Normally, the volume of the expanded sorghum will be more than eight times that of the original grain. It is a low-cost technology and the capital investment for the popping unit is also very low.

Popping of sorghum is still done using this sand roasting method in India. Undoubtedly, hot sand acts as a very effective media for heat transfer

Table 5.15. Nutritional composition of sorghum flakes. (Source: Dayakar *et al.*, 2014.)

Nutrient	Amount (/100 g)	Amino acid	Amount (g/100 g of protein)
Moisture (g)	10.55	Aspartic	6.32
Protein (g)	7.23	Threonine	3.17
Fat (g)	1.79	Serine	4.86
Total dietary fibre (g)	5.97	Glutamic acid	24.8
Insoluble (g)	5.43	Proline	3.64
Soluble (g)	0.54	Glycine	2.89
Carbohydrates (g)	73.8	Alanine	11.36
Calcium (mg)	10.94	Cysteine	0.48
Magnesium (mg)	68.9	Valine	5.44
Copper (mg)	0.1	Methionine	0.6
Manganese (mg)	0.53	Isoleucine	3.88
Iron (mg)	8.77	Leucine	14.6
Zinc (mg)	0.88	Tryosine	4
Phosphorus (mg)	110	Phenylalanine	5.64
Thiamine (mg)	0.45	Histidine	2.22
Niacin (mg)	1.93	Lysine	2.03
Riboflavin (mg)	0.15	Arginine	3.68
Total carotenoids (mg)	ND		
Β-carotene (mg)	ND		
Total folic acid (μg)	1569		
Vitamin C (mg)	ND		

Note: ND, not detectable.

but the popped grains generally get contaminated with fine sand particles and this affects the eating quality. This also increases the silica content of the product (Mishra *et al.*, 2014). Moreover, the process is highly drudgerous and workers are exposed to very high levels of heat and dust. To overcome the drawbacks, popping of sorghum using hot air and contact heat as the heat transfer medium has been explored. In these methods, the grains, equilibrated to about 16–18% moisture content, are subjected to HTST treatment as before. However, in the case of hot air popping, the grains are exposed to a hot air stream; the resulting popped grain comes out of the popping machine (Fig. 5.36) and is lighter due to moisture absorption as a result of hot air stream injection; this produces a product with desirable taste and texture similar to popped corn. In the case of the contact heat popper, a rotary cone heated from the outside with a continuous agitation lever facilitates popping.

Popped sorghum produced using these methods will a have milky white colour, crunchy texture and highly desirable and characteristic aroma. The gelatinized starchy endosperm exhibits a spongy texture with continuous air cells (Pawar *et al.*, 2014; Sharma *et al.*, 2014).

Gun popping is another traditional method of cereal popping, extensively practised in East Asian countries but to date not popular in India. The process requires decorticated grains and the resulting product has highly desirable functional and sensory qualities. The equipment comprises a heavy duty rotary barrel-shaped machine with an opening on one end with provision to firmly close it with a lid on the other end (Fig. 5.37). It is positioned horizontally and heated either by liquefied petroleum gas (LPG), furnace oil, electricity or fire wood. The decorticated sorghum, equilibrated to about 21% moisture content, is dropped in the drum and heated while it rotates. The moisture content of the sorghum forms steam due to heat and exerts very high pressure in the barrel. Heating is continued until the pressure inside the drum attains about 0.7 bar, after which the burner is turned off and the lid is opened quickly. Suddenly the material, which is under very high pressure, is exposed to atmospheric pressure, which causes expansion in all directions of the grain (Lai and Cheng, 2004). As with sand popping, this is a batch process. Normally, the expansion ratio will be more than 6.0. The pop sorghum (Fig. 5.38) will have an attractive milky white colour and a desirable aroma. However, it

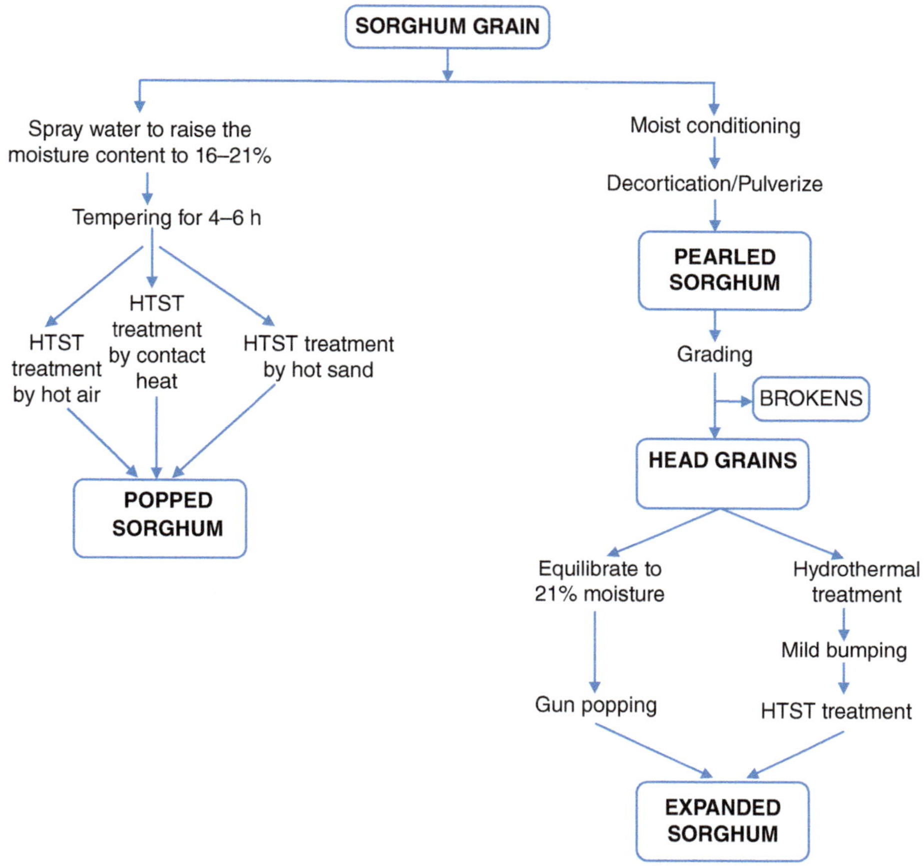

Fig. 5.34. Flow chart preparation of popped and expanded sorghum.

will be slightly soggy. In order to improve the crispness, pops may be deep fried or toasted in salt medium. Since it has gelatinized starch, the product will be hygroscopic and so becomes soggy when exposed to high humidity and therefore requires storage in a closed container soon after preparation. The sorghum pop may be seasoned and garnished with spices and condiments for snacks. It may also be used for the preparation of RTE supplementary foods and as an adjunct to brewing.

The nutritional qualities of popped sorghum using native sorghum following the sand popping and that followed by gun popping, which requires decorticated sorghum, differ considerably. The former contains the seed coat matter and as result it will contain almost all the nutrients of the native sorghum used for popping. In contrast, the gun popped sorghum contains slightly lower levels of all the nutrients except starch (Saravanabavan *et al.*, 2013).

Expanded, or 'puffed' rice is a popular product, called *murmura* in India. It is also an RTE product, whereby the shape of the grain bulges in all directions without bursting open like popped grain. Expanded sorghum can be prepared by HTST treatment of parboiled, dehulled and bumped sorghum. The expansion may be only double, but it is a delicious product and can be coated with spices and condiments or with sugar syrup. This is used as an ingredient in health bars and similar products.

To conclude, in general the traditional method of sand popping of sorghum should be avoided and alternative methods such as air popping, contact heat popping and gun popping should be encouraged. Sorghum cultivars that yield better expansion and more than 90% yield of popped grains should be identified and popularized. Diversification of the use of popped sorghum for value added products such as supplementary foods, snacks and allied uses should be encouraged.

Fig. 5.35. Traditional method of (*bhatti*) popping of sorghum.

Fig. 5.36. Hot air popper.

Fig. 5.37. Gun puffing machine.

Fig. 5.38. Sorghum puffs.

5.6.8 Sorghum extruded products

Food extrusion is one of the traditional technologies followed in almost all countries. In China, Japan, Korea and many of the East Asian countries rice is used for extrusion, whereas in most of the other countries it is wheat. Nowadays food extrusion is practised on an industrial scale. Mainly cold extrusion was practised until recently, while hot extrusion or extrusion cooking started to gain popularity. Although these methods are not extensively practised for sorghum, there are a few reports of this (Llopart *et al.*, 2014).

Sorghum is one of the starchy cereals that exhibit good extrusion cooking characteristics. Fine grits or flour is mixed with additional water to raise the moisture content to about 20% and then extruded in a single or twin screw extruder (Fig. 5.39) at the appropriate temperature, rpm of the screws and feed rate. The extrudate is normally a fully cooked product (Fig. 5.40) containing gelatinized starch. Hence, it is an RTE product. During extrusion, the material undergoes the glass transition state and as a result the biomolecules lose their cellular integrity, and the proteins, carbohydrates and lipids intermix forming a homogenous mass under high pressure and temperature. Some proportion of the insoluble dietary fibre undergoes rupturing and gets converted into soluble fibre. Due to the high temperature, pressure and constant kneading effect, the enzymes get inactivated and the products also become microbiologically safe. The product

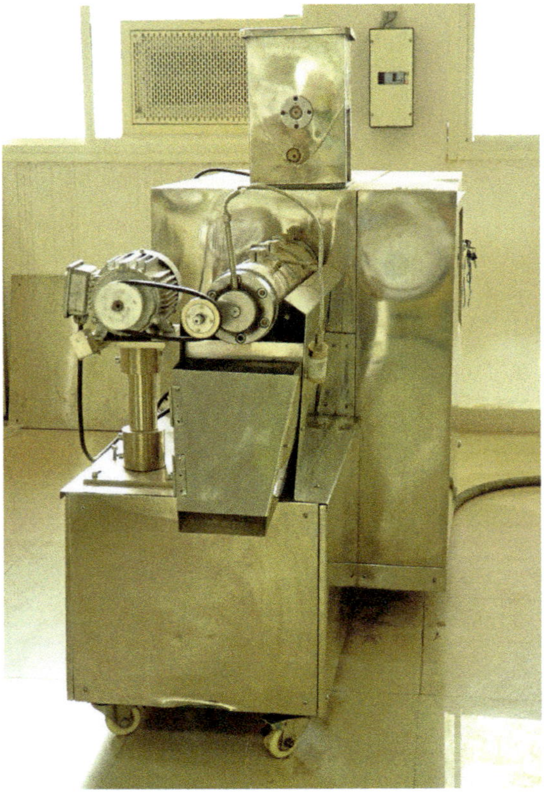

Fig. 5.39. Extrusion cooker.

Fig. 5.40. Sorghum extrusion product (*masala*-coated snacks).

coming out of the extrusion cooker die takes the shape of the die but gets inflated, forming a spongy texture (Hamaker *et al.*, 1994). The products can be made into different shapes and sizes as desired and can be seasoned with spice to prepare snacks or could be cooked with sugar syrup. Sorghum blended with soya, bengal gram, green gram or milk powder in appropriate proportion can be extrusion cooked to prepare nutritious weaning and supplementary foods (Malleshi *et al.*, 1996). For preparation of these kinds of products, normally dehulled sorghum is preferred. Extrusion cooking may also be followed to prepare breakfast cereals. Sorghum blended with cereals such as wheat and rice, as well as legumes, may be used for extrusion cooking to prepare multigrain RTE products. Extrusion cooking is therefore a very good technology for value addition to sorghum.

Breakfast cereals

Normally, maize is processed to prepare flakes for breakfast cereals; it has technological advantages in terms of high starch content and a poor rigid protein matrix. However, sorghum can also be processed to prepare breakfast cereals in the form of flakes (Fig. 5.31B). The process involves pulverizing decorticated sorghum to finer grits or flour and mixing it with water to hydrate it to about 28–30% moisture content. The hydrated flour is then cooked and extruded using a twin screw extrusion cooker, where it undergoes a glass transition stage and extrudes out in the form of a gelatinized mass. Next the extrudate is cut into small pieces (now having about 20% moisture); these are then flattened in a heavy duty roller flaker machine to an appropriate thickness (depending on the gap between the rolls, but <1 mm), and finally dried and blistered to prepare flakes. The flakes, which are of near-uniform size and nearly circular in shape, are normally coated with sugar syrup, cocoa or some other flavour, and dried. They are also normally fortified with the necessary vitamins and minerals to enhance their nutritional value. Sorghum-based breakfast cereals can be prepared using sorghum blended with soy or grain legumes in appropriate proportions; such a cereal would be nutritionally superior to sorghum flakes alone. This type of flake is of high value and a convenience food; it is RTE and is generally taken as a breakfast cereal along with cold or warm milk.

Vermicelli and pasta

Pasta and vermicelli are prepared by cold extrusion of wheat worldwide, including in most of the Asian countries, in cold extruder machines (Fig. 5.41). Sorghum and millet or even green gram is also used to prepare such products. Vermicelli, commonly called *semiya* in India, comprises strands of about 2–3 mm diameter that are normally circular in cross-section (Fig. 5.42), while pasta is prepared in different shapes, such as shells (Fig. 5.43), cones, rings, tubes, stars and alphabets. Vermicelli used to be prepared by hand-rolling the cold-water dough on wheat flour, relying on the skill of the person, but these days it is mostly prepared using a vermicelli/pasta machine. The flour or fine grits of wheat flour is mixed with cold water to raise the moisture content (to about 30%), allowed to stand for about 30 min and subjected to cold extrusion. The machine is fitted with dies of different sized perforations or of different shapes that determine the diameter of the vermicelli strands or shape/size of pasta products, respectively (Dayakar *et al.*, 2015). The cold-extruded products are normally steamed for a few minutes to stabilize the vermicelli/pasta and then dried.

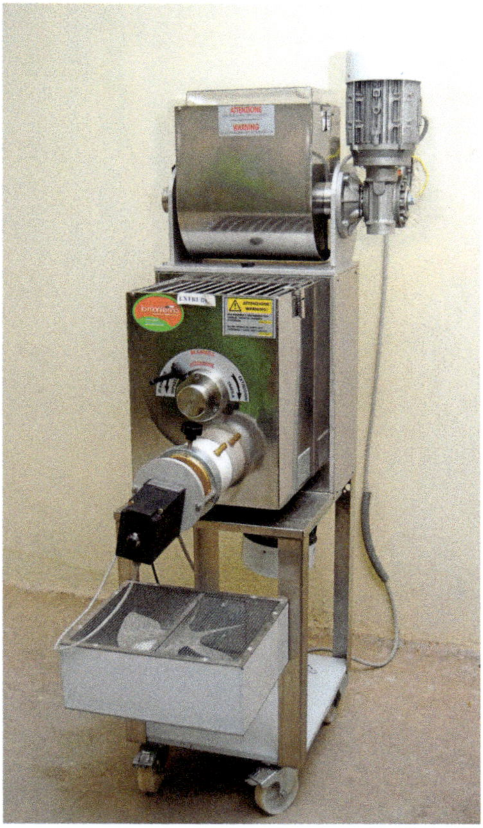

Fig. 5.41. Cold extruder.

Fig. 5.42. Sorghum vermicelli.

However, in the case of sorghum the process needs different treatments to facilitate starch to act as a binder to form a cohesive dough and also to impart strength to the product. This is essential as sorghum proteins do not form gluten

Fig. 5.43. Sorghum pasta.

and so the flour does not form a sticky dough on mixing with cold water. In view of this, the starch content of sorghum needs to be partially gelatinized to enable it to form dough that can be extruded. This is generally done by steaming the dough to about 30% moisture content. During extrusion the material experiences high pressure and as a result the starch gets gelatinized without swelling and forms a homogenous mass that is able to take the shape of the die. The extrudates retain their shape after drying but will lose their integrity on cooking. Hence, soon after extrusion the extrudates require steaming, preferably under pressure, to strengthen them; the vermicelli/pasta will then retain their shape on cooking without disintegration. Leaching losses are negligible. After cooking in excess water, the product should not give away more than 10% of solubles and will cook to a soft-textured product.

Normally, fine grits obtained from hard endosperm of sorghum make better quality vermicelli and pasta. Very often a small percentage of wheat gluten or a functional ingredient such as gum is added to sorghum flour, for improving the cohesiveness of the dough. Alternatively, a blend of sorghum and wheat flour may be used to prepare sorghum vermicelli and pasta. Whole grain sorghum pasta offers the advantage of being a non-gluten product that also provides high fibre and complex carbohydrates, although pasta/vermicelli made from dehulled sorghum has greater acceptability.

DEVELOPMENT AND STANDARDIZATION OF SORGHUM PASTA. IIMR, Hyderabad, has conducted extensive studies on the preparation of pasta from sorghum, optimizing various parameters for its production from sorghum as well as from a composite blend of sorghum with wheat, soy, flax seed etc. and is successfully producing it on a pilot scale and establishing a marketing channel. ANGRAU also developed a gluten-free pasta. The formulations used and nutrient composition of sorghum pasta developed using sorghum and wheat as the main ingredients are given in Table 5.16.

The moisture level was highest with the 100% sorghum pasta extrudate, followed by the 80, 70, 60 and 50% sorghum pasta extrudates. However, the optimum moisture level (<14%) was seen in all the extrudates. The protein content was highest in the sorghum pasta with the highest wheat content (50:50 sorghum:wheat) and declines according to the decline in wheat content; the trend is therefore likely to be due to the addition of wheat. The fat and fibre content was highest in the 80% sorghum pasta and 100% sorghum pasta (control) extrudates. The amylose content was highest in the control and among the experimental samples it was highest in the 80% sorghum pasta extrudate, followed by 70, 60 and 50% sorghum pasta extrudates.

The amylose content gradually decreased from the 100% (control) to the 50% sorghum pasta, i.e. with an increase in wheat composition; this is likely to be due to the higher amylose content of sorghum in relation to wheat. The energy values of all products were similar, ranging from 347.5 to 349 kcal/100 g, and met the nutritional requirements (regardless of age), i.e. 300 kcal/100 g and nearly 10 g protein/100 g.

The cooking time of different sorghum pasta samples ranged from 6 to 9 min (Table 5.17). The 100% sorghum pasta sample (control) showed a significantly higher ($P<0.05$) cooking time of 9 min compared to the other blends (average of 6 min). The percentage water absorbed was in the range of 86.20 to 199.74 g/100 g, with sorghum/wheat pasta showing a significantly higher percentage of water absorbed than the control (100% sorghum pasta) ($P<0.05$). Cooking loss indicates the ability of the pasta to maintain its structural integrity during the cooking process. The solid gruel loss was observed in the range of 7.41 to 10.12 g/100 g (Table 5.17), being highest for 100% sorghum pasta and lowest for the sample with the lowest sorghum content (i.e. the 50:50 sorghum:wheat sample), the difference between them being significant statistically ($P<0.05$). Moreover, with an increase in cooking time in sorghum pasta blends, the amylose networks were degraded, thus increasing the cooking losses. In the case of blends, the solid loss in gruel was intermediate and with the 50:50 blend the solid loss in percentage terms was lowest and was significantly different from the other four blends ($P<0.05$). The water absorption index was found to be more for the control (36.44±0.08) followed by the 80:20 sample (38.79±0.03). The water absorption index of the extrudates increased with the increase of sorghum semolina percentage in the composite mixes.

The water solubility index was highest for the sample with most wheat (the 50:50 sample) followed by the 60:40 sorghum:wheat sample and was lowest for the sorghum-only sample, i.e. the water solubility index of the extrudates increased when wheat semolina incorporation increased from 20% to 50% in the composite mix samples. The SEM microphotograph shows that the microstructure of sample T1 (Fig. 5.44) had the least morphological changes of the food components and showed the best results. This is useful when preparing sorghum pasta which retains its structure intact, without breaking even after cooking; this is similar to wheat pasta and could be used for commercial adaption.

The mean scores of sensory evaluation showed that all the extruded products prepared from composite blends were within the acceptable range, while the 50:50 blend product

Table 5.16. Formulations and nutrient composition of sorghum pasta. (Source: Dayakar *et al.*, 2014.)

	Control	T1	T2	T3	T4
Formulation					
Sorghum semolina (%)	100	50	60	70	80
Wheat semolina (%)	0	50	40	30	20
Nutrient composition					
Moisture (%)	10.02	8.25	8.32	9.2	9.41
Protein (g/100 g)	9.6	9.7	9.5	9.3	9
Fat (g/100 g)	0.84	0.44	0.73	0.7	0.82
Fibre (g/100 g)	1.02	0.9	0.76	0.77	0.77
Amylase (%)	18.48	13.78	13.98	15.33	16.59
Carbohydrate (g/100 g)	74.7	74.2	72.5	71.5	71.3
Energy (kcal/100 g)	349	348.4	348.1	347.8	347.5

Table 5.17. Quality characteristics of sorghum pasta. (Source: Dayakar *et al.*, 2014.)

Treatment	Cooking time (min)	Cooked wt (g/100 g)	Water absorbed (g/100 g)	Solid loss (g/100 g)	Water solubility index	Water absorption index
Control	9.0±0.40	18.25±0.43	86.20±3.50	10.12±0.14	8.38±0.45	36.44±0.08
T1	6.0±0.53	17.75±0.20	180.48±5.12	7.41±0.52	11.28±0.12	21.27±1.56
T2	7.0±0.12	16.25±0.36	199.74±7.15	8.24±0.43	9.43±0.01	31.42±1.24
T3	6.0±0.22	16.75±0.23	177.56±4.40	9.36±0.26	9.32±0.52	31.39±0.27
T4	8.0±0.29	15.75±0.12	120.35±6.22	8.25±0.22	8.50±0.56	38.79±0.03

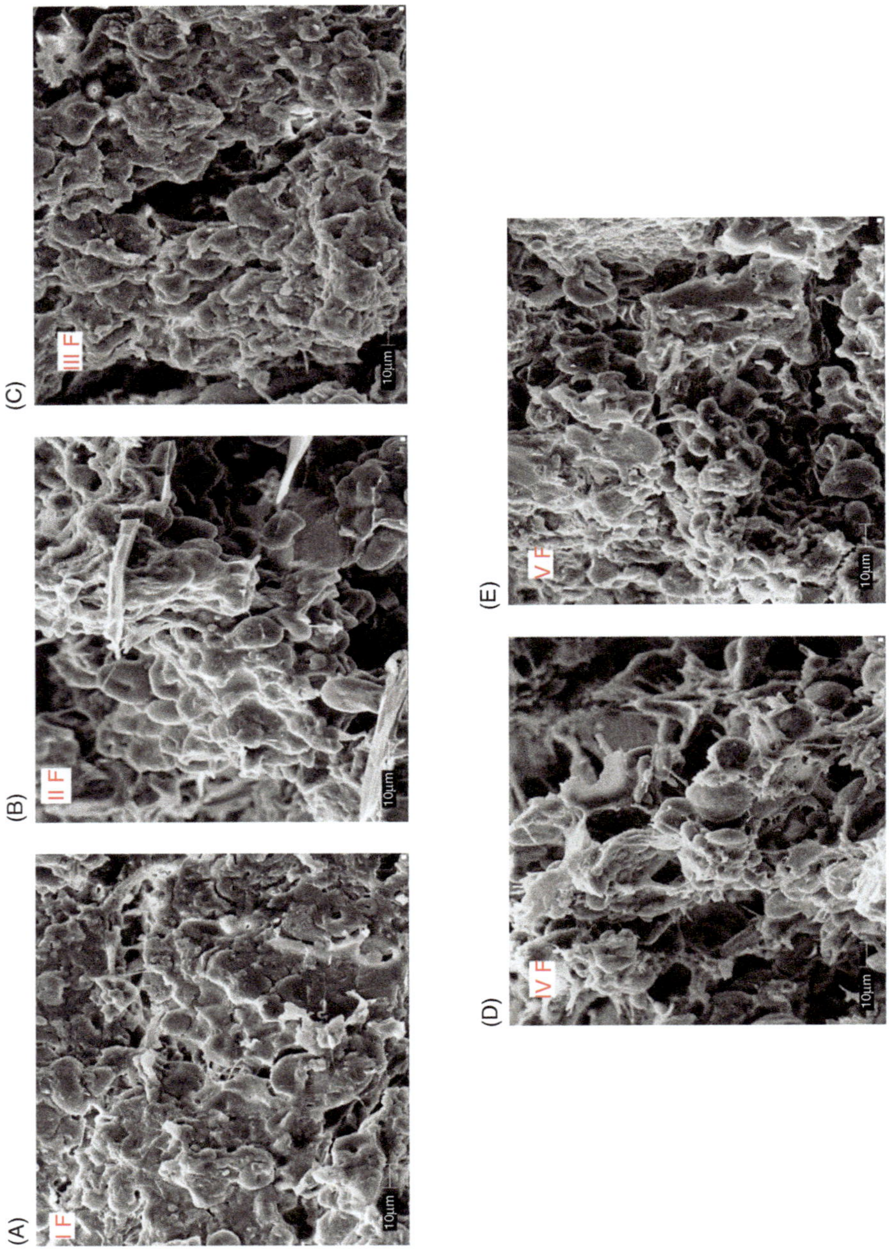

Fig. 5.44. SEM microphotograph of different pasta blends, showing (A) T1, (B) T2, (C) T3, (D) T4 and (E) T5 formulations of pasta.

prepared had significantly better colour, texture, taste, stickiness, firmness, bulkiness and overall acceptability.

SORGHUM-RICH PASTA WITH SOYA, CHANNA *AND GUAR GUM.* Sorghum pasta was further developed by blending sorghum flour separately with soya protein concentrate (SPC) and *channa* flour (a flour made from chickpeas and used in many Asian cuisines) along with guar gum. The control pasta was prepared using 100% durum wheat to compare the quality parameters (Table 5.18). Pasta was analysed for the proximate composition, dietary fibre, protein digestibility, carbohydrate digestibility, cooking loss and gel consistency (Table 5.19). The pasting profile, texture, colour and sensory values of pasta were also studied. Cooking quality of the pastas revealed that sorghum pasta with different blends exhibited more cooking loss (6–14%) than that of durum wheat (control) pasta (5%) ('cooking loss' indicating the ability of pasta to maintain structural integrity during the cooking process). Protein, ash and fat content in pasta with SPC and *channa* increased compared to control wheat and control sorghum pasta (Table 5.18). The gel consistency of cooked sorghum pasta (113–152 mm) showed a significantly higher value compared to blends as such (30–115 mm). Pasting indices, i.e. peak viscosity, hot paste viscosity and cold paste viscosity, of sorghum pasta ranged from 160–337 BU, 118–250 BU and 254–536 BU, respectively, which are relatively high compared to that of traditional wheat pasta (120–150 BU), thus developing a sticky texture, but the formulation T1 had pasting indices similar to the control wheat pasta.

The colour of pasta, i.e. whiteness (L*) and average colour value (ΔE), ranged from 53.2 to 59.2 and from 44.9 to 49.9, respectively. The texture value, in terms of the shear force required to cut the cooked pasta (N), ranged from 0.5 (sorghum control) to 1.3 (sorghum pasta blended with *channa* and guar gum), with the latter similar to that of control wheat pasta. The addition of *channa* and guar gum thus gave firmness to the pasta; as soggy pasta is not acceptable to the consumer, this T5 formulation would be useful for commercial adaptation.

SOY-INCORPORATED SORGHUM PASTA. Formulation of sorghum pasta is usually more exigent than that of gluten-containing products, as the major structure-forming component, namely gluten, is absent. However, pasta from millets has a great future since it is rich in fibre and low in fat. In the present study, two sets of formulations (Table 5.20) were used for the development of gluten-free pasta that possesses a texture and taste similar to that of control wheat pasta: one incorporating defatted soya flour, the other

Table 5.18. Formulation and proximate composition of sorghum-rich pasta. (Source: Dayakar *et al.*, 2014.)

Sample	Formulation	Protein (g/100 g)	Moisture (g/100 g)	Fat (g/100 g)	Ash (g/100 g)
T1	Control wheat	12.5±0.35	7.7±0.49	1±0.12	0.6±0.05
T2	Control sorghum	9.3± 0.28	7.8±0.57	3.3±0.25	1±0.17
T3	Sorghum + SPC	13.5±0.44	8.1±0.15	4.4±0.40	1.3±0.14
T4	Sorghum + *channa*	12.7±0.43	7.7±0.19	5.1±0.13	1.3±0.20
T5	Sorghum + *channa* + guar gum	13.6±0.38	7.8±0.20	6±0.14	0.7±0.47

Table 5.19. Quality characteristics of sorghum-rich pasta. (Source: Dayakar *et al.*, 2014.)

Sample	Cooking time (min)	Cooked wt (g)	Cooking loss (g/100 g)	IVSD (g/100 g)	IVPD (g/100 g)	Shear force (N)
T1	4.7±0.3	51.6±0.8	5.5±0.1	18.8±0.2	91.3±0.7	1.5±0.1
T2	3.1±0.1	50.1±2.5	14.2±0.4	22±0.5	81.2±0.6	0.5±0.2
T3	3.2±0.1	47.6±1.5	6.4±0.6	20.1±0.16	93.1±0.8	1±0.01
T4	3.1±0.2	52.2±2.5	5.8±0.1	19.6±0.5	94.7±0.5	1.1±0.2
T5	3.3±0.2	52.3±0.9	5.8±0.2	18.8±0.3	95.4±0.8	1.3±0.2

Note: IVSD, *in vitro* starch digestibility; IVPD, *in vitro* protein digestibility.

incorporating caseinates (caseinates are proteins produced from casein in skimmed and sometimes 1% milk, termed hydrocolloids) and using carboxymethyl cellulose (CMC).

Among the hydrocolloid-containing pasta samples, the T4 combination with wheat semolina at 20% and hydrocolloid-1 at 2% scored best in terms of colour, taste and texture. The overall acceptability was similar to control pasta. Among the standardized formulations of gluten-free soy-incorporated pasta, T5 with defatted soy flour (36%), CMC (2%) and SG1 (2%) scored highly and were most acceptable.

Formulation of soy-incorporated gluten-free pasta was usually more exigent and the swelling index of pasta products was directly proportional to the water absorption. The swelling index of wheat pasta was higher than the gluten-free pasta blends. Higher swelling indices were recorded for both T1 and T2 (1.12%) followed by T3 and T4 (0.97%) and was lowest in blends of gluten-free soy- incorporated pasta (Table 5.21).

FLAXSEED-INCORPORATED SORGHUM PASTA. Flaxseed is a rich source of polyunsaturated fatty acids (at about 80% of total oil content) with relatively high proportions of omega-3 and omega-6 fatty acids, both of which are essential and should be consumed since the body cannot manufacture them from another source. In the study, flaxseed flour was incorporated at different levels in wheat- and sorghum-based pasta and the cooking quality was assessed (Table 5.22). Even though the control wheat pasta had the highest acceptability among all the formulations (using organoleptic studies), the next highest score was for the combination of wheat/sorghum semolina with 5% flaxseed (T6), which had high acceptability. However, more than 5% incorporation of flaxseed in sorghum-based pasta was not acceptable and, similar to wheat pasta, the interaction of sorghum grain components with flaxseed resulted in an undesirable taste and flavour.

DEVELOPMENT AND STANDARDIZATION OF SORGHUM VERMICELLI. Sorghum vermicelli was developed using sorghum and wheat as the main ingredients (Table 5.23). Organoleptic characteristics indicated that all the products developed with sorghum and wheat had good acceptability (see Section 5.6.13). However, the overall acceptability

Table 5.20. Soy-incorporated sorghum pasta formulations. (Source: Dayakar *et al.*, 2014.)

	Composition (%)					
	Sorghum semolina	Wheat semolina	Defatted soy flour	Hydrocolloid-1	Hydrocolloid-2	CMC
Control	–	100	–	–	–	–
T1	60	28	10	2	–	–
T2	60	28	10	–	2	–
T3	60	20	18	–	2	–
T4	60	20	18	2	–	–
T5	60	–	36	2	–	2
T6	71	–	25	2	–	2

Note: CMC, carboxymethyl cellulose.

Table 5.21. Cooking quality of different soy-incorporated pasta formulations. (Source: Dayakar *et al.*, 2014.)

Formulation	Swelling index (%)	Dry matter (%)	Cooking losses (%)	Cooking time (min)
Control	1.94	90	5.06	14
T1	1.12	83	7.95	12
T2	1.12	80	7.95	12
T3	0.97	85	7.32	13
T4	0.97	87	7.32	14
T5	0.64	67	7.08	14
T6	0.62	76	6.91	15

Table 5.22. Formulation and cooking quality of flaxseed incorporated sorghum pasta formulations. (Source: Dayakar *et al.*, 2014.)

Formu-lation	Wheat semolina (%)	Sorghum semolina (%)	Flax seed flour (%)	Swelling index (g of water/g of dry pasta)	Dry matter (g /100g of raw pasta)	Cooking losses (g/100 g of raw pasta)	Cooking time (min)	Water absorption (%)
T1	100	0	0	2.58	89	4.97	12	28
T2	60	40	0	4.21	86	5.67	13	31
T3	95	0	5	3.21	91	6.1	14	34.5
T4	90	0	10	2.89	89	7.92	13	39
T5	85	0	15	2.71	92	9.67	14	35
T6	38	57	5	4.56	89	7.12	13	39
T7	36	54	10	4.42	90	9.45	15	41

of formulation T1 was 4.6, which was higher than other formulations. It was revealed from the cooking time that the 50:50 formulation took the least time to cook and thus could be successfully used to develop an acceptable product. The effect of drying temperature on the moisture content of sorghum vermicelli was also assessed, indicating that the moisture content decreased with an increase in drying temperature and drying time.

Hot-extruded sorghum snack

Hot extrusion combines heating with extrusion to create a puff-shaped cooked product. Extrusion is accomplished by single screw and twin-screw extruders. Commercially, most of the extruded snacks are prepared from maize; in the current study the extruded snack is produced using sorghum grits, rice flour and soya flour. The snack can be coated with spices to create variations in taste and flavour.

In order to determine a consistent sensory quality formulation used in the RTE snack, several combinations were tried, such as sorghum flour ranging from 30 to 70% and soy flour ranging from 10 to 20%. The resulting ingredient composition used was: dehulled sorghum flour (50%), maize flour (30%), defatted soy flour (15%) and roasted bengal gram flour (5%). The grits of all the raw material were conditioned overnight and then extruded in a twin screw extruder at 120°C and with a cutter speed of 350 rpm. The extrudates were then dried in a hot air dryer, coated with spice mix and finally packed in nitrogen flushed pouches. The nutrient composition for this sorghum hot-extruded snack are tabulated in Table 5.24.

5.6.9 Fat-free sorghum crunches

Fat-free sorghum crunches were made using whole/dehulled sorghum flour combined with spice mixes, green chilles, *jeera* powder, sesame seeds, curry leaves, coriander leaves and salt. All the mixture was conditioned for initiating gelatinization and it was then terminated by the application of dry heat at high temperature. The nutrient composition for fat-free sorghum crunches are given in Table 5.25.

Table 5.23. Formulations used for the preparation and cooking quality of sorghum vermicelli. (Source: Dayakar *et al.*, 2014.)

Formulation	Ingredients (%)		Water solubility index	Water absorption index	Cooking time (min)
	Sorghum semolina	Wheat semolina			
Control	100	0	7.38±0.45	34.44±0.08	8.0±0.40
T1	50	50	10.28±0.12	20.27±1.56	5.0±0.53
T2	60	40	9.53±0.01	29.42±1.24	6.0±0.12
T3	70	30	9.12±0.52	30.39±0.27	6.5±0.22
T4	80	20	8.45±0.56	35.79±0.03	7.0±0.29

Table 5.24. Nutrient composition for sorghum hot-extruded snack per 100 g. (Source: Dayakar *et al.*, 2014.)

Nutrient	Value
Energy (kcal)	356
Protein (g)	12.9
Fat (g)	4.08
Dietary fibre (g)	12.88
Calcium (mg)	43.5
Iron (mg)	1.73
Zinc (mg)	0.3

Table 5.25. Nutrient profile of fat-free sorghum crunches per 100 g. (Source: Dayakar *et al.*, 2014.)

Nutrient	Value
Energy (kcal)	343
Protein (g)	10
Fat (g)	1.3
Dietary fibre (g)	21.48
Calcium (mg)	182
Iron (mg)	4.5
Zinc (mg)	2.4

5.6.10 Instant mixes

Instant and ready-to-reconstitute foods have become well-established products in western countries. In a busy and fast moving life where the woman is often also a bread-winner along with the emergence of new technologies and increased per capita income of the Indian population, instant products have their own importance; it is the need of the hour to develop traditional foods as convenience foods.

Although there are a large number of food products that are either RTE or just require warming, such as retort-pouched products, many consumers prefer freshly cooked food and hence opt for convenience mixes that require minimal preparation. Accordingly, instant mixes for products such as *dosa*, *idli* and *upma* are marketed extensively and termed 'ready mixes'. To use, they are mixed with water and left for a few minutes to hydrate and form a batter. Thus, wet grinding and overnight fermentation is avoided. The mixes not only offer convenience but can also be used at the desired level as and when required. Most of these ready mixes are prepared from rice or wheat as the base material. But nowadays, due to the fear of diabetes, consumption of rice-based products is decreasing, and because of coeliac disease/gluten intolerance, wheat-based products are also being side-tracked. In view of these, millet-based convenience foods are gaining prominence.

To prepare instant *idli* or *dosa* mixes, sorghum flour or fine grits are normally given a mild toasting to develop an aroma, reduce stickiness and to reduce the moisture content, which helps in their storage stability. The flour or fine grits are mixed with black gram flour, leavening agents (such as acidulates) and salt. For the preparation of sorghum *upma* mix and similar products, sorghum semolina of an appropriate grade is used. It is given a mild toasting to reduce the stickiness, and then garnished with spice, curry leaves, coriander leaves, onion, garlic and also sometimes leavening agents. Ready mixes for *bisibele bath* and for sweet dishes using sorghum flakes are similarly prepared.

Since millet offers several health benefits, there exists scope for its use in the production of traditional and contemporary convenience mixes. IIMR has carried out research in developing sorghum-based instant and ready-to-reconstitute foods, the results of which are discussed here.

Instant sorghum idli and dosa mix

To prepare the sorghum *idli* mix, sorghum fine semolina, black gram *dal*, salt, citric acid and sodium bicarbonate were used as the main ingredients. Sorghum fine semolina (180 μm) and black gram *dal* flour (2:1), citric acid (5 g), salt (6 g) and sodium bicarbonate (6 g) were mixed uniformly in a blender. The formulated mix was packed in MPET packing material. The shelf life of this *idli*-mix was found to be 3 months. IIMR's results (Table 5.26) show that this instant *idli*-mix has high levels of calcium, iron, zinc and riboflavin in comparison to the control (traditional sorghum *idli*), while protein, magnesium and folic acid content are similar to that of the control.

To prepare sorghum *dosa*-mix, sorghum flour and black gram *dal* (2:1), salt (6 g), citric acid (5 g) and sodium bicarbonate (6 g) were used as main ingredients and mixed uniformly in a blender, then packed in MPET. The shelf life of *dosa* mix was found to be 6 months.

Instant dhokla mix

To prepare sorghum *dhokla* mix, sorghum fine semolina (180 μm; 50 g) and black gram *dal* (180 μm; 50 g), (i.e. at a 1:1 ratio), Bengal gram *dal* (22 g), salt (4.6 g), citric acid (2.8 g), and sodium bicarbonate (4 g) were used as the main ingredients. They were mixed uniformly in a blender and packed in MPET. Its shelf life was found to be 3 months. Instant *dhokla* mix made from sorghum is rich in riboflavin, calcium, iron, zinc and

Table 5.26. Nutritional composition of instant sorghum *idli* and control (traditional sorghum *idli*). (Source: Dayakar *et al.*, 2014.)

Nutrient	Instant sorghum *idli* (/100 g)	Control (traditional sorghum *idli*) (/100 g)
Energy (kcal)	364.0	346.0
Carbohydrates (g)	71.7	72.0
Protein (g)	12.4	12.5
Fat (g)	1.6	0.8
Riboflavin (mg)	1.5	0.1
Folic acid (μg)	45.7	49.3
Calcium (mg)	10.2	5.8
Iron (mg)	7.2	0.2
Zinc (mg)	0.9	0.2
Magnesium (mg)	102.3	103.3

magnesium in comparison to the control (Bengal gram flour *dhokla)* while carbohydrate, protein, fat and folic acid levels are similar to those of the control (Table 5.27).

Instant upma-mix

To prepare instant sorghum *upma* mix, sorghum semolina (500 μm; 100 g), bengal gram *dal* (2 g), mustard seeds (0.5 g), curry leaves (0.3 g), dried green chillies (0.8 g), dried ginger (0.6 g), salt (3 g), citric acid (0.2 g) and oil (10 g) were used as ingredients. Semolina, mustard seeds and bengal gram *dal* were roasted separately. To the semolina, roasted mustard seeds, bengal gram *dal*, dehydrated curry leaves, ginger, salt and citric acid were added and mixed. The formulated mix was packed in MPET. The shelf life of *upma* mix was found to be 6 months.

Instant vermicelli kheer mix

To prepare instant sorghum vermicelli *kheer* mix, sugar (61%), sorghum vermicelli (30%), raisins (1%), edible vegetable oil (2%), cashew nuts (1%), edible starch (4.5%) and cardamom (0.5%) were used as ingredients. The organoleptic evaluation showed the acceptability of sorghum instant vermicelli *kheer* mix is far better than that of wheat vermicelli *kheer* mix. It was also found to have a shelf life of 6 weeks.

Instant flakes kheer mix

To prepare instant sorghum flake *kheer* mix, sugar (61%), sorghum flakes (30%), raisins (1%), cooking vegetable oil (2%), cashew nuts (1%), edible starch (4.5%) and cardamom (0.5%) were used as ingredients. The organoleptic evaluation showed the acceptability of sorghum instant flakes *kheer* mix is far better than that of the control *kheer* mix. It was also found to have a shelf life of 6 weeks.

Table 5.27. Nutritional composition of instant sorghum *dhokla*-mix and control *dhokla* (Bengal gram flour *dhokla*). (Source: Dayakar *et al.*, 2014.)

Nutrient	Instant *dhokla*-mix (/100 g)	Control *dhokla* (Bengal gram flour *dhokla*) (/100 g)
Energy (kcal)	352	350
Carbohydrates (g)	67.8	66.5
Protein (g)	15.7	17.5
Fat (g)	1.9	1.5
Riboflavin (mg)	1.1	0.1
Folic acid (μg)	75.2	74.1
Calcium (mg)	10.9	8.2
Iron (mg)	5	0.3
Zinc (mg)	0.7	0.1
Magnesium (mg)	116.7	71.4

5.6.11 Formulation of *masala* for the preparation of sorghum flakes and pops

A *masala* is a mixture of the ground spices used in Indian cookery; it is used as seasoning, adding flavour and enhancing palatability of food. *Aam chat* (mango tangy spice-mix) *masala*, *pudina* (mint) seasoning and coriander *masala* were developed and standardized for *masala* sorghum flakes and pops preparation. The ingredients of the *aam chat masala* and *pudina* seasoning are given in Table 5.28; and those for coriander *masala* are given in Table 5.29. The flow chart for the preparation of *masala* and *masala* flakes/pops is given in Fig. 5.45.

Table 5.28. Formulations used for the preparation of *aam chat masala* and *pudina* seasoning (per 100 g). (Source: Dayakar *et al.*, 2014.)

Ingredient	*Aam chat masala* (g)	*Pudina* seasoning (g)
Maltodextrin	29.4	29.4
Salt	12	12
Black salt	4	4
Sugar	18	18
Green chilli	5	5
Cumin	2.5	2.5
Coriander	8.5	8.5
Ginger	1.2	1.2
Black pepper	1	1
Turmeric	1	1
Asafoetida	0.02	0.02
Cardamom	0.7	0.7
Cinnamon	1.2	1.2
Clove	0.28	0.28
Citric acid	1.5	1.5
Amchur powder	9.5	–
Malic acid	0.6	0.6
Pudina oil	0.5	–
Calcium carbonate	0.1	0.1
Paprika oleoresin	0.3	0.3

Table 5.29. Formulation of coriander *masala* for the preparation of sorghum flakes and pops. (Source: Dayakar *et al.*, 2014.)

Ingredient	Amount (%)	Ingredient	Amount (%)
Chilli	3.7	Nutmeg	0.08
Onion powder	9	Cardamom	0.1
Cumin	2.1	Paprika extract	0.3
Garlic	5	Iodized salt	16.7
Coriander	15.64	Sugar	13
Turmeric	2.9	Sorghum flour	7.4
Aniseed	0.05	Edible vegetable oil	3
Black pepper	0.3	Potassium chloride	1
Fenugreek	0.03	MSG	0.8
Ginger	1.8	Citric acid	0.5
Clove	0.2	Maltodextrin	16.4

Note: MSG, monosodium glutamate.

Weighing of ingredients

↓

Light roasting spices in the edible oil

↓

Cooling of roasted material

↓

Addition of other ingredients

↓

Grinding of the above all weighed ingredients

↓

Vacuum packaging in LDPE bags

↓

Storage at room condition

↓

Roasted flakes or puffed pops (about 80 g)

↓

Heat oil and add 9 g of *aam chat masala* to make it into slurry

↓

Add flakes or pops to slurry and mix well

Fig. 5.45. Flow chart for the preparation of *masala* flakes/pops.

5.6.12 Standardization of flavourings for sorghum pasta

Tomato and *masala* flavourings were standardized for sorghum pasta (Table 5.30), to be used as a seasoning after the pasta is cooked. The procedure involved the roasting and grinding of spices and condiments and the addition of ingredients such as onion powder, garlic, sorghum flour, potassium chloride, sugar, citric acid, maltodextrin and salt. Organoleptic evaluation was conducted, using an equivalent commercial pasta/noodle flavouring as control. Sensory scores show that the IIMR formulations had higher acceptability than the commercial ones.

5.6.13 Sensory properties of cooked sorghum foods for daily consumption

At IIMR, traditional foods that are cooked from scratch (i.e. not instant mixes) and consumed daily – *roti, upma, dosa* and *idli* – were prepared using sorghum, and evaluated for their sensory properties. The sorghum cultivar M35-1 was used for the test recipes. The grain was procured and primary processed (winnowed, sieved, cleaned, pearled and dehulled) and products such as flour, coarse grits and semolina were prepared at IIMR (according to the requirement). The other ingredients required were brought from local retail markets. The recipes were carefully prepared according to the procedures outlined in *Delicious Sorghum Foods* (Ratnavathi *et al.*, 2008). The fuel used was LPG. Control foods were prepared using traditional standard recipes with wheat or rice and legumes wherever applicable and the test recipes were prepared with sorghum or with a sorghum/legume combination. All the recipes were subjected to sensory evaluation.

Sensory evaluation

This test was carried out with semi-trained panellists ($n = 15$), both male and female, aged 20–50 years. Care was taken that the panellists had no food allergies and were apparently healthy. The

Table 5.30. Formulation of tastemakers for pasta. (Source: Dayakar *et al.*, 2014.)

Tomato flavouring		*Masala* flavouring	
Ingredient	Amount (%)	Ingredient	Amount (%)
Tomato powder	22	SMP	22
Sugar	17	Sweet whey powder	14.2
Salt	12	Onion powder	11.5
Milk solids	14	Coriander	3.3
Sweet whey powder	4.4	Coriander leaves	1.8
Onion powder	3	Garlic	3.1
Onion flakes	4.3	Turmeric	2.3
Green bell pepper	7.1	Red chilli powder	0.5
Maltodextin	1	Cumin	0.1
Garlic granules	2.8	Aniseed	0.05
Edible vegetable oil	1.5	Black pepper	0.35
Corn flour	2	Fenugreek	0.03
Potassium chloride	1	Ginger	0.07
Cumin	1.2	Paprika extract	0.4
Paprika extract	0.5	Clove	0.1
Black pepper	1	Nutmeg	0.03
Green chilli	1.6	Green cardamom	0.1
Oregano	0.9	Sugar	9.5
I+G	1.2	Sorghum flour	5.1
Malic acid	0.4	Corn flour	3.1
$CaCO_3$	0.8	Salt	11.2
Cream plus butter buds	0.3	Maltodextrin	4
		I+G	1.2
		Edible vegetable oil	2
		Potassium chloride	1
		Citric acid	0.3
		Calcium carbonate	0.8
		Butter buds supreme	0.3
		Butter buds Asia	0.5
		Mustard	0.07
		Soya bean	1

Note: SMP, skimmed milk powder.

tests were conducted in a large hall illuminated with natural light. The questionnaire was prepared in English. The responses are recorded on a five-point Hedonic scale (1, very bad; 2, bad; 3, fair; 4, good; 5, very good) for appearance, colour, shape and size, odour, texture, taste and overall acceptability. General demographic data (name, age, gender and date of evaluation) were collected (Mahony, 1986). The products for panellists to sample were coded sample A and sample B. Sample A was prepared with sorghum as the main ingredient and sample B (the control) was prepared from rice or wheat. The panellists were given sample A followed by sample B. Drinking water was provided to the panellist to minimize any residual effects between samples.

Statistical analysis

Mean, standard deviation and frequency distribution were computed, and the Student's *t*-test was used to check the mean differences between control and test samples.

Results

The sorghum foods were compared with respective controls prepared with traditional ingredients, thus: sorghum *roti* was compared with wheat *roti*; sorghum *upma* was compared with wheat semolina *upma*; sorghum *dosa* was compared with *dosa* prepared from rice and black gram *dal*; sorghum *idli* was compared with *idli* prepared from rice semolina and black gram *dal*. Sorghum products were well accepted by 80% of the panellists. The results of the study clearly show the overall acceptability score of wheat *roti* was 100% and sorghum *roti* was 86.7%. Sorghum *roti* was prepared from whole millet flour. These variations may be due to the bland taste of sorghum, which could be enhanced by making *roti* with a combination of sorghum and wheat flours. This product can be consumed daily by those who have the habit of taking wheat *rotis*.

The overall palatability of sorghum products were: *dosa* 83.4%, *idli* 100% and *upma* 88.9% (Table 5.31). The sorghum semolina/grits have to be soaked in water for some time before cooking to make the finished product as soft as wheat *upma*. The soaking may be for as long as 1 h. According to House *et al.* (2000), some of the traits that affect the food quality and acceptability of sorghum grain are colour, shape, size and texture. The sensory scores revealed that the sorghum *dosa* was liked by the panellists. The mean value of percentage acceptability of colour, shape, size, texture, taste and overall palatability of sorghum *dosa* and control *dosa* are significantly different due to marked variations in colour, odour, appearance and taste. As studies state, colour is one of the traits that affects the acceptability of sorghum because of the presence of bran (Desikachar and Malleshi, 1987).

The mean scores of taste and overall palatability of sorghum *idli* were found to be more or less similar to those of the control. Virtually all the studies assure the sorghum quality and have found a significant association between the colour of the grain and the quality of the food prepared from them. The results of the percentage acceptability of sorghum products are given in Table 5.31.

Products like sorghum *dosa* and *idli* can be considered more nutritious than *roti* and *upma* as fermentation increases the nutrient content and bio-availability of various nutrients. The fermentation of sorghum improves its nutrient availability and protein digestibility. Sorghum is not a source of vitamin C but on germination, some amount of vitamin C is synthesized in the grain and on fermentation there is a further rise in vitamin content (Taur *et al.*, 1984).

5.7 Conclusion

Through recent initiatives, various aspects of millet processing interventions and development of end product-specific production have been developed, with the aim of increasing demand for the grain. Sorghum-based product formulations compare well with other cereal-based products and hold great promise. The marketing of sorghum/millet, cluster analysis of some of the developed sorghum products, entrepreneurship development and the resulting strengthening of millet as a commodity through the value chain model is discussed in detail in the following chapter.

Table 5.31. Comparison of percentage acceptability of sorghum recipes and control recipes. (Source: Dayakar *et al.*, 2014.)

Recipe		Sensory attribute (%)							
		Appearance	Colour	Shape	Size	Odour	Texture	Taste	Overall acceptability
Roti	Sorghum	100.0	100	100	99.9	86.7	94	86.7	86.7
	Control (wheat)	100	100	100	100	100	93.3	100	100
Upma	Sorghum	100	100	61.1	61.1	88.9	88.9	88.9	88.9
	Control (wheat)	100	100	66.7	66.7	84.1	75	91.7	91.7
Dosa	Sorghum	83.3	83.3	75.0	75.0	58.4	83.3	75	83.4
	Control (rice)	100	100	66.7	66.7	84.1	75.0	91.7	91.7
Idli	Sorghum	100	100	100	100	100	92.3	100	100
	Control (rice)	100	100	100	100	84.5	100	94.5	100

References

Abd Elmoneim, O., Elkhalifa, Schiffler, B. and Bernhardt, R. (2005) Effect of fermentation on the functional properties of sorghum flour. *Food Chemistry* 92(1), 1–5.

Ali, S.Z., Meera, M.S. and Malleshi, N.G. (2003) Processing of sorghum and pearl millet for promoting wider utilization for food. In: *Proceedings of an Expert Meeting*. ICRISAT, Patancheru, India, pp. 169–187.

Chavan, U.D., Chavan, J.K. and Kadam, S.S. (1988) Effects of fermentation on soluble proteins and in vitro digestibility of sorghum, green grain and sorghum green gram blends. *Journal of Food Science* 53, 1574–1575.

Chavan, U.D., Pansare, S.S., Patil, J.V. and Shinde, M.S. (2015) Preparation and nutritional quality of sorghum papads. *International Journal of Current Microbiology and Applied Science* 4(5), 806–823.

Dayakar, R.B., Patil, J.V., Hymavathi, T.V., Nirmal Reddy, K. and Rajendra Prasad, M.P. (2014) Terminal report of NAIP sub-project on *Creation of Demand for Millet Foods through PCS Value Chain*. Directorate of Sorghum Research, Rajendranagar, Hyderabad, AP- 500 030.

Dayakar, R.B., Bhargavi, G., Kalpana, K., Vishala, A.D., Ganapathy, K.N. and Patil, J.V. (2015) Development and standardization of sorghum pasta using extrusion technology. *Journal of Food Science and Technology* 52(10), 6828–6833.

Desikachar, H.S.R. and Malleshi, N.G. (1987) Milling of jowar for culinary uses. In: Ingle, U.M., Kulkarni, D.M. and Thorat, S.S. (eds) *Proceedings of National Seminar on Technology and Applications for Alternative Uses of Sorghum*. Parbhani, India, pp. 37–44.

Devaraju, B., Begum, M., Begum, S.S. and Vidya, K. (2003) Nutritional quality, sensory attributes and storage study of finger millet (*Eleusine coracana*) composite flour pasta. *Proceedings of the 5th International Food Convention*, Mysore, 5–8 December.

Dlamini, N.R., Taylor, J. and Rooney, L.W. (2007) The effect of sorghum type and processing on the antioxidant properties of African sorghum-based foods. *Food Chemistry* 105(4), 1412–1419.

Eastman, P. (1980) *An End to Pounding: a New Mechanical Flour Milling System in Use in Africa*. IDRC, Ottawa, ON, Canada.

Elmaki, H.B., Babiker, E.E. and El Tinay, A.H. (1999) Changes in chemical composition, grain malting, starch and tannin contents and protein digestibility during germination of sorghum cultivars. *Food Chemistry* 64(3), 331–336.

FAO (1995) *Sorghum and Millets in Human Nutrition*. Food and Nutrition Series 27, pp. 23–26. FAO, Rome.

Fast, R.B., Lanhoff, G.H., Taylor, D.D. and Getgood, S.J. (1990) Flaking ready-to-eat breakfast cereals. *Cereal Foods World* 35(3), 295–296

Geervani, P. and Vimala, V. (1993) *Rural Food Enterprises: Operational Research on Development of Sorghum Food Enterprises for Alternate Uses and Supplementary Feeding*. ANGRAU, Hyderabad, India.

Gopaldas, T., Inamdar, F. and Patel, J. (1982) Malted versus roasted, young child mixes: viscosity, storage and acceptability trials. *Indian Journal of Nutrition and Dietetics* 19, 327–336.

Hallgren, L., Rexen, F., Petersen, P.B. and Munck, L. (1992) Industrial utilization of whole crop sorghum for food and industry. *Proceedings of the International Workshop on Policy, Practice, and Potential Relation to Uses of Sorghum and Millets, February 8–12 1988, ICRISAT, Bulawayo, Zimbabwe*. ICRISAT, Patancheru, pp. 121–130.

Hamaker, B.R., Mertz, E. and Axtell, J.D. (1994) Effect of extrusion on sorghum kafirins solubility. *Cereal Chemistry* 71, 515–517.

Hoseney, R.C., Zeleznak, K. and Abdelrahman, A. (1983) Mechanism of popcorn popping. *Journal of Cereal Science* 1, 43–52.

House, L.R., Gomez, M., Murty, D.S., Sun, Y., Verma, B.N. (2000) Development of some agricultural industries in several African and Asian countries. In: Smith, C.W. and Frederiksen, R.A. (eds) *Sorghum: Origin, History, Technology, and Production*. Wiley, New York, pp. 131–190.

Hugo, L. F., Rooney, L. W. and. Taylor, J.R.N. (2003) Fermented sorghum as a functional ingredient in composite breads. *Cereal Chemistry* 80(5), 495–499.

Hulse, J.H., Laing, E.M. and Pearson, O.E. (1980) *Sorghum and the Millets: their Composition and Nutritive Value*. Academic Press, USA.

Itagi, K.S. (2003) Development and evaluation of millet based composite food for diabetes II. Ph.D. thesis, University of Agricultural Science, Dharwad, India.

Katti, S.V., Kumar, S. and Malleshi, N.G. (2008) Functional properties of finger millet flours prepared in different cereal pulverizers. *Journal of Food Science Technology* 45(5), 398–405.

Kazanas, N. and Fields, M.L. (1981) Nutritional improvement of sorghum by fermentation. *Journal of Food Science* 46, 81–821.

Kebakile, M.M., Taylor, J.R.N. and Rooney, L.W. (2008) Roller milling: an alternative dry milling process for sorghum. *Proceedings of the Workshop on Traditional Grains,* National Food Technology Research Centre Lobatse, Botswana, pp. 26–32.

Khedker, M.B., Bakane, P.H. and Kambale, A.K. (2008) Effect of grain moisture on popping quality of sorghum. *International Journal of Agricultural Engineering* 1(2), 71–73.

Krishna Kumari, S. and Thayumanavan, B. (1997) Comparative study of resistant starch from minor millets on intestinal responses, blood glucose, serum cholesterol and triglycerides in rats. *Journal of Food Sciences and Agriculture* 75, 296–302.

Kulkarni, K.D., Parlikar, D. and Bhagwat, S.M. (1987) Product application studies on sorghum malt: cookies, weaning food and beer wort. In: Ingle, U.M., Kulkarni, D.N. and Thorat, S.S. (eds) *Proceedings of the National Seminar on Technology and Applications for Alternative Uses of Sorghum*. Parbhani, Maharashtra, India, pp. 130–143.

Lai, H. and Cheng, H.H. (2004) Properties of pre-gelatinized rice flour made by hot air or gun puffing. *International Journal of Food Science and Technology,* 39, 201–212.

Llopart, E.E., Drago, S.R., De Greef, D.M., Torres, R.L. and González, R.J. (2014) Effects of extrusion conditions on physical and nutritional properties of extruded whole grain red sorghum. *International Journal of Food Science and Nutrition* 65(1), 34–41.

Lu, S. and Walker, C.E. (1988) Laboratory preparation of ready to eat breakfast flakes from grain sorghum flour. *Cereal Chemistry* 65(4), 377–378.

Mahgoub, S.E. and Elhag, S.A. (1998) Effect of milling, soaking, malting, heat-treatment and fermentation on phytate level of four Sudanese sorghum cultivars. *Food Chemistry* 61(1): 77–80.

Mahony, O.M. (1986) *Sensory Evaluation of Food – Hedonic Scale*. Marcel Dekker, Inc., New York.

Malleshi, N.G. (1987) Malting of sorghum for food uses. In: Ingle, U.M., Kulkarni, D.N. and Thorat, S.S. (eds) *Proceedings of the National Seminar on Technology and Applications for Alternative Uses of Sorghum*. Parbhani, Maharashtra, India, pp. 144–152.

Malleshi, N.G. and Klopfenstein, C.F. (1998) Nutrient composition, amino acid and vitamin contents of malted sorghum, pearl millet, finger millet and their rootlets. *International Journal of Food Science and Nutrition* 49, 412–442.

Malleshi, N.G., Hadimani, N.A., Chinnaswamy, R. and Klopfenstein, C.F. (1996) Physical and nutritional qualities of extruded weaning foods containing sorghum, pearl millet, or finger millet blended with *mung* beans and nonfat dried milk. *Plant Foods for Human Nutrition* 49, 181–189.

Martin, G. (2007) E-Course development – Module VI: Post production management, Grain Quality, Nutrition, and Postharvest Center. IRRI Germany, pp. 55–57. Available at: www.knowledgebank.irri.org/ericeproduction/bodydefault.htm#VI.D.2_Milling_systems.htm (accessed 27 August 2015).

Mishra, G., Joshi, D.C. and Pandaa, B.K. (2014) Popping and puffing of cereal grains: a review. *Journal of Grain Processing and Storage* 1(2), 34–46.

Munck, L. (1995) New milling technologies and products; whole plant utilization by milling and separation of the botanical and chemical components. In: Dendy, D. (ed.) *Sorghum and Millet for Human Food. Proceedings of a Symposium on Sorghum and Millet for Human Food, Vienna*. Tropical Products Institute, London, pp. 223–281.

Murthy, D.S., Subramanian, V., Suryaprakash, S., Patil, H.D. and House, L.R. (1984) Amylase activity and sprout damage in sorghum (*Sorghum bicolor* L. Moench). *Cereal Chemistry* 61(5), 415–418.

Murty D.S. and Kumar, K.A. (1995) Traditional uses of sorghum and millets. In: Dendy, D.A.V. (ed.) *Sorghum and Millets: Chemistry and Technology*. American Association of Cereal Chemists Press, St. Paul, Minnesota, pp. 185–221.

Osman, M.A. (2004) Changes in sorghum enzyme inhibitors, phytic acid, tannins and in vitro protein digestibility occurring during Khamir (local bread) fermentation. *Food Chemistry* 88(1), 129–134.

Panasiuk, O. and Donald, D. (1984) Cyanide content of sorghum sprouts. *Journal of Food Science* 49(3), 791–793.

Pathak, P., Srivastava, S. and Grover, S. (2000) Development of food products based on millets, legumes and fenugreek seeds and their suitability in the diabetic diet. *International Journal of Food Sciences and Nutrition* 51(5), 409–414.

Pawar, S.G., Pardeshi, I.L., Borkar, P.A. and Rajput, M.R. (2014) Optimization of process parameters of microwave puffed sorghum based ready to eat (RTE) foods. *Journal of Ready to Eat Food* 1(2), 59–68.

Radha, R. and Vijayalakshmi, P. (2007) Hypolipidemic potential of *Panicum miliare* on selected cardiovascular subjects. *Ancient Science of Life* 26(4), 29–32.

Rao, P.H. and Shurpalekar, S. R. (1976) Utilization of milo in bakery products. *Journal of Food Science and Technology* 13(6), 293–299.

Rasper, V.F. (1977) Palyi's compact system for debranning of sorghum and millet. In: Dendy, D.A.V. (ed.) *Proceedings of a Symposium on Sorghum and Millet for Human Food*. Tropical Products Institute, London, pp. 59–72.

Ratnavathi, C.V. and Patil, J.V. (2013) Sorghum utilization as food. *Journal of Nutrition and Food Science.* 4, 42–47.

Ratnavathi, C.V., Biswas, P.K., Pallavi, M., Maheshwari, M., Vijaykumar, B.S. and Seetharama, N. (2003) Alternative uses of sorghum-methods and feasibility: Indian perspective in alternate uses of sorghum and pearl millet in Asia. In: *Proceedings of the Expert Meeting, ICRISAT*, Patancheru, India, pp. 188–99.

Ratnavathi, C.V., Sailaja, V., Komala, V.V. and Dayakar, R.B. (2008) *Delicious Sorghum Foods*. National Research Centre for Sorghum, Hyderabad, India, pp. 34–45.

Rooney, L.W. and Taylor, J.R.N. (2000) Malted sorghum as a functional ingredient in composite bread. *Cereal Chemistry* 77(4), 428–432.

Saravanabavan, S.N., Shivanna, M.M. and Bhattacharya, S. (2013) Effect of popping on sorghum starch digestibility and predicted glycemic index. *Journal of Food Science and Technology* 50(2), 387–392.

Sautier, D. and O'Deye, M. (1989) *Mil, Mais, Sorgho – Techniques et Alimentation au Sahel*. Harmattan. Paris.

Serna-Saldivar, S.O., Clegg, C. and Rooney, L.W. (1994) Effects of parboiling and decortication on the nutritional value of sorghum (*Sorghum bicolor* L. Moench) and pearl millet (*Pennisetum glaucum* L.). *Journal of Cereal Science* 19(1), 83–89.

Shankara, R., Ananthachar, T.K., Narasimha, H.V., Krishna Murthy, H. and Desikachar, H.S.R. (1984) Improvement of the traditional edge runner process for rice flake production. *Journal of Food Science and Technology* 21(3), 121–122.

Shankara, R., Malleshi, N.G., Krishnamurthy, H., Narayana, M.N. and Desikachar, H.S.R. (1985) Development of a mini grain mill for dehusking and grinding of cereal. *Journal of Food Science and Technology.* 22, 91–93.

Sharma, V., Champawat, P.S. and Mudgal, V.D. (2014) Process development for puffing of sorghum. *International Journal of Current Research and Academic Review* 2(1), 164–170.

Subramaniam, V., Jambunatha, R. and Sambasiva Rao, W. (1983) Textural properties of sorghum dough. *Cereal Chemistry* 48, 1650–1653.

Taur, A.T., Pawar, V.D. and Ingle, U.M. (1984) Effect of fermentation on nutritional improvement of grain sorghum (*Sorghum bicolor* (L.) Moench). *Indian Journal of Nutrition and Dietetics* 21, 129–136.

Viraktamath, C.S., Raghavendra, G. and Desikachar, H.S.R. (1971) Use of rice milling machinery for commercial pearling of grain sorghum (*jowar*) and culinary uses for pearled sorghum products. *Journal of Food Science and Technology* 8, 11–13.

Waniska, R.D. and Rooney, L.W. (2000) Structure and chemistry of the sorghum caryopsis. In: Smith, C.W. and Frederiksen, R.A. (eds) *Sorghum: Origin, History, Technology, and Production*. John Wiley and Sons, Inc., New York, pp. 649–687.

Watson, S.A. (1970) Wet-milling and products. In: Wall, J.S. and Ross, W.M. (eds) *Sorghum Production and Utilization*. AVI, Westport, CT. pp. 602–626.

Yang, P. and Seib, P.A. (1996) Wet milling of grain sorghum using a short steeping period. *Cereal Chemistry* 73(6), 751–755.

6

Value Chain: Interventions Related to Consumer Acceptability, Entrepreneurship, Commercialization, Promotion and Policymaking

6.1 Consumer Acceptability and Market Feasibility of Newly Developed Sorghum Products

A test was conducted by IIMR in collaboration with expert consultants with the objective of assessing consumer acceptability and price preferences, and to further suggest suitable marketing strategies for five sorghum-based food products, namely sorghum-rich multigrain *atta* (flour), sorghum vermicelli, sorghum semolina, sorghum flakes and sorghum-rich biscuits. The core objective of the study was to understand consumer behaviour on their acceptance of products and to determine consumer profiling, with resulting delineation of specific target segments in selected cities, and advice on positioning and provision of inputs on branding, communication, pricing and distribution strategies. It was found that prior knowledge and awareness of the health and nutritional benefits of sorghum has emerged as one of the most important factors for acceptance of new products and purchase decisions.

6.1.1 Sampling and methodology

The cities chosen for the study were Hyderabad, New Delhi and Pune. Based on the statistical significance level, over 400 household respondents/samples were covered in each city. Except for sorghum-rich biscuits, all other products have to be cooked prior to consumption. Each household was supplied with a sample kit containing sorghum-rich multigrain *atta*, sorghum vermicelli, sorghum semolina, sorghum flakes and sorghum-rich biscuits.

Information was collected from respondents in two stages on various factors such as packaging, product attributes before and after cooking, and experiences encountered during preparation and consumption, captured through a structured questionnaire. In stage one, information pertaining to demographic profiles (age, family size, dietary preference, income level, number employed, number of children, number of elders and occupation of the lady member of the household) was collected. In stage two, the consumers provided their acceptability rating on product attributes, such as packaging, appearance, colour, smell, texture, preparation process and taste, in both raw and cooked forms. For this they used a five-point hedonic scale (5, liked very much; 4, liked moderately; 3, neither liked nor disliked; 2, disliked moderately; 1, not liked at all). Consumers rated information on prior awareness of sorghum and its benefits that led to the decision to purchase or not to purchase the product, based on a binary (yes/no) scale.

The final outcome of the consumers' responses for each of the products are presented in this chapter. Other details/outputs of the analysis are included in Appendix 9.

Factors/dimensions used for the analysis

Factors were grouped and titled on the basis of the characteristics of their composite variables:

- **Organoleptic** ('factor 1'). This is the most important dimension identified by the respondent. This deals with the physical attributes and organoleptic characteristics of the product. Physical attributes include product appearance in its raw form as well as during the preparation (cooking) form and finished form. Organoleptic attributes include sensory aspects (cooked), appearance, test, aroma, texture etc.
- **Prior awareness on health and other benefits of sorghum** ('factor 2'). This dimension includes awareness about the consumption of sorghum and the benefits or perceptions attached to it. Awareness of nutritional benefits, ease in digestion, energy provision for work, modernity etc., are the parameters grouped in this dimension.
- **Comparison of product with traditional product** ('factor 3'). The comparison of the selected sorghum-rich product with the traditional/currently consumed equivalent on the basis of preparation process difficulties and taste are the major attributes included under this dimension.
- **Socio-economic aspects and value for money** ('factor 4'). Respondents' income, expenditure on food and product pricing were some of the important aspects covered under this dimension.
- **Demography** ('factor 5'). Family size, number of children, family background and occupations of the respondents are the attributes included in this dimension.
- **Myths related to sorghum** ('factor 6'). In many areas consumers have the misconception that sorghum should be restricted only to the old, ill and adult members of the family. This prejudice is another dimension included.

Cluster analysis

Cluster analysis was carried out to classify the different consumer profiles based on their willingness to purchase the product or not. Following this, the clusters were further grouped as 'good', 'average' and 'bad' in terms of clarity in the output. Each cluster is compared and classified based on the analysis of the consumer feedback through three parameters: percentage of purchase vs non-purchase out of the total; attribute likewise ranking; and perceptions in comparison with the traditional/current products consumed.

6.1.2 Sorghum-rich multigrain *atta*

The important dimensions of the consumers' responses on sorghum-rich multigrain *atta* are listed in Table 6.1. A common rule of thumb for dropping the least important factors from the analysis is the K1 rule. Factors with eigenvalues >1.0 were selected. Based on this, demography was the last factor selected, having an eigenvalue of 1.340. The reliability of the data and consistency in the responses received for factor 1 ('organoleptic') was 0.944, which is much higher than the measuring point (the cut-off point) of 0.600. Similarly, Cronbach's alpha for factors 2 ('prior awareness') and 3 ('comparison with traditional products') were 0.926 and 0.768, respectively. The test could not be conducted for factors 4 and 5 as it consists of independent variables, and responses are on a different scale. The percent variance explained by these five factors is 62%.

The organoleptic factor, with an eigenvalue of 9.191 (Fig. 6.1), emerged as the most important property of sorghum-rich multigrain *atta*

Table 6.1. Important factors/dimensions in the consumer's response to sorghum-rich multigrain *atta*. (Source: Dayakar *et al.*, 2014a.)

Dimensions/(factor)	Cronbach's alpha	Eigenvalue	Variance explained (%)
Organoleptic (1)	0.944	9.191	27.85
Prior awareness (2)	0.926	5.169	15.66
Comparison with traditional products (3)	0.768	2.143	6.5
Socio-economic and value for money (4)	N.A.		
Demography (5)	N.A.		

perceived by the consumers. In this, attributes of *chapathi* or other final products of sorghum-rich multigrain *atta* are more important than the preparation and appearance of the flour in its raw form. Prior awareness of the sorghum products and their nutritional and health benefits is the second most important attribute, followed by comparison with traditional wheat *atta*.

Clusters of respondents for sorghum-rich multigrain atta

The graphical representation of the major characteristics and features of the clusters are presented below (Table 6.2, Figs 6.2–6.4). The output of the cluster is further statistically tested (chi-square test and probability (*P*-value)) to ascertain significant similarities and differences between both clusters.

The results of the chi-square and probability tests on socio-economic characters show that there are significant differences ($P<0.05$) between the clusters in terms of family size, children, food habit, household income and willingness to pay the purchase price. On the other hand, the clusters do not have significant differences in terms of occupation of the female respondents and household expenditure on food from the total income. A detailed table of the consumer profile from both good and average clusters is given in Appendix 9. A similar technique was followed for analysing the clusters for the other four products, discussed later in this chapter.

Attributes influencing the purchase of sorghum-rich multigrain atta

Factor analysis in Table 6.2 enabled the identification of important dimensions from the response of the consumers with no dependent variable (purchase/not purchase). The ultimate aim is to find out the attributes that have an impact on the purchasing decision of the consumer. A logistic regression analysis (LRA) model was developed and stepwise analysis carried out using various statistical tools.

The odds ratio score (ORS) estimates indicate the probability that the decision to purchase would increase if there is a one-unit increase in the Likert scale (Likert scale scores 1 through 5) and is tabulated in Table 6.3. Statement 1 ('*Jowar* [sorghum] is wholesome nutrition for the entire family') emerged as the most important attribute that could increase the chances of purchase. The response collected was a non-numeric scale (agree/disagree) so that increasing awareness of sorghum's benefit to the consumer would certainly increase the chances of purchase. This could be the interpretation of the odds ratio score.

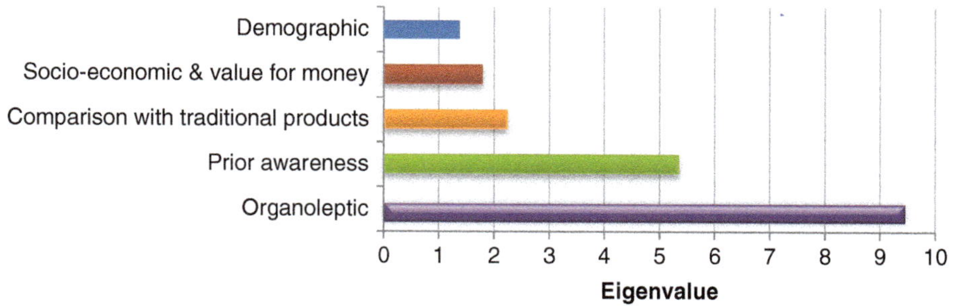

Fig. 6.1. Factors for consideration in the purchase of sorghum-rich multigrain *atta*.

Table 6.2. Cluster-wise consumer response on purchase decision (*atta*). (Source: Dayakar *et al.*, 2014b.)

Cluster	Would purchase	Would not purchase	Total respondents	Cluster output
1	206	29	235	Average
2	6	23	29	Bad
3	24	21	45	Bad
4	4	24	28	Bad
5	817	19	836	Good
6	41	0	41	Good

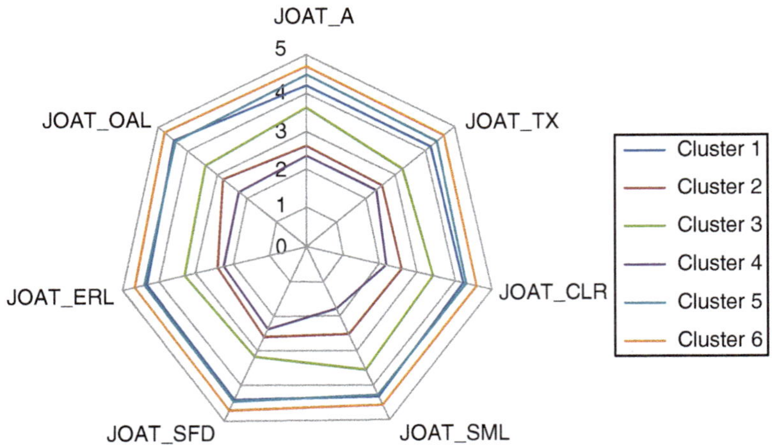

Scale: 5, liked very much; 4, liked moderately; 3, neither liked nor disliked; 2, disliked moderately; 1, disliked very much.

Fig. 6.2. Cluster profile based on product attributes (*atta*). See list on p. xxv for definitions of cluster analysis abbreviations.

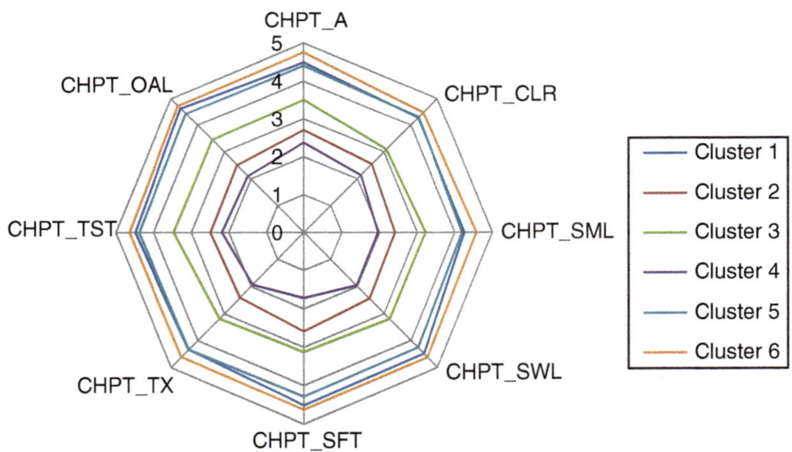

Scale: 5, liked very much; 4, liked moderately; 3, neither liked nor disliked; 2, disliked moderately; 1, disliked very much.

Fig. 6.3. Cluster profile based on product attributes (*chapathi*). See list on p. xxv for definitions of cluster analysis abbreviations.

On the other hand, *chapathi* texture and taste are the most influential variables, estimated to increase the likelihood of purchase by almost 1.9 times each. *Chapathi* swelling and ease of rolling could increase the purchasing chance by 1.61 times and 1.58 times, respectively, when there is an increase in one unit in the liking Likert scale.

Prediction of correct purchase decision

The results (Table 6.4) indicate that 88% of those who are willing to purchase and 5% of those who are not willing to purchase the sorghum-rich multigrain *atta* are observed and predicted to be correctly classified into their group. On the other hand, 5% of those who are willing to purchase and 2% of those who are not

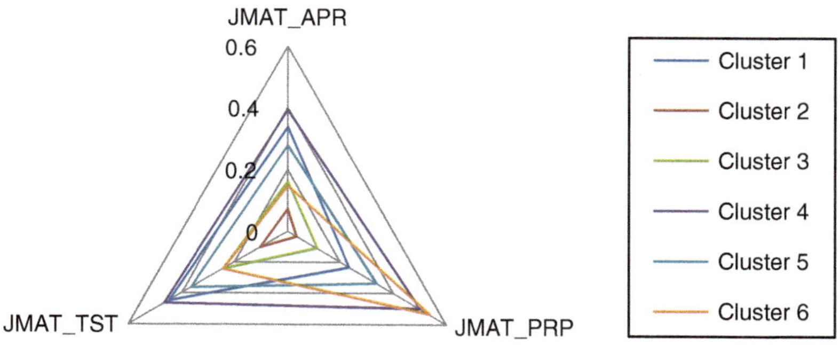

Scores: 0, traditional product is better; 1, *jowar* product is better or both are the same.

Fig. 6.4. Cluster profile based on comparison with wheat *atta*. See list on p. xxv for definitions of cluster analysis abbreviations.

Table 6.3. Variable/attribute influential in purchasing decision. (Source: Dayakar *et al.*, 2014b.)

Factor	MLE estimate	Odds ratio estimate
Intercept	−9.2045	N.A.
Stat_1	2.4959	12.133
JOAT_ERL	0.4605	1.585
CHPT_CLR	0.3395	1.404
CHPT_SWL	0.4801	1.616
CHPT_TX	0.6327	1.883
CHPT_TST	0.6285	1.875

Note: The analysis of maximum likelihood estimates (MLE) was used to obtain parameter estimates. Significance of parameter estimates was based on the Wald chi-square value at $P<0.05$. See list on p. xxv for definitions of cluster analysis abbreviations.

willing to purchase the sorghum-rich multigrain *atta* are observed and predicted to be incorrectly classified. In conclusion, the purchasing decision of sorghum-rich multigrain *atta* could be predicted with 93% accuracy and 7% inaccuracy.

Purchase prediction accuracy based on individual attributes (hit ratio)

Based on the logistic regression analysis (LRA) at *P*-value of 0.05, hit rate (%) is the correct classification of an unknown unit (product) into a group (purchased vs not purchased). The hit ratio of purchase prediction accuracy based on individual attributes for sorghum-rich multigrain *atta* is given in Fig. 6.5. From this, it can be seen that *chapathi* texture alone can be used for predicting the *atta* purchase decision with an accuracy of 76.5%. Good swelling of *chapathi* (roasting on direct fire or after putting on oil drops) is another important attribute/factor that can be used as a single predictor, with a purchase prediction accuracy of 73.3%. Ease in rolling of *chapathi* and colour can also provide accuracy in predicting the correct purchase decision, at 71.3 and 71%, respectively.

Similarities and dissimilarities in the response of respondents on important attributes of sorghum-rich multigrain atta

Surveyed cities were grouped in pairs and analysed for attributes influencing purchase (Fig. 6.6). The objective was to find out which pairs were similar or different in their level of response, with results helping to decide the communication strategy required to market sorghum-rich multigrain *atta* in particular cities. For this, a paired comparison test was conducted using ANOVA techniques, with a confidence level set at 95% ($P<0.05$). Hyderabad and Delhi do not have a significant difference in response received for *chapathi* taste, texture and ease of rolling. Hyderabad and Delhi, and Hyderabad and Pune are very similar in their responses for *chapathi* texture; however, the differences are more marked when compared with the former pairs. Delhi and Pune is the only combination of the three cities that has significant differences among all the variables.

Respondents in Hyderabad liked the overall attributes of the sorghum-rich multigrain *atta* compared to the respondents of Pune and Delhi, and are close to liking all attributes very much. In Delhi, respondents are satisfied with *chapathi*'s ability to swell, its texture and taste, but colour and

Table 6.4. Correct classification of actual purchase decision and prediction on correctness of purchase for sorghum-rich multigrain *atta*. (Source: Dayakar *et al.*, 2014b.)

	No. of respondents			% of respondents		
	Would not purchase	Would purchase	Total	Would not purchase	Would Purchase	Total
Would not purchase	59	57	116	5	5	10
Would purchase	21	1078	1098	2	88	90
Total	80	1135	1215	7	93	100
	Correct prediction		1137	Correct prediction		93
	Incorrect prediction		78	Incorrect prediction		7

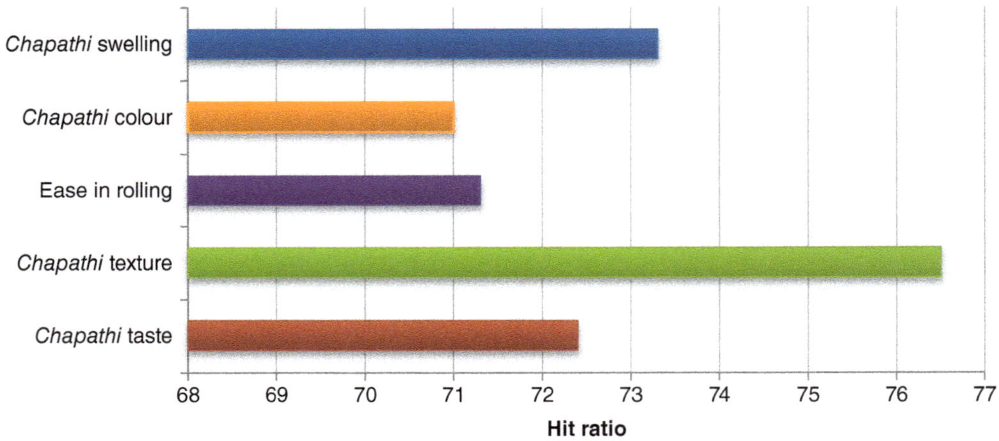

Fig. 6.5. Purchase prediction accuracy based on individual attributes for sorghum-rich multigrain *atta*.

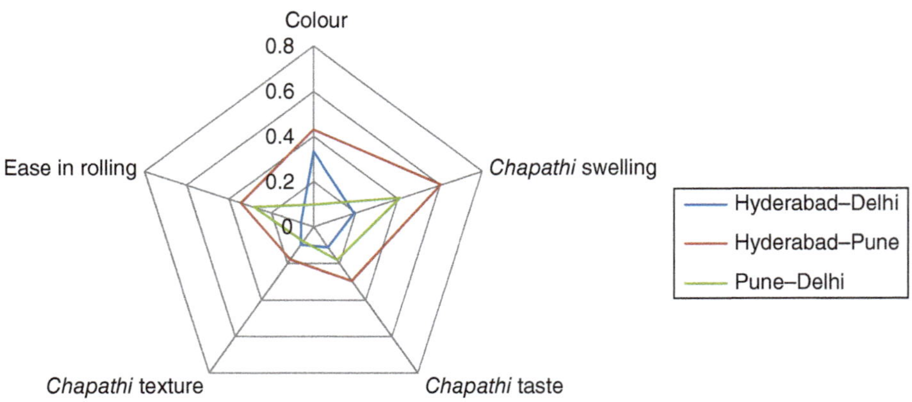

Scores: 0, traditional product is better; 1, *jowar* product is better or both are the same.

Fig. 6.6. Cities' pairwise responses on influential attributes of *atta*.

ease in rolling have an average performance on the Likert scale. In Pune, colour, taste, texture and swelling of *chapathi* has been liked moderately by the respondents, while ease in its rolling is close to being liked very much (Fig. 6.7).

6.1.3 Sorghum biscuits

The most important factors influencing the consumer's response to sorghum biscuits are given in Table 6.5. Demography, with an eigenvalue of 1.27, was the last variable to be selected. The reliability of the data and consistency in responses received (shown as the value for Cronback's alpha; Table 6.5) is 0.896 for factor 1 (organoleptic attributes) – which is much higher than the measuring point of 0.600 – however, the value is higher (0.913) for factor 2 ('prior awareness'), which can be interpreted as the attribute having the most reliable response. The percent variance explained by these six factors is 58.5%.

A sorghum biscuit's organoleptic properties is the most important consideration for the consumer, followed by prior awareness of the health and other benefits of sorghum. Comparison with other biscuits that they are currently consuming is the third most important factor in evaluating the sorghum biscuits. Ultimately, the final experience of the consumer in terms of sensory and comparative attributes is going to be the major dimension for accepting or rejecting the biscuit (Fig. 6.8).

Clusters of respondents for sorghum biscuits

The representation of the major characteristics and features of the clusters are presented in Table 6.6 and Figs 6.9 and 6.10.

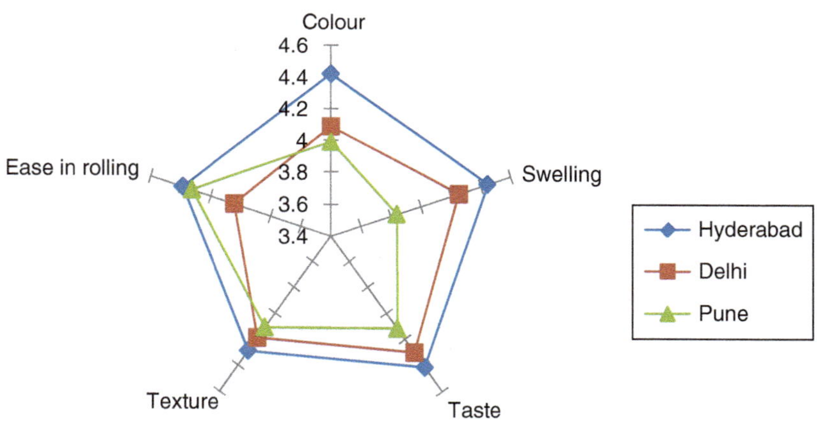

Scale: 5, liked very much; 4, liked moderately; 3, neither liked nor disliked; 2, disliked moderately; 1, disliked very much.

Fig. 6.7. Impact of important attributes of *atta* on individual city's respondents.

Table 6.5. Important factors/dimensions in the consumer's response to sorghum biscuits. (Source: Dayakar *et al.*, 2014b.)

Dimensions/(factors)	Cronbach's alpha	Eigenvalue	Variance explained (%)
Organoleptic (1)	0.896	5.91	21
Prior awareness (2)	0.913	4.08	15
Comparison with traditional products (3)	0.748	1.99	7
Socio-economic and value for money (4)	N.A.	1.66	6
Myth of sorghum to the decision maker and income (6)	N.A.	1.43	5
Demography (5)	N.A.	1.27	4.5

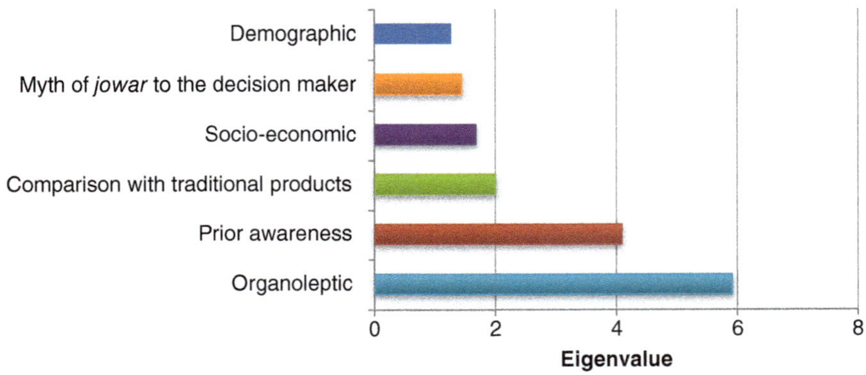

Fig. 6.8. Factors for consideration in the purchase of sorghum biscuits.

Table 6.6. Clusters of consumers based on purchase decision for sorghum biscuits. (Source: Dayakar *et al.*, 2014b.)

Cluster	Would purchase	Would not purchase	Total respondents	% purchase	Cluster output
1	6	40	46	13	Bad
2	43	0	43	100	Good
3	153	11	164	93	Average
4	199	4	203	98	Good
5	731	28	759	96	Good

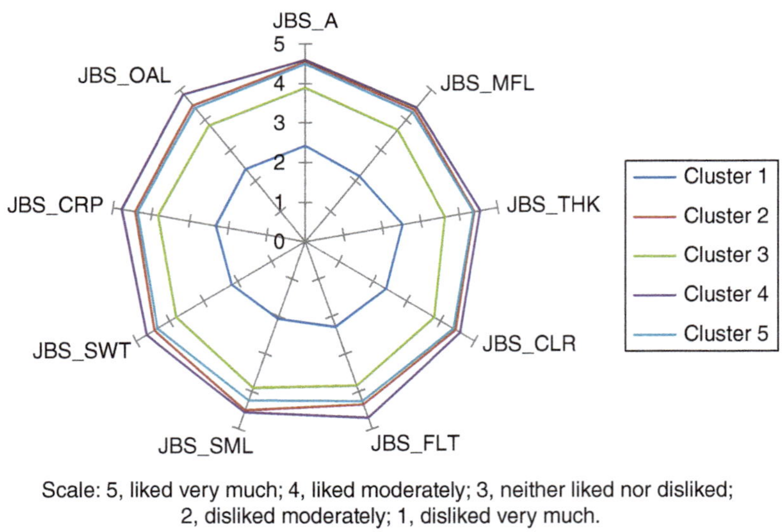

Fig. 6.9. Cluster-wise score on sorghum biscuit attributes. See list on p. xxv for definitions of cluster analysis abbreviations.

The total respondents were formed into five clusters. Each cluster was compared and classified based on the analysis of the consumer feedback through three parameters: percentage that would purchase vs those that would not purchase out of the total, attribute likewise ranking, and perceptions on how the product compared with traditional/currently consumed biscuits.

The biscuit cluster differs widely from that of the multigrain *atta*. The clusters are significantly different in terms of price level, buying preference, quantity of biscuits consumed in a month, percent

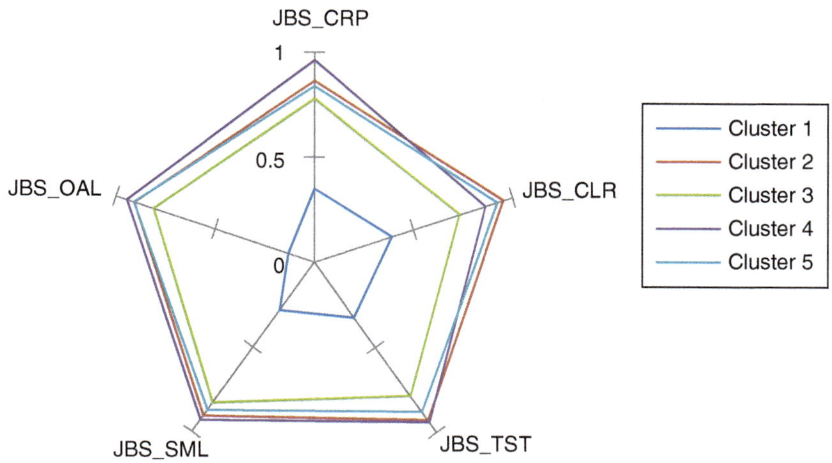

Score: 0, traditional product is better; 1, *jowar* product is better or both are the same.

Fig. 6.10. Cluster-wise score on sorghum biscuit attributes in comparison with traditional biscuits. See list on p. xxv for definitions of cluster analysis abbreviations.

expenditure on food from household income, monthly household income, food habit and number of children. A detailed table of the consumer profile for both good and average clusters is given in Appendix 9.

Attributes influencing the purchase of sorghum biscuits

The most important attribute that could influence the purchasing decision of the consumer is the smell of the sorghum biscuit in comparison with traditional biscuits, followed by its sweetness. As shown in Table 6.7, these two attributes could increase the probability of purchasing the sorghum biscuit by 4.10 and 2.67 times, respectively, when there is an increase in one unit in the Likert scale. Mouth feel of the biscuit is another important attribute, followed by its crispiness; these factors increase the purchasing likelihood by 1.677 and 1.68 times, respectively. Awareness of the benefits of sorghum does not have a significant influence on purchasing decision.

Prediction of correct purchase decision

To draw an inference (Table 6.8), 92% of those who are willing to purchase and 3% of those who are not willing to purchase the sorghum biscuits are observed and predicted to be correctly classified into their group. On the other hand, 4% of those who are willing to purchase and 1% of those who are not willing to purchase the sorghum biscuits are observed and predicted to be incorrectly classified. In conclusion, the purchasing decision of sorghum biscuits could be predicted with 95% accuracy and 5% inaccuracy.

Purchase prediction accuracy based on individual attributes (hit ratio)

Sweetness of the biscuit can be used as a single predictor, with a purchase decision prediction accuracy of 71.5% (Fig. 6.11). Similarly, mouth feel and crispiness of the biscuits can also be used for predicting the purchase decision with an accuracy of 65.8% and 64.9%, respectively.

Similarities and dissimilarities in the response of respondents on important attributes of sorghum biscuits

Hyderabad and Delhi respondents do not show a significant difference in terms of response to sorghum biscuit crispiness and sweetness attributes. Similarly, Delhi and Pune respondents have similar responses to mouth feel attribute. All cities have significant differences in response to the attribute of sorghum biscuit smell/aroma in comparison with traditional biscuits (Fig. 6.12). Hyderabad respondents very much liked all sorghum biscuit attributes compared to respondents from Delhi and Pune. Mouth feel and sweetness of the sorghum-rich biscuit is not much liked by the Pune respondents. Delhi and

Table 6.7. Odds ratio estimate to predict purchase influencing attributes of sorghum biscuits. (Source: Dayakar *et al.*, 2014b.)

Factor	MLE estimate	Odds ratio estimate
Intercept	−6.2408	N.A.
Sweetness	0.9833	2.673
Mouth feel	0.5173	1.677
Smell	1.4114	4.102
Crispiness	0.5185	1.68

Note: The analysis of maximum likelihood estimates (MLE) was used to obtain parameter estimates. Significance of parameter estimates was based on the Wald chi-square value at $P<0.05$.

Table 6.8. Correct classification of actual purchase decision and prediction on correctness of purchase of sorghum biscuits. (Source: Dayakar *et al.*, 2014b.)

	No. of respondents			% of respondents		
	Would not purchase	Would purchase	Total	Would not purchase	Would purchase	Total
Would not purchase	37	46	83	3	4	7
Would purchase	18	1114	1132	1	92	93
Total	55	1160	1215	5	95	100
	Correct prediction		1151	Correct prediction		95
	Incorrect prediction		64	Incorrect prediction		5

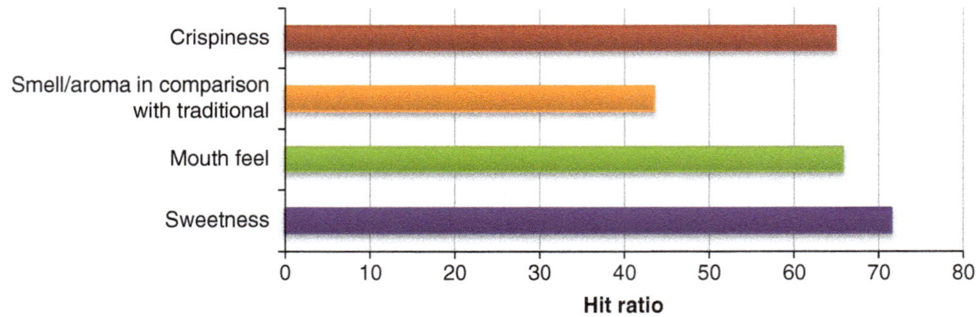

Fig. 6.11. Purchase prediction accuracy based on individual attributes for sorghum biscuits.

Table 6.9. City-wise responses on influential attributes of sorghum biscuits. (Source: Dayakar *et al.*, 2014b.)

	Hyderabad		Delhi		Pune			
	0	1	0	1	0	1	Chi sq.	*P*
Smell/aroma in comparison with traditional	30	405	74	301	72	333	32.1	<0.0001

Note: 0, traditional product is better; 1, sorghum product is better or both are the same.

Hyderabad respondents more than moderately liked the crispiness and sweetness of the sorghum biscuit. On the other hand, respondents of Pune are nearly satisfied with all three attributes of the sorghum biscuit (Table 6.9, Fig. 6.13).

6.1.4 Sorghum vermicelli

The most important factors influencing the consumer's response to sorghum vermicelli are given in Table 6.10. Demography, with an eigenvalue of

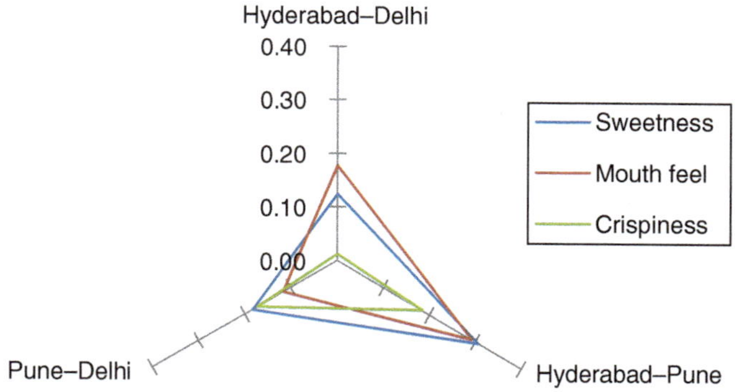

Score: 0, traditional product is better; 1, *jowar* product is better or both are the same.

Fig. 6.12. Cities' pairwise responses on influential attributes of sorghum biscuit.

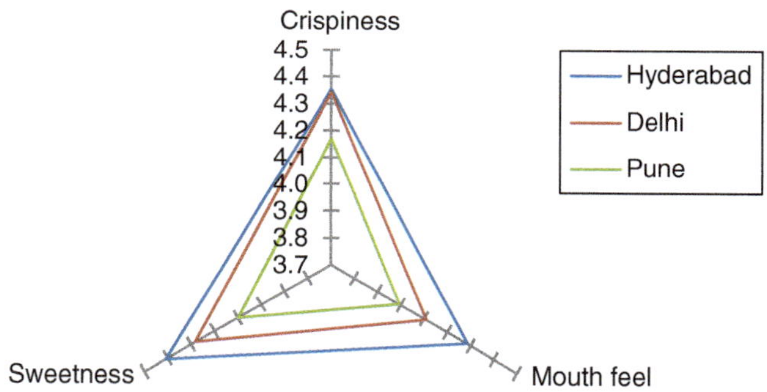

Scale: 5, liked very much; 4, liked moderately; 3, neither liked nor disliked; 2, disliked moderately; 1, disliked very much.

Fig. 6.13. Impact of important attributes of sorghum biscuits on individual city's respondents.

Table 6.10. Important factors/dimensions in the consumer's response to sorghum vermicelli. (Source: Dayakar *et al.*, 2014b.)

Dimensions/(factor)	Cronbach's alpha	Eigenvalue	Variance explained (%)
Organoleptic (1)	0.950	8.82	28
Prior awareness (2)	0.914	4.02	13
Comparison with traditional products (3)	0.699	1.81	6
Socio-economic and value for money (4)	N.A.	1.65	5
Myth of sorghum to the decision maker and income (6)	N.A.	1.36	4.5
Demography (5)	N.A.	1.28	4

1.28, was the last variable to be selected. The reliability of the data and consistency in the responses received for factor 1 ('organoleptic' attribute) is 0.950, which is much higher than the measuring point of 0.600. Cronbach's alpha for factors 1, 2 and 3 is 0.950, 0.914 and 0.699, respectively (Table 6.10, Fig. 6.14). The percent variance explained by the six most important factors is 59.5%.

Clusters of respondents for sorghum vermicelli

The sorghum vermicelli clusters are significantly different in terms of the area from which the majority of respondents purchase (i.e. purchase place of the majority), buying preferences (branded or unbranded), occupation of the male member of the family and size of the family (total members in a household) (Table 6.11 and Figs 6.15–6.17). A detailed table of consumer profiles from both good and average clusters is given in Appendix 9.

Attributes influencing the purchase of sorghum vermicelli

As shown in Table 6.12, the taste of sorghum vermicelli in comparison with wheat vermicelli is the most influential attribute; this attribute alone can increase chances of the purchase of sorghum vermicelli by 6.26 times when there is an increase in one unit in the Likert scale. Ease of preparation of sorghum vermicelli is similar to that of traditional vermicelli, resulting in a potential increase in likelihood of purchase by 3.24 times, and overall appearance of the sorghum vermicelli as traditional vermicelli could influence the purchasing likelihood by 2.45 times if there is a one-unit increase in the Likert scale or if the appearance is similar to that of wheat vermicelli. Smell/aroma and stickiness of the cooked vermicelli could also enhance the purchasing likelihood by 2.26 and 1.49 times, respectively.

Prediction of correct purchase decision

Table 6.13 shows that 85% of those who are willing to purchase and 7% of those who are not willing to purchase the sorghum vermicelli are observed and predicted to be correctly classified. On the other hand, 6% of those who are willing to purchase and 2% those who are not willing to purchase the sorghum vermicelli are observed and predicted to be incorrectly classified. In conclusion, the purchasing decision of sorghum vermicelli could be predicted with 92% accuracy and 8% inaccuracy.

Purchase prediction accuracy based on individual attributes (hit ratio)

The purchase prediction accuracy based on individual attributes for sorghum vermicelli

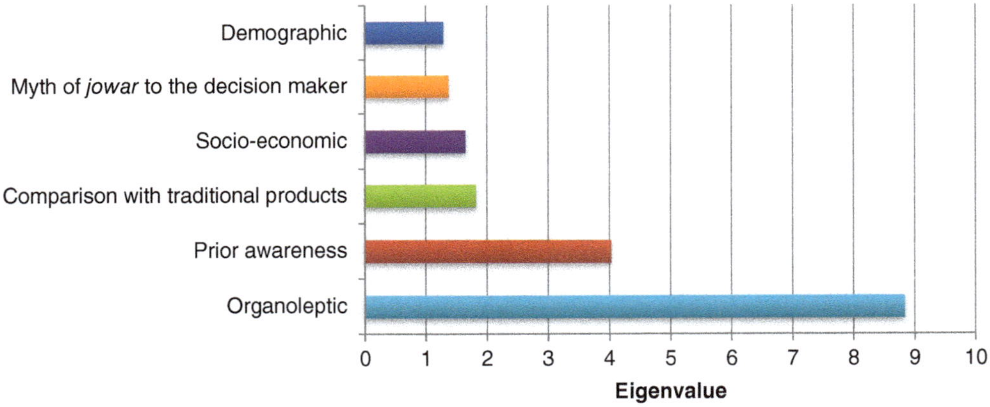

Fig. 6.14. Factors for consideration in the purchase of sorghum vermicelli.

Table 6.11. Clusters of consumers based on purchase decision for sorghum vermicelli. (Source: Dayakar *et al*., 2014b.)

Cluster	Would purchase	Would not purchase	Total respondents	% purchase	Cluster output
1	555	29	584	95	Good
2	7	50	57	12	Bad
3	166	29	195	85	Average
4	325	51	376	86	Average

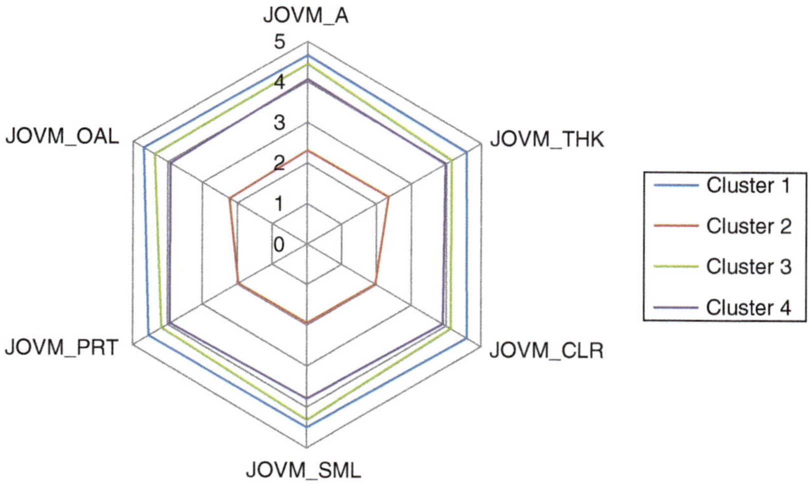

Scale: 5, liked very much; 4, liked moderately; 3, neither liked nor disliked; 2, disliked moderately; 1, disliked very much.

Fig. 6.15. Cluster-wise score on raw vermicelli's attributes. See list on p. xxv for definitions of cluster analysis abbreviations.

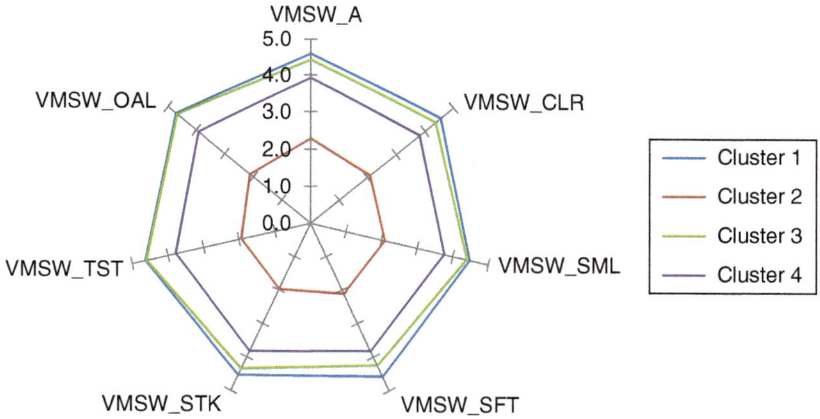

Scale: 5, liked very much; 4, liked moderately; 3, neither liked nor disliked; 2, disliked moderately; 1, disliked very much.

Fig. 6.16. Cluster-wise score on cooked (sweet) vermicelli's attributes. See list on p. xxv for definitions of cluster analysis abbreviations.

is given in Fig 6.18. Sorghum vermicelli smell, sweet vermicelli stickiness and sweet vermicelli overall can be used as predictors, for which the decision to purchase vermicelli can be predicted with an accuracy of 65.9, 64.4 and 63.5%, respectively. Taste in comparison with wheat vermicelli is a slightly less accurate predictor, with an accuracy of 56.7%.

Similarities and dissimilarities in the response of respondents on important attributes of sorghum vermicelli

Delhi and Pune have similar responses to the stickiness of vermicelli. In terms of overall attributes, Hyderabad and Delhi have similar responses. Hyderabad–Delhi and Delhi–Pune

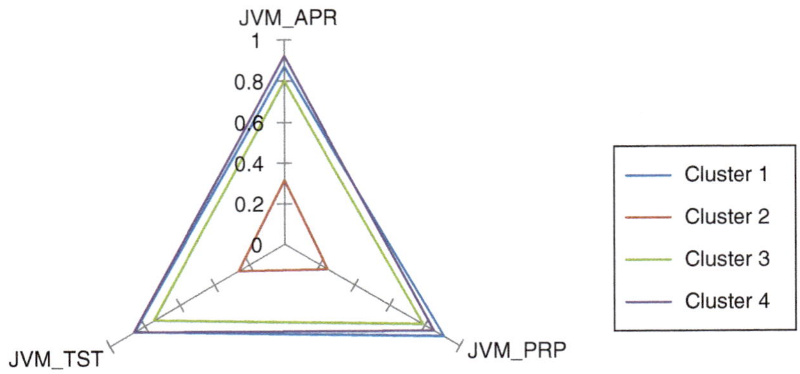

Score: 0, traditional product is better; 1, *jowar* product is better, or both are the same.

Fig. 6.17. Cluster-wise score on sorghum vermicelli's attributes in comparison with traditional vermicelli.

Table 6.12. Odds ratio estimate to predict purchase influencing attributes of sorghum vermicelli. (Source: Dayakar *et al.*, 2014b.)

Factor	MLE estimate	Odds ratio estimate
Intercept	0.8243	NA
Smell	0.8181	2.266
Stickiness (sweet vermicelli)	0.3982	1.489
Overall (sweet vermicelli)	0.5142	1.672
Appearance	0.8984	2.456
Ease of preparation	1.1763	3.242
Taste	1.8347	6.263

Note: The analysis of maximum likelihood estimates (MLE) was used to obtain parameter estimates.

Table 6.13. Correct classification of actual purchase decision and prediction on correctness of purchase for sorghum vermicelli. (Source: Dayakar *et al.*, 2014b.)

	No. of respondents			% of respondents		
	Would not purchase	Would purchase	Total	Would not purchase	Would purchase	Total
Would not purchase	84	75	159	7	6	13
Would purchase	28	1028	1056	2	85	87
Total	112	1103	1215	9	91	100
	Correct prediction		1112	Correct prediction		92
	Incorrect prediction		103	Incorrect prediction		8

have similar responses in the smell/aroma of sorghum vermicelli. The comparative attributes of sorghum vermicelli in comparison to wheat vermicelli are also significantly different among the three cities, although the majority have responded that sorghum vermicelli is better or no different to wheat vermicelli (Table 6.14, Fig. 6.19).

Pune respondents did not much like the smell of sorghum during the preparation and consumption of sweet vermicelli. Hyderabad again emerged as the city to most like all attributes of sorghum vermicelli. On the Likert scale, Hyderabad and Delhi respondents liked it above moderate level. Pune again remained conservative, only averagely liking the vermicelli attributes (Fig. 6.20).

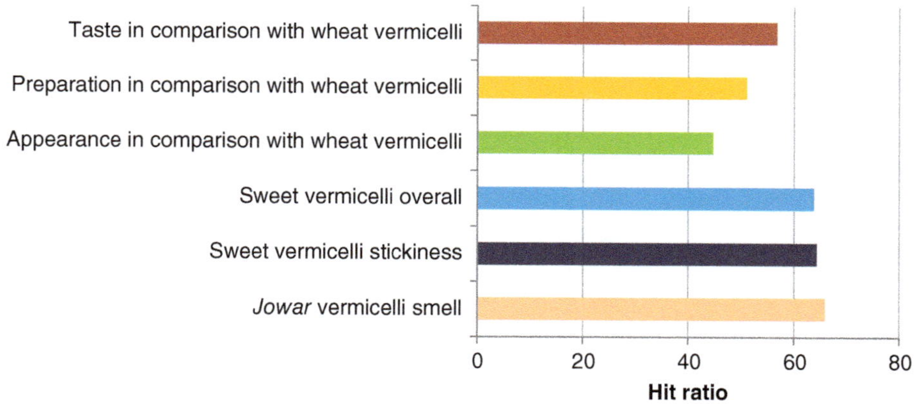

Fig. 6.18. Purchase prediction accuracy based on individual attributes for sorghum vermicelli.

Table 6.14. City-wise responses on influential attributes of sorghum vermicelli. (Source: Dayakar *et al.*, 2014b.)

	Hyderabad		Delhi		Pune			
	0	1	0	1	0	1	Chi sq.	*P*
Taste in comparison with wheat vermicelli	60	375	113	262	52	353	48.62	<0.0001
Appearance in comparison with wheat vermicelli	48	387	99	276	37	368	54.06	<0.0001
Preparation in comparison with wheat vermicelli	48	387	95	280	51	354	35.85	<0.0001

Note: 0, traditional product is better; 1, sorghum product is better or both are the same.

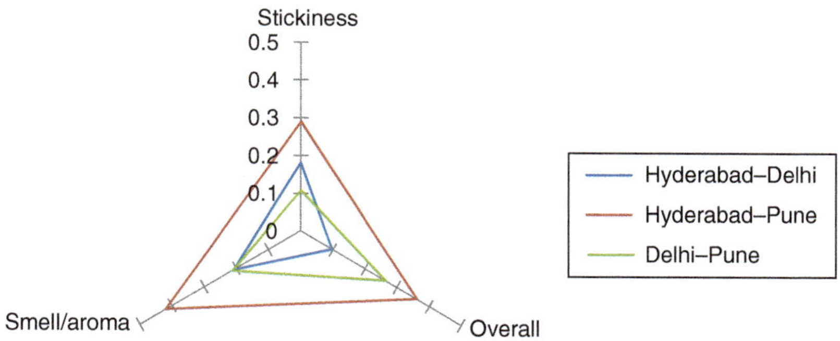

Score: 0, traditional product is better; 1, *jowar* product is better or both are the same.

Fig. 6.19. Cities' pairwise responses on influential attributes of sorghum vermicelli.

6.1.5 Sorghum semolina (for *uppit*)

Semolina (*suji*, or *rawa*) is used for making different types of food items in the areas of Pune, Delhi and Hyderabad. In Hyderabad, respondents have tried to make *uppit* (*upma*) from the semolina, whereas Delhi respondents have prepared *sira* (*halwa*, or 'sweet'). In Pune, respondents have tried to prepare both *sira* and *uppit*. Both types of foods have different attributes in terms of preparation and taste. To map the attributes of the respondents correctly and for improved clarity, the sorghum semolina has been divided into two groups – *uppit* and *sira* – before analysis.

The most important factors influencing the consumer's response to sorghum semolina for *uppit* are given in Table 6.15. Demography, with an eigenvalue of 1.4, was the last variable to be selected. The reliability of the data and consistency in the responses received for factor 1 ('organoleptic') is 0.946, which is much higher than the measuring point of 0.600; the value is lower for factor 2 ('prior awareness') and 3 ('comparison with traditional products'), these attributes being less influential on consumers' response. The percent variance explained by these six factors is 58.25% (Table 6.15). The myth that sorghum should be restricted to ill and old members of the family is the new dimension which appeared to be an important factor for the consumer to allow or avoid consumption of sorghum in their household (Fig. 6.21).

Clusters of respondents for sorghum semolina (for uppit)

The clusters based on purchase decisions are presented in Table 6.16. The sorghum semolina for *uppit* clusters are significantly different in terms of price level, the purchase place of the majority, buying preferences (branded or unbranded), percent expenditure on food, monthly household income, occupation of the female member of the family and number of children (Figs 6.22–6.24). A detailed table of consumer profiles from both good and average clusters is given in Appendix 9.

Attributes influencing the purchase of sorghum semolina (for uppit)

Ease of or similarity in preparation in comparison with wheat semolina is a very important factor/attribute influencing purchase (Table 6.17). The respondents seek no variation in preparation process from their regular semolina. The taste of sorghum semolina in comparison with wheat semolina is the second most important attribute influencing purchase decision. As shown in Table 6.17, these two attributes could increase

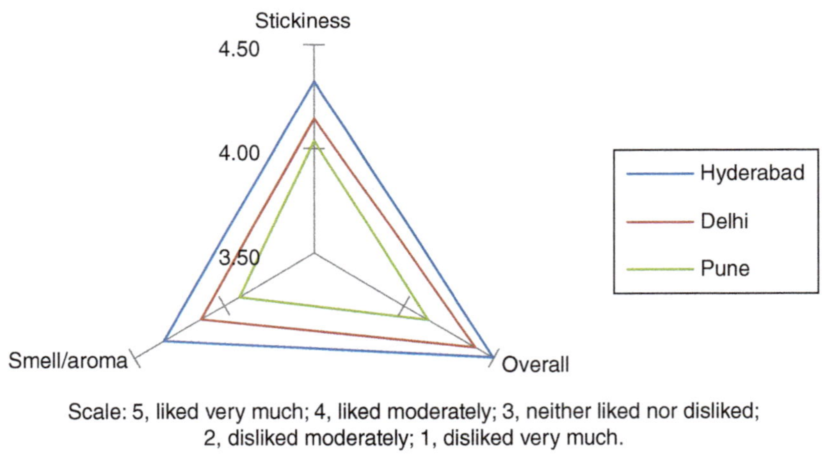

Fig. 6.20. Analysis of the response of members of individual cities with respect to important attributes.

Table 6.15. Important factors/dimensions in the consumer's response to sorghum semolina (for *uppit*). (Source: Dayakar *et al*., 2014b.)

Dimension/(factor)	Cronbach's alpha	Eigenvalue	Variance explained (%)
Organoleptic (1)	0.946	9.38	29
Prior awareness (2)	0.72	2.58	8
Comparison with traditional products (3)	0.697	2.45	7.5
Socio-economic and value for money (4)	N.A.	1.57	5
Myth of sorghum to the decision maker and income (6)	N.A.	1.5	4.75
Demography (5)	N.A.	1.4	4

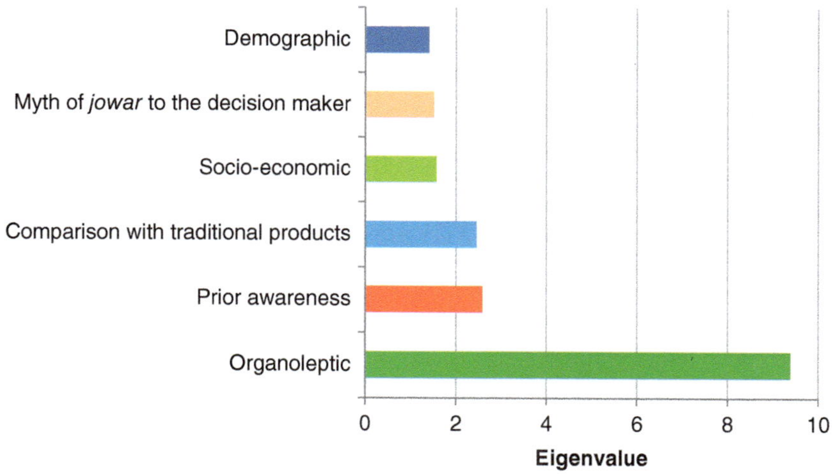

Fig. 6.21. Factors for consideration in the purchase of sorghum semolina (for *uppit*).

Table 6.16. Clusters of consumers based on purchase decision for sorghum semolina (for *uppit)*.

Cluster	Would purchase	Would not purchase	Total respondents	% purchase	Cluster output
1	250	28	278	90	Average
2	206	3	209	98	Good
3	11	41	52	21	Bad
4	6	18	24	25	Bad
5	167	10	177	90	Good

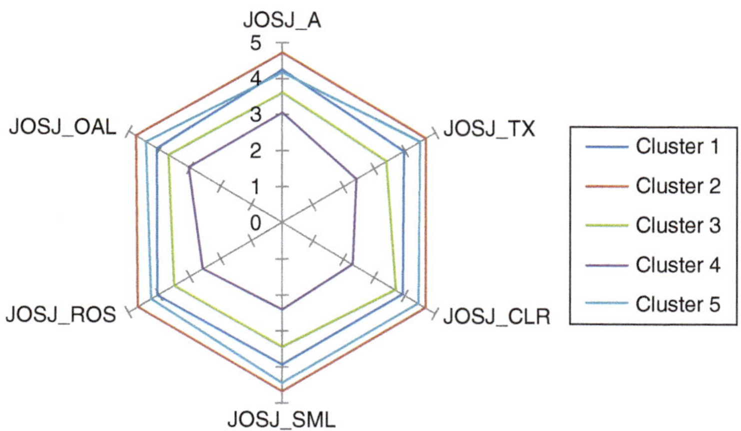

Score: 5, liked very much; 4, liked moderately; 3, neither liked nor disliked; 2, disliked moderately; 1, disliked very much.

Fig. 6.22. Cluster-wise score on sorghum semolina (for *uppit*). See list on p. xxv for definitions of cluster analysis abbreviations.

the likelihood of purchasing sorghum semolina by 3.775 and 2.671 times, respectively, when there is an increase of one unit in the Likert scale or when it is the same as regular semolina. The taste of the final product (*uppit*) is another important attribute that could increase the chance of purchase by 1.997 times, followed by *uppit* appearance (1.844 times).

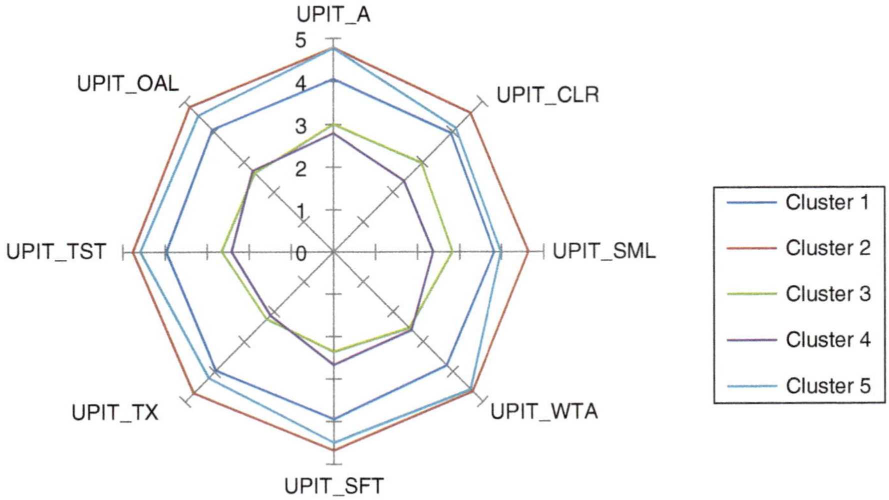

Score: 5, liked very much; 4, liked moderately; 3, neither liked nor disliked; 2, disliked moderately; 1, disliked very much.

Fig. 6.23. Cluster-wise score on sorghum *uppit*. See list on p. xxv for definitions of cluster analysis abbreviations.

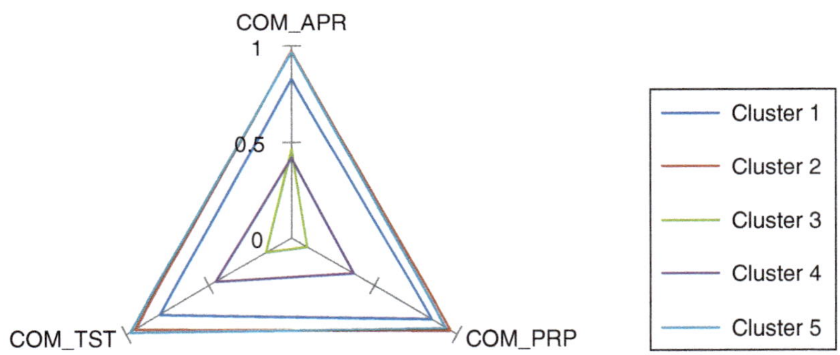

Score: 0, traditional product is better; 1, *jowar* product is better or both are the same.

Fig. 6.24. Cluster-wise score on sorghum semolina's attributes in comparison to traditional semolina (for *uppit*). See list on p. xxv for definitions of cluster analysis abbreviations.

Table 6.17. Odds ratio estimate to predict purchase influencing attributes of sorghum semolina (for *uppit*). (Source: Dayakar *et al.*, 2014b.)

Factor	MLE estimate	Odds ratio estimate
Uppit appearance	0.6117	1.844
Uppit smell	0.4439	1.559
Uppit taste	0.6917	1.997
Preparation in comparison with wheat semolina	1.3283	3.775
Taste in comparison with wheat semolina	0.9826	2.671

Note: The analysis of maximum likelihood estimates (MLE) was used to obtain parameter estimates. Significance of parameter estimates was based on the Wald chi-square value at $P<0.05$.

Prediction of correct purchase decision

To draw an inference from Table 6.18, 85% of those who are willing to purchase and 8% of those who are not willing to purchase the sorghum semolina are observed and predicted as being correctly classified. On the other hand, 5% of those who are willing to purchase and 1% of those who are not willing to purchase the sorghum semolina are observed and predicted as being in the incorrect classification group. Thus, the purchasing decision of sorghum semolina could be predicted with 94% accuracy and 6% inaccuracy (Table 6.18).

Purchase prediction accuracy based on individual attributes (hit ratio)

The taste of the final product of sorghum semolina (for *uppit*) can be used as a single predictor, with an accuracy of 73.7% (Fig. 6.25). The smell and appearance of the final product (*uppit*) are also important attributes that predict purchasing decision, which can be used with 72.5 and 70.7% accuracy, respectively. Similarity in the preparation process and taste in comparison with traditional semolina can be used with 59.9 and 57.5% accuracy, respectively, for predicting the purchase decision.

Similarities and dissimilarities in the response of respondents on important attributes of sorghum semolina for uppit

The comparison here is made between the two cities of Hyderabad and Pune as Delhi's respondents did not prepare *uppit* from the sorghum semolina and ANOVA analysis could not be performed due to the lack of pair-forming variables. Hyderabad and Pune (Tables 6.19, 6.20) have significant differences in the responses for appearance, smell and taste of the final semolina product (*uppit*). The smell of the *uppit* made from sorghum is not much liked by the respondents of Pune.

6.1.6 Sorghum semolina (for *sira*)

Here, value for money and food habit was the last factor included, having an eigenvalue of 1.12 (Table 6.21). The reliability of the data and consistency in the responses received for factor 1 ('organoleptic') is 0.940, which is much higher than the measuring point of 0.600. It is also higher than the next most important factors, those of 2 ('prior awareness') and 3 ('comparison with traditional products'), suggesting that factor 1 is the most reliable response. The percent variance explained by these six factors is 57.5% (Fig. 6.26).

Clusters of respondents for sorghum semolina (for sira)

The sorghum semolina (for *sira*) clusters are markedly different from the *uppit* clusters. The clusters have more similarities among the

Table 6.18. Correct classification of actual purchase decision and prediction on correctness of purchase for sorghum semolina (for *uppit*). (Source: Dayakar *et al.*, 2014b.)

	No. of respondents			% of respondents		
	Would not purchase	Would Purchase	Total	Would not purchase	Would Purchase	Total
Would not purchase	61	39	100	80	5	14
Would purchase	9	631	640	1	85	86
Total	70	670	740	9	91	100
	Correct prediction		692	Correct prediction		94
	Incorrect prediction		48	Incorrect prediction		6

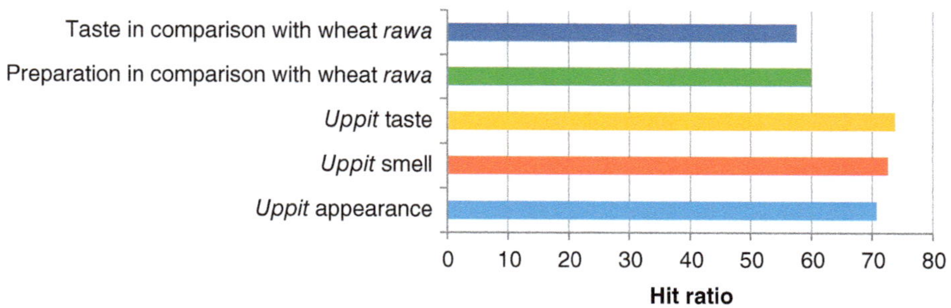

Fig. 6.25. Purchase prediction accuracy based on individual attributes for sorghum semolina (for *uppit*).

Table 6.19. City-wise responses on influential attributes of sorghum semolina (for *uppit*). (Source: Dayakar *et al.*, 2014b.)

	Hyderabad	Pune	*t*-value	*Ps*>(*t*)
Sample size	434	305		
Mean UPIT_A	4.4309	4.1508	4.76	<0.0001
Mean UPIT_SML	4.0624	3.8492	3.03	0.0025
Mean UPIT_TST	4.3341	4.0197	5.15	<0.0001

Note: *Ps*>(*t*) indicates probability value of sample is more than the *t*-value. See list on p. xxv for definitions of cluster analysis abbreviations.

Table 6.20. City-wise responses on influential attributes of sorghum semolina compared with wheat semolina. (Source: Dayakar *et al.*, 2014b.)

	Hyderabad		Pune			
	0	1	0	1	Chi sq.	*P*
Taste in comparison with wheat semolina	60	375	69	236	9.71	0.0018
Preparation in comparison with wheat semolina	49	386	76	229	23.8	<0.0001

Note: 0, traditional product is better; 1, sorghum product is better or both are the same.

Table 6.21. Important factors/dimensions in the consumer's response to sorghum semolina (for *sira*). (Source: Dayakar *et al.*, 2014b.)

Dimension/(factor)	Cronbach's alpha	Eigenvalue	Variance explained (%)
Organoleptic (1)	0.940	8.13	25
Prior awareness (2)	0.917	4.91	15
Comparison with traditional products (3)	0.641	1.68	5
Socio-economic and value for money (4)	N.A.	1.64	5
Demography (5)	N.A.	1.38	4
Myth of sorghum related to food habit and value of money (6)	N.A.	1.12	3.5

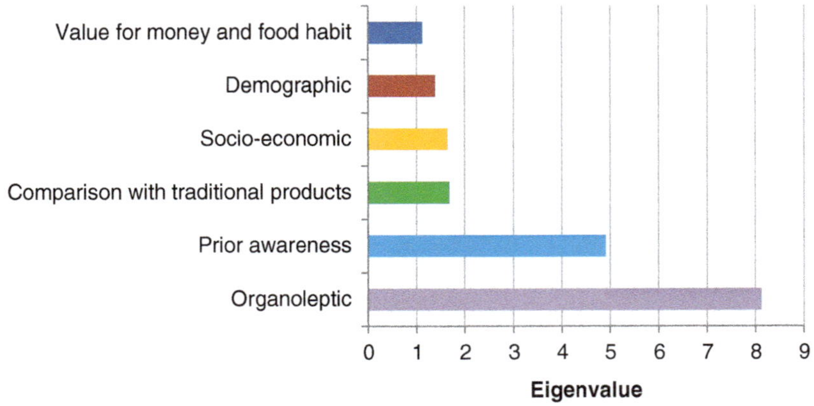

Fig. 6.26. Factors for consideration in the purchase of sorghum semolina (for *sira*).

respondents. Purchase place of the majority, buying preferences (branded or unbranded), quantity of semolina consumed in a month and number of children in a family or household are the major parameters showing significant differences between the good and average cluster (Table 6.22, Figs 6.27–6.29). A detailed table of consumer profiles from both good and average clusters is given in Appendix 9.

Attributes influencing the purchase of sorghum semolina (for sira)

An analysis of maximum likelihood estimates (MLE) was used to obtain parameter estimates. Significance of parameter estimates was based on the Wald chi-square value at $P<0.05$. The factors having the greatest influence on the purchasing decision by the consumers when they compare it to wheat semolina are appearance, preparation process and taste, which could increase the likelihood of purchasing sorghum semolina by 4.21, 3.19 and 4.57 times, respectively, when there is an increase in one unit in the Likert scale or when the attributes are the same as wheat semolina (Table 6.23). Smell of the final product (*sira*) is another important attribute that could increase the likelihood of purchase by 1.7 times. Colour (influencing first-sight purchasing decision) and texture of sorghum semolina of the final product (*sira*) are other important attributes, increasing the likelihood of purchase by 2.247 and 1.467 times, respectively (Table 6.23).

Prediction of correct purchase decision

Results (Table 6.24) show that 67% of those who are willing and 15% those who are not

Table 6.22. Clusters of consumers based on purchase decision for *sira*. (Source: Dayakar *et al.*, 2014b.)

Cluster	Would purchase	Would not purchase	Total respondents	% purchase	Cluster output
1	1	27	28	3.57	Bad
2	81	56	138	59	Bad
3	159	35	194	81	Average
4	101	13	114	89	Good

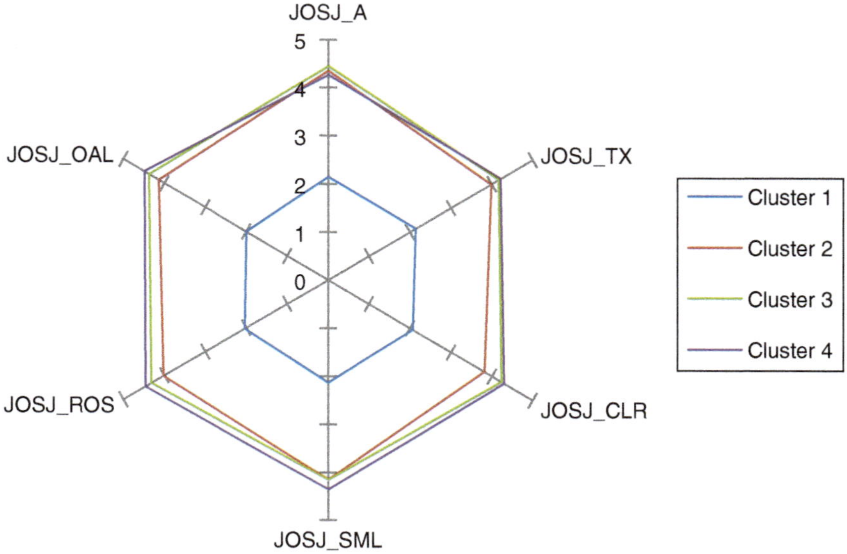

Score: 5, liked very much; 4, liked moderately; 3, neither liked nor disliked; 2, disliked moderately; 1, disliked very much.

Fig. 6.27. Cluster-wise score on sorghum *rawa*'s attributes (raw). See list on p. xxv for definitions of cluster analysis abbreviations.

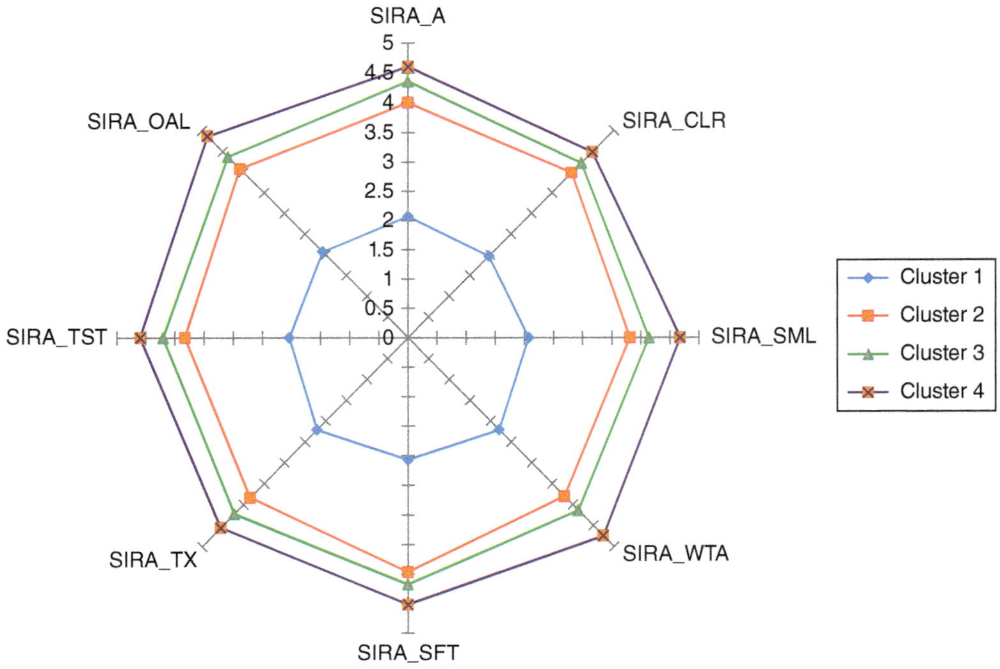

Score: 5, liked very much; 4, liked moderately; 3, neither liked nor disliked; 2, disliked moderately; 1, disliked very much.

Fig. 6.28. Cluster-wise score on sorghum *rawa*'s attributes (cooked). See list on p. xxv for definitions of cluster analysis abbreviations.

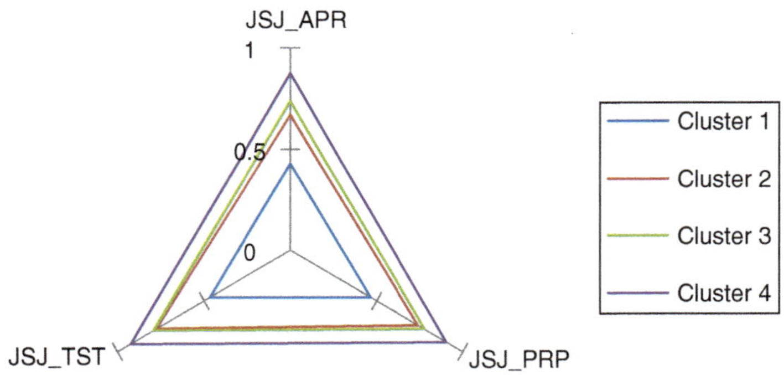

Score: 0, traditional product is better; 1, *jowar* product is better or both are the same.

Fig. 6.29. Cluster-wise preference on sorghum semolina's attributes in comparison with those of traditional semolina. See list on p. xxv for definitions of cluster analysis abbreviations.

willing to purchase the sorghum semolina (for *sira*) are observed and predicted as being correctly classified. On the other hand, 12% of those who are willing and 6% of those who are not willing to purchase the sorghum semolina are observed and predicted as being incorrectly classified in that group. In conclusion, the purchasing decision of sorghum semolina can be predicted with 82% accuracy and 18% inaccuracy.

Purchase prediction accuracy based on individual attributes (hit ratio)

Purchase prediction accuracy based on individual attributes for sorghum semolina (for *sira*)

Table 6.23. Odds ratio estimate to predict purchase influencing attributes of sorghum semolina (for *sira*). (Source: Dayakar *et al.*, 2014b.)

Factor	MLE estimate	Odds ratio estimate
JOSJ_TX	0.3834	1.467
JOSJ_CLR	0.8098	2.247
SIRA_SML	0.5719	1.772
COM_APR	1.4385	4.214
COM_PRP	1.1588	3.186
COM_TST	1.5201	4.573

Note: The analysis of maximum likelihood estimates (MLE) was used to obtain parameter estimates. Significance of parameter estimates was based on the Wald chi-square value at $P<0.05$. See list on p. xxv for definitions of cluster analysis abbreviations.

Table 6.24. Correct classification of actual purchase decision and prediction on correctness of purchase for sorghum semolina (for *sira*). (Source: Dayakar *et al.*, 2014b.)

	No. of respondents			% of respondents		
	Would not purchase	Would purchase	Total	Would not purchase	Would purchase	Total
Would not purchase	71	58	129	15	12	27
Would purchase	29	316	345	6	67	73
Total	100	374	474	21	79	100
	Correct prediction		387	Correct prediction		82
	Incorrect prediction		87	Incorrect prediction		18

are given in Fig. 6.30. The smell of sorghum semolina can be used as the single predictor and the decision to purchase sorghum semolina can be predicted with an accuracy of 52%. Similarly, the colour and raw appearance of semolina can be used to predict the purchase decision with an accuracy of 49% and 46%, respectively.

Similarities and dissimilarities in the response of respondents to important attributes of sorghum semolina (for sira)

Delhi respondents have moderately liked the important attributes of sorghum semolina when preparing *sira* from it, but the liking of respondents from Pune was average (Table 6.25). Delhi and Pune respondents are very close in terms of their perception of sorghum semolina's colour and texture, whereas the smell of the final product (*sira*) in both sets of responses is significantly different. Similarly, in comparison with traditional semolina, responses of both the cities are not significantly different (Table 6.26).

6.1.7 Sorghum flakes

The reliability of the data and consistency in the responses received for factor 1 is 0.952, which is much higher than the measuring point of 0.600. Cronbach's alpha for factor 2, 6 and 3 is 0.914, 0.732 and 0.572, respectively. The percent variance explained by these six factors is 61.5% (Table 6.27, Fig. 6.31).

Clusters of respondents for sorghum flakes

Socio-economic profiles of the respondents grouped into good and average clusters is presented in Table 6.28 and Figs 6.32 and 6.34. Most of the parameters, except for occupation of the male member of the family and family size, have significant differences between good and average clusters. A detailed table of consumer profiles

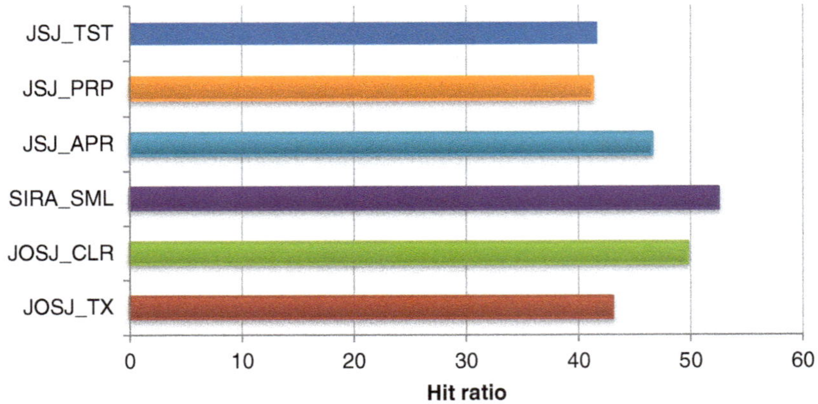

Fig. 6.30. Purchase prediction accuracy based on individual attributes for sorghum semolina (for *sira*). See list on p. xxv for definitions of cluster analysis abbreviations.

Table 6.25. City-wise responses on influential attributes of sorghum semolina (for *sira*). (Source: Dayakar *et al*., 2014b.)

	Delhi	Pune	*t*-value	*P*>(*t*)
Sample	373	100	N.A.	N.A.
JOSJ_CLR	4.0188	3.96	0.65	0.514
Texture	4.0241	3.93	1.13	0.2587
SIRA_SML	4.131	3.75	3.99	<.0001

Note: *P*>(*t*) indicates probability value of sample is more than the *t*-value. See list on p. xxv for definitions of cluster analysis abbreviations.

Table 6.26. Response of influential attributes of sorghum *rawa (sira)* by city. (Source: Dayakar *et al*., 2014b.)

	Delhi		Pune			
	0	1	0	1	Chi sq.	*P*
Appearance in comparison with wheat semolina	109	265	20	80	3.33	0.068
Preparation in comparison with wheat semolina	77	297	28	72	2.51	0.1129
Taste in comparison with wheat semolina	76	298	20	80	0.005	0.943

Note: 0, traditional product is better; 1, sorghum product is better or both are the same.

Table 6.27. Important factors/dimensions in the consumer's response to sorghum flakes. (Source: Dayakar *et al*., 2014b.)

Dimensions/(factor)	Cronbach's alpha	Eigenvalue	Variance explained (%)
Organoleptic (1)	0.952	8.44	28
Prior awareness (2)	0.914	4.2	14
Myths of sorghum to the decision maker (6)	0.732	1.89	6
Comparison with traditional products (3)	0.572	1.53	5
Socio-economic and value for money (4)	N.A.	1.4	4.5
Demography (5)	N.A.	1.27	4

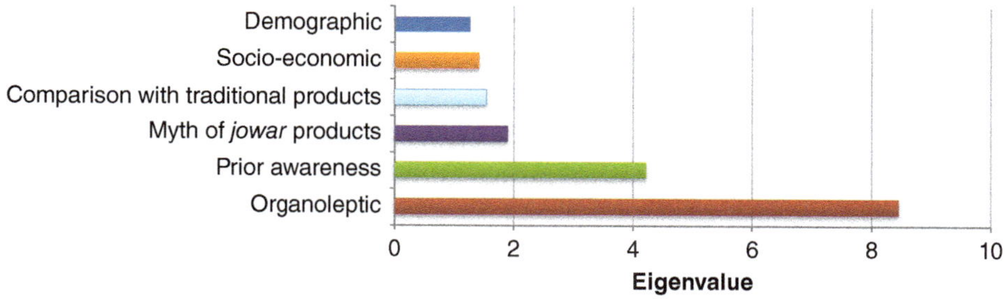

Fig. 6.31. Factors for consideration in the purchase of sorghum flakes.

Table 6.28. Clusters of consumers based on purchase decision for sorghum flakes. (Source: Dayakar *et al.*, 2014b.)

Cluster	Would purchase	Would not purchase	Total respondents	% purchase	Cluster output
1	105	4	109	96	Good
2	521	63	584	89	Good
3	173	57	230	75	Average
4	0	34	34	0	Bad
5	146	112	258	56	Bad

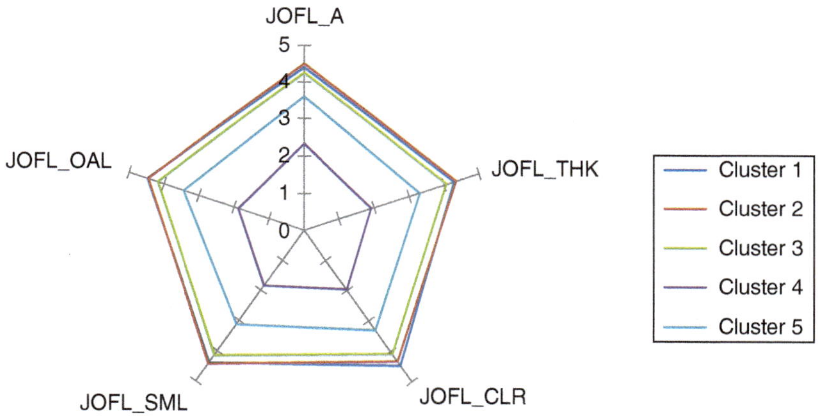

Score: 5, liked very much; 4, liked moderately; 3, neither liked nor disliked; 2, disliked moderately; 1, disliked very much.

Fig. 6.32. Cluster-wise score on raw sorghum flakes' attributes. See list on p. xxv for definitions of cluster analysis abbreviations.

from both good and average clusters is given in Appendix 9.

Purchase influencing attributes of sorghum flakes

Results (Table 6.29) show that perception regarding the sorghum product's ability to release energy in a sustainable manner could drive the likelihood of purchase by 2.74 times. Similarity of sorghum flake and rice flake preparation processes is a very influential attribute; it alone can increase the chances of purchasing sorghum flakes by 2.20 times with an increase in one unit of the liking scale. Also, but to a lesser extent, appearance, water absorption, softness and overall experience of cooked flakes *(poha)* could increase the likelihood of purchase by 1.75, 1.79,

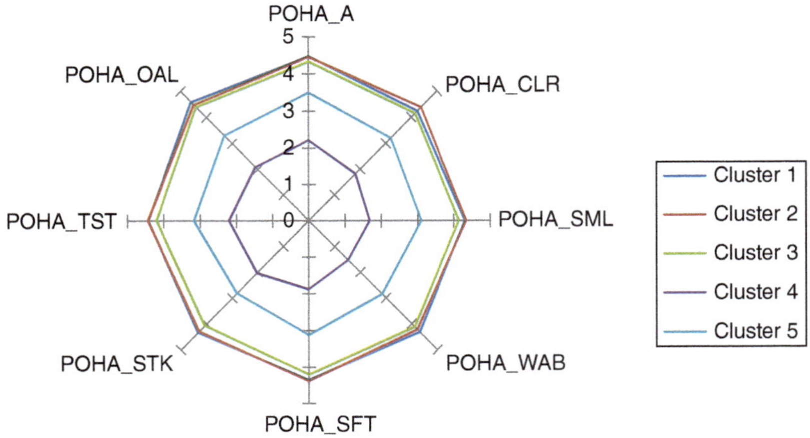

Score: 5, liked very much; 4, liked moderately; 3, neither liked nor disliked; 2, disliked moderately; 1, disliked very much.

Fig. 6.33. Cluster-wise score on cooked sorghum flakes' (*poha*) attributes. See list on p. xxv for definitions of cluster analysis abbreviations.

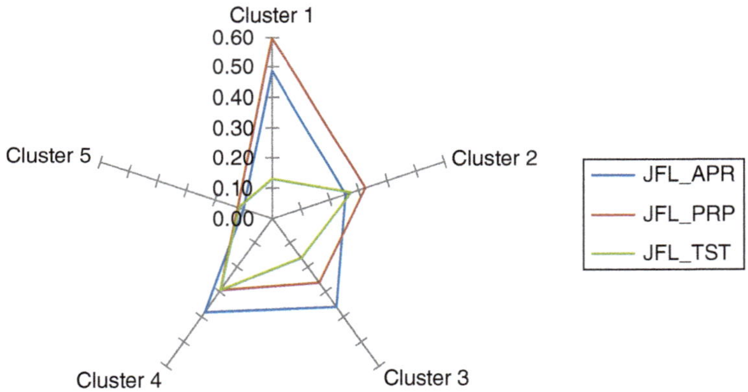

Score: 0, traditional product is better; 1, *jowar* product is better or both are the same.

Fig. 6.34. Cluster-wise score on sorghum flakes' attributes in comparison with traditional (maize) flakes. See list on p. xxv for definitions of cluster analysis abbreviations.

1.58 and 1.63 times, respectively, if there is a one-unit increase in the Likert scale (Table 6.29).

Prediction of correct purchase decision

As shown in Table 6.30, 73% of those who are willing to purchase and 12% of those who are not willing to purchase the sorghum flakes are observed and predicted as being correctly classified in their group. On the other hand, 11% of those who are willing and 5% those who are not willing to purchase the sorghum flakes are observed and predicted as being incorrectly classified. In conclusion, the purchasing decision for sorghum flakes could be predicted with 84% accuracy and 16% inaccuracy.

Purchase prediction accuracy based on individual attributes (hit ratio)

The purchase prediction accuracy based on individual attributes for sorghum flakes is given in Fig. 6.35, showing prediction accuracy for all of >60% except for the preparation process, which

Table 6.29. Odds ratio estimate to predict purchase influencing attributes of sorghum flakes. (Source: Dayakar *et al.*, 2014b.)

Factor	MLE estimate	Odds ratio estimate
Stat_7	1.0094	2.744
POHA_A	0.5612	1.753
POHA_WAB	0.5829	1.791
POHA_SFT	0.458	1.581
POHA_OAL	0.49	1.632
COM_PRP	0.7902	2.204

Note: The analysis of maximum likelihood estimates (MLE) was used to obtain parameter estimates. The significance of parameter estimates was based on the Wald chi-square value at $P<0.05$. See list on p. xxv for definitions of cluster analysis abbreviations.

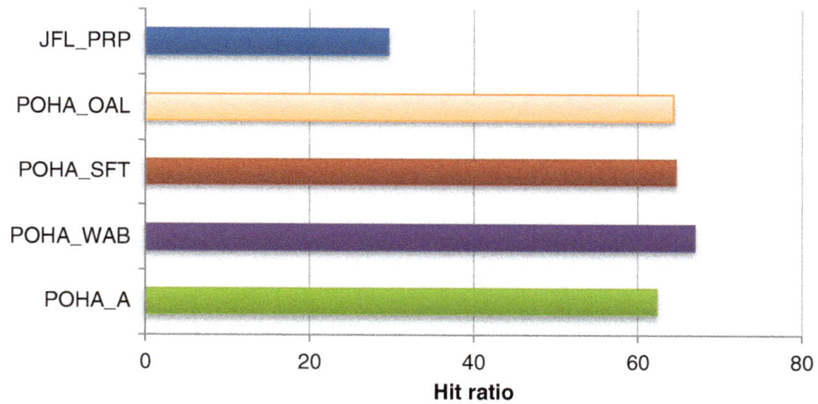

Fig. 6.35. Purchase prediction accuracy based on individual attributes for sorghum flakes. See list on p. xxv for definitions of cluster analysis abbreviations.

has much lower accuracy (29.6%). The water absorption attribute has the highest purchase prediction accuracy of 66.9%, followed by softness of sorghum flakes (64.6%).

Similarities and dissimilarities in the response of respondents to important attributes of sorghum flakes

Delhi and Hyderabad have similar responses in terms of the appearance of the cooked sorghum flakes (*poha*), overall experience, water absorption of *poha* and its softness attributes. But, in terms of these attributes, the responses of the other two city pairings are significantly different (Fig. 6.36). In Delhi and Hyderabad, respondents have very much liked the attributes of sorghum flakes that most influence purchase decision; the average score for these attributes is above 4.15, which falls near to extreme liking of the product. In Pune, respondents are still conservative in replacing the rice flakes in terms of those purchase-influencing attributes mentioned above. The average score assigned by the respondents falls close to moderately liking the product (Fig. 6.37).

6.1.8 Conclusion

According to the cluster analysis, all the processed sorghum products studied were acceptable to the consumer. Organoleptic attributes and prior awareness of heath/other benefits have the highest eigenvalues for all five sorghum products tested for consumer acceptability. The clusters are significantly different in terms of price level, buying preference and quantity, and the prices of *eatrite* sorghum products are, in general, the same or lower than existing alternative cereal products available in the market. Sorghum products are promoted as health beneficial, with superior nutritional profiles over existing *maida*-/rice-/

Table 6.30. Correct classification of actual purchase decision and prediction on correctness of purchase for sorghum flakes. (Source: Dayakar *et al.*, 2014b.)

	No. of respondents			% of respondents		
	Would not purchase	Would purchase	Total	Would not purchase	Would purchase	Total
Would not purchase	140	130	270	12	11	22
Would purchase	60	885	945	5	73	78
Total	200	1015	1215	16	84	100
	Correct prediction		1025	Correct prediction		84
	Incorrect prediction		190	Incorrect prediction		16

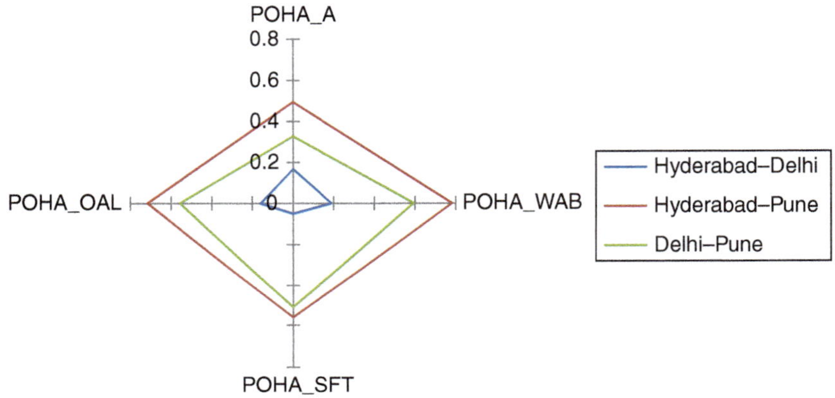

Score: 0, traditional product is better; 1, *johar* product is better or both are the same.

Fig. 6.36. Cities' pairwise responses on influential attributes of sorghum flakes. See list on p. xxv for definitions of cluster analysis abbreviations.

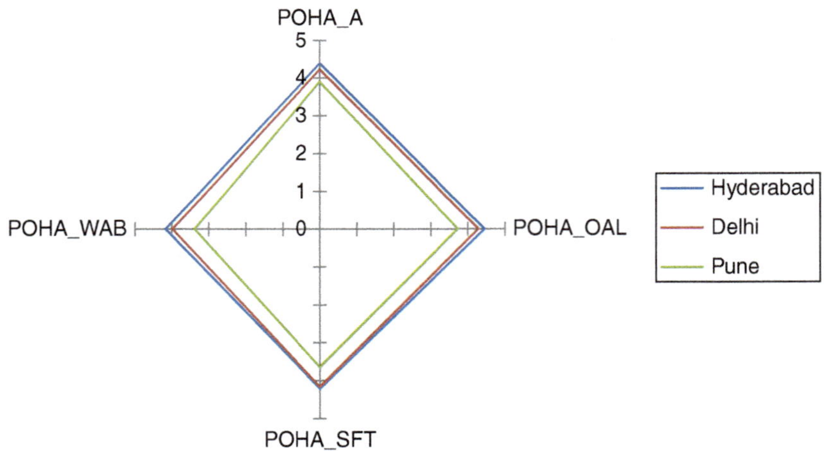

Score: 5, liked very much; 4, liked moderately; 3, neither liked nor disliked; 2, disliked moderately; 1, disliked very much.

Fig. 6.37. Impact of important attributes on individual city's respondents. See list on p. xxv for definitions of cluster analysis abbreviations.

maize-based products, and consumers who had prior awareness of health benefits of these products preferred them. The most influential attributes on the decision to purchase sorghum multigrain *atta* and flakes was the perception of the product as an energy source/health food. But for sorghum biscuits and vermicelli, smell and taste, respectively, are the most important attributes influencing the purchasing decision. For sorghum semolina, ease and similarity in preparation as compared to wheat semolina were the most important factors.

6.2 Entrepreneurship Development in Millet Product Technologies

In view of decreasing millet consumption (despite its nutritional superiority) and its potential to bolster dryland agriculture, efforts have been made to provide all stakeholders with a common platform, so as to increase participation and demand within the value chain of millet. An integral part of this is developing the entrepreneurship of farmers and processors.

Farmers usually follow old practices of subsistence farming for sorghum cultivation, and entrepreneurs are not usually aware of the prospects of sorghum commercialization, nor are public servants and people in general aware of its nutritional benefits. In 2008, under the National Agricultural Innovation Project (NAIP), the Indian Institute of Millets Research (IIMR) and partners instigated a pilot-scale commercialization project in sorghum, in order to identify its feasibility and sustainability in value addition. With a view to promoting and initiating commercialization of sorghum as different health food products, interventions were made on processing by developing novel sorghum ready-to-eat/-cook (RTE/RTC) products, such as multigrain *atta*, flakes, *rawa* and pasta. The interventions and new technology are being transferred to farmers, technical assistants and entrepreneurs through various training programmes.

6.2.1 Entrepreneurship development at farm level

New or fine-tuned value-added processing technologies were disseminated through rural entrepreneurship programmes to various farmers' groups, rural women and NGOs in the district of Parbhani, Maharashtra, particularly to those farmers who had participated in on-farm production of *rabi* sorghum via the successful model of e-Choupal within the previous 2 years. The processing and entrepreneurship development programmes were simultaneously conducted in the target area and at IIMR, Hyderabad. Table 6.31 reveals that over eight seasons INR 1,833,342 has been spent on 4000 farmers, that is INR 229 per farmer per season. The salient features of entrepreneurship development programmes that have been undertaken under the millet value chain subproject are given below:

- **Field days.** Field days have been conducted off-season in Parbhani, before the sowing starts, wherein participating farmers are educated in improved methods of cultivation and the various developments of sorghum products and associated technology. Farmers also interact and discuss the roles of both parties (farmer and processor) in the integrated value chain of sorghum.
- **Literature and creating awareness.** Sorghum recipe books and recommended packages of practice (PoP) for *rabi*/*kharif* sorghum cultivation in local dialects/languages were distributed among the farmers, including videos on *roti*-making in sorghum and CDs. The progamme has also been covered on TV shows (in popular local Telugu channels in Andhra Pradesh such as TV 5, Sakshi TV, Gemini TV, HM TV and ETV-2), news and print media (six articles in newspapers were published).
- **Processing training.** The farmers were exposed to various sorghum processing technologies and were able to learn about them practically at the processing laboratory at IIMR, Hyderabad.
- ***Choupal haat* programme.** Twelve *Choupal haat* programmes were conducted in the Parbhani and Nanded districts. Here, stalls were set up in rural markets exhibiting various sorghum processed and semi-processed products developed at IIMR, with distribution of accompanying literature. This was an effective way of interacting with farmers and motivating them to invest in sorghum for greater income.

6.2.2 Entrepreneurship Development Programmes for women, rural entrepreneurs and urban start-up entrepreneurs

The ultimate goal of Entrepreneurship Development Programmes (Fig. 6.38) is to disseminate a thorough knowledge of post-harvest management, which includes linking farmers with markets, processing, the nutritional importance of sorghum, and marketing. The topics discussed with the stakeholders (progressive farmers, rural entrepreneurs, NGOs, SHGs, small- and medium-scale processors, women and entrepreneurs) are given below.

- **Value chain experience in sorghum cultivation.** The trainees under this programme were educated on intensive sorghum cultivation through public–private partnerships (PPPs) whereby sorghum cultivation was introduced and given orientation as a commercial proposition to change the mindset of the participating farmers, including technological backstopping for the participating farmers with improved cultivars, buy-back facilitation and value addition at the farm level. All these experiences were disseminated in such a way that it could be replicated in the context of other millets too.

Table 6.31. Expenditure incurred in entrepreneurship trainings and related services (average for four seasons; four each of *kharif* and *rabi*). (Source: Dayakar *et al.*, 2014b.)

Particulars	Amount (INR)	% share
Entrepreneurship Development Programme		
Training	701,150	38
Literature and creating awareness	1,009,062	55
Choupal haat programme	123,490	7
Total expenditure (A)	1,833,342	100
Factorization of expenditure per farmer (A/1000)	1,833.34	
Factorization of expenditure per farmer per season	229.17	

Fig. 6.38. Entrepreneurship development training at IIMR.

- **Nutritional importance of sorghum.** This programme highlighted the problem of malnutrition and over-nutrition among the populace of the country and how the various diseases that have been prevailing can be prevented by consuming sorghum. The nutritional values of sorghum and the comparative effects on sorghum vs non-sorghum consumers (from National Institute of Nutrition studies) were disseminated to the trainees. The superiority of sorghum over other cereals such as rice and wheat in terms of nutrition, minerals and vitamins were discussed, and the advantage of sorghum's low glycaemic index and glycaemic load, its gluten-free nature and its high fibre content were thoroughly discussed to impart a clear understanding of its nutritional benefits.
- **Post-harvest technologies of sorghum/millets.** The entire scope of processing and enhancing the values of farmers' produce at optimum intervention was dealt with in this training. Processing interventions in sorghum has led to the removal of inconveniences in preparation of traditional *rotis* and offers the health-conscious urban consumer a wide range of choices for healthy and convenient RTC/RTE foods. Food technologies, especially in the instant foods sector where big scope exists for small-scale processors (for those who can invest small capital), were discussed.
- **Branding, packaging and labelling.** The issue of capacity building was addressed, a relatively new subject but crucial for commercialization in this context. Creating a brand, the advantage of innovative packaging and labelling was tackled. The most common question that arose was whether it is possible to begin commercializing under their own brand; so guidelines were given, and the issue of financial flexibility (capital requirement) was discussed. Economic feasibility plans were drawn and business plan formats were worked out for entrepreneurship development.

6.2.3 Entrepreneurship Development Programmes in various states

Technology-based Entrepreneurship Development Programmes have been conducted by IIMR four times per year in the states of Maharastra, Tamil Nadu, Andhra Pradesh and Telangana. The topics discussed in these programmes are:

- Changing RTE patterns: a case of value chain development in millets. Under this topic the participants are informed of the prevailing RTE food products available on-shelf in the market. Subsequently, the scope of millets, with regard to nutritional and health benefits for the consumers and its prospects in food processing and their commercialization are taught, based on the first-hand experience of IIMR and partner institutes.
- Scope/opportunities for RTE foods in India.
- Indian scenario: the nutritional divide, consumer trends.
- Prospects for RTE millets and 'health and wellness' foods.
- Value chain development on millet RTE foods.

In addition, entrepreneurship development at IIMR, Hyderabad, invites experts to give lectures and hands-on training on subjects ranging from on-farm improved sorghum production to extending the final products. The scope and various opportunities that could emanate from sorghum processing at various capacities by the trainees are taken into consideration while planning the subject matter.

6.3 Commercialization of Sorghum Product Technologies on a Pilot Scale

In 2015, the market for health foods in India was estimated to be worth US$7.5–10 billion according to a study undertaken by the Tata Strategic Management group. The demand for fortified, whole grain, high fibre, low cholesterol, natural and organic foods is expected to grow (Live mint, 2013). The recent efforts of the IIMR-led intervention in a PPP under NAIP have led to commercialization of some millet products, though up-scaling is a challenge. However, they have shown that it is possible for millet to be used in convenience foods similar to those of rice and wheat, so as to offer choices to urban health-conscious consumers. Similarly, efforts from major private players such as Britannia Industries Ltd (through their NutriChoice brand) and ITC Ltd (multigrain biscuits), show that millets can play a major role. In recent times Britannia Industries have entered a research Memorandum of Understanding (MoU) with IIMR on sorghum biscuits and bakery products, which is another landmark. Apart from big players there are many small-scale players who produce extruded snacks from sorghum, as well as from pearl millet in Mumbai, Rajasthan, Tamil Nadu and Andhra Pradesh. Among the millets, finger millet-based malt is a more acceptable product both in traditional and non-traditional consuming areas.

Some of these products have premium prices and are acceptable as up-market products that cater to the high income groups. Gluten-free sorghum products are commercialized by an entrepreneur, based in Kolkata, for catering to coeliac patients. Processed small millets are commercialized in a big way by entrepreneurs from Nasik, Maharashtra, and are being exported. Although substantial studies have not been attempted on the bioavailability of minerals in millets, the general awareness that is being created across the country and the world over, is that millets are nutri-cereals and offer promise to health-seeking populations who desire to keep lifestyle diseases at bay. IIMR alone has entered more than 15 MoUs with private partners. However, advocacy and policy prescription would take the consumption to greater heights, which would generate not only demand for millets but also raise standards of living for the poor dryland farmers growing these crops.

6.3.1 Branding, labelling and packaging

Branding

A brand can be understood simply as a name termed for a product, service or commodity. The most vital and important part of a brand is its name and logo that distinguishes one seller's product from those of others.

Effective branding can result in higher sales not only of one product, but also of other products associated with that brand. For example, if a customer loves sorghum biscuits and trusts the brand, he or she is more likely to try other sorghum products offered by the company, such as sorghum cookies and other sorghum RTE products. Brand is the personality that identifies a product, service or company (name, term, sign, symbol, or design, or a combination of them) and how it relates to key constituencies, such as customers, staff, partners and investors.

The IIMR brand *eatrite* (Fig. 6.39) was launched for the commercialization of sorghum product technologies in the twin cities of Secunderabad and Hyderabad on a pilot scale with the tag lines 'eat *jowar* – stay healthy' and 'health meets taste'. The name itself refers to its health benefits and nutritional superiority over alternative products available on-shelf in the market. IIMR, as a government agency, received wide support from across different organizations and individuals and incidentally attracted lots of entrepreneurs from this initiative. Many firms entered MoUs with the institute for marketing or technological collaborations as they became confident of the demand for these health products.

Labelling

In labelling, it is of upmost importance to mark the product with the standard classifications. For example, if the product is a vegetarian product in India it should be marked green, as is seen in the *eatrite* logo. Generally, for encryption, EAN-13 codes such as 126 formats, are used. Other necessary requirements are credentials, and contact and mandatory packing information, such as quantity, net weight, Maximum retail price (MRP), expiry date, batch number and barcode, which encrypts a product's details (Fig. 6.40) and helps in maintaining stock inventory, faster billing at point of sale and product searches on the internet using smartphones.

Good layout is essential. The brand name should always be at the centre with maximum prominence; the product identity should get equal prominence so as to describe to the

Fig. 6.39. The *eatrite* logo.

Fig. 6.40. Barcode for *eatrite* products.

customer what is inside the pack. It is also important to mention how to use/make the product and to highlight its benefits, so that people can understand the product thoroughly.

Packaging

Packaging involves the science, art and technology of enclosing or protecting products for distribution, storage, sale and end use. It also refers to the process of design, evaluation and production of packages. Packaging depends upon the need and end-product use. There are various packaging types, such as poly packaging, paperboard and foil. The product and the package are two different things. The product is to be further processed and consumed, but its packaging is thrown away after removing the product. However, packaging adds value to the product, with respect to the customer's perception. A well-designed package can complement or even enhance the product if it is designed to help in transit of the product. Good package design is not just about the product – it is about convenience as well.

The choice of material needs to be considered, depending upon the shelf life, product type and the finish/look. If a product is fragile and might get crushed easily, then it is advisable to pack it in, for example, a paperboard box packing or a plastic container. Packaging experts should be consulted to obtain a perfect understanding of the product packing needs. The different packaging types for *eatrite* products are shown in Fig. 6.41.

6.3.2 Launching of *eatrite* brand by IIMR

The *eatrite* brand has been promoted by IIMR in India for the commercialization and promotion of millet (mainly sorghum) processed products. This brand has been initiated under NAIP, a unique recognition, as an institutional brand promotion is uncommon. The *eatrite* name was chosen to reflect the characteristics of carbohydrates present in sorghum, which are complex in nature and are released slowly during digestion, regulating hunger and helping prevent over-eating. Eating sorghum is also 'right' because it is high in fibre, minerals, vitamins and antioxidants and has manifold health benefits (see Chapter 3).

The range of products (Fig. 6.42) under this brand includes: sorghum-rich multigrain flour, sorghum semolina, sorghum pasta, sorghum vermicelli, sorghum flakes, sorghum roasted flakes and sorghum biscuits. Five different business plans were developed and commercialized for the sorghum products; these evolved according to their relative merits in terms of the farmer's share in the consumer rupee. IIMR has now signed an MoU with more than ten entrepreneurs; that with the multinational company Britannia Industries Ltd is a most prominent achievement. Other terms of reference include adopting processing technologies (cold extrusion) at the pilot scale for commercialization (GUV Foods), technology liaising on multigrain *atta* and biscuits (Hope Blessing Enterprise Ltd, Delhi) and taking up *eatrite* products for commercialization. Branding, labelling with

Fig. 6.41. Different types of packaging.

Fig. 6.42. A range of different *eatrite* products.

nutritional profiles, recipes, mandatory information and packing were unique in sorghum as, up until that point, there had been no other sorghum products available on the market except multigrain *atta*.

This has had a huge impact on consumer perception of product diversification and flagging nutritional superiority. The acceptability among health-conscious urban consumers has been very high, and the response of entrepreneurs to produce and market these products as future alternative nutri-cereals has been overwhelming. As a result, MoUs with IIMR and ANGRAU have been made.

6.3.3 Commercialization of the sorghum product technologies

The IIMR-led consortium developed and standardized more than 25 sorghum products aimed at removing the inconveniences associated with its preparation and reviving demand for these traditional cereals. Based on the assessment of the market potential, consumer acceptability and aspects of convenience, they were shortlisted to a total of nine products, which were subsequently labelled and packaged (with packs designed through professional agencies, containing nutritional profiles and other mandatory information).

Salient features of commercialization and scaling-up

- The *eatrite* brand name was registered and launched by IIMR with the unique selling proposition 'Eat *jowar* – stay healthy' (Fig. 6.43), used for commercialization of sorghum product technologies developed under NAIP.
- Successfully commercialized *eatrite* products at the pilot scale in Hyderabad through Heritage Retail Stores across the city and through local retail shops.

- For sustainability, IIMR has signed MoUs with 17 prospective entrepreneurs from Hyderabad, New Delhi and Bangalore, with varying emphasis on production, promotion and marketing of sorghum product technologies.
- An MoU was signed between Britannia and IIMR (the first time the Indian Council of Agricultural Research signed with a multinational company) (Fig. 6.44) to work jointly on R&D of bakery and biscuit products.
- The commercialization was solidly backed with popularization measures by adopting various communication strategies.
- Horizontal expansion is envisaged in Bangalore, Chennai, Pune and Delhi.

The *eatrite* products, being convenient and healthy, were targeted primarily at the urban population. The products were aggressively promoted across Hyderabad city: through roadshows, nutritional campaigns, and in print and electronic media. They were marketed by retail

Fig. 6.43. Advertisement for IIMR brand *eatrite*.

Fig. 6.44. IIMR and Britannia Industries Ltd signing the MoU.

chain stores such as Heritage Fresh and Big Bazaar Future Retail and were explored for their market elsewhere in local retail stores. The response has been very good and there is an increasing demand for their availability across the city and other major metropolises; multigrain *atta*, pure sorghum biscuits and flakes in particular have had good demand across all age groups. IIMR signed MoUs with private entrepreneurs so as to make these products available on-shelf across Hyderabad. These products were tested for their nutrient profiles and nutritional suitability by NIN, Hyderabad, who were then able to recommend the products for those at risk of, or suffering from, lifestyle diseases such as diabetes and obesity, for all age groups.

Initially, aggressive promotion through roadshows and health awareness campaigns in, for example, public parks and malls, displayed product technologies and distributed samples. The resulting impact of popularization and increase of awareness among the masses then stimulated entrepreneurs, who began to either approach IIMR for exclusive marketing or to manufacture and market the product technologies in their own brand.

6.4 Sensitization of Policymakers

With the consumption of millets – especially sorghum – drastically reduced in India and the realization of their importance in nutritional security of the country, it became necessary to make policymakers aware of the problem. IIMR has taken up policy sensitization through various national and international conferences as discussed below.

6.4.1 Organizing national and international conferences

Under the subproject of NAIP, the IIMR and partner organizations have tried to sensitize the policymakers towards promotion of millets and flagging them as crops of the future in the event of climate change. They have done this through conducting various national and international seminars (Fig. 6.45).

A national seminar on millets was jointly organized by IIMR and the Directorate of Agriculture and Cooperation (DAC) in 2010, which was followed by a brainstorming session in November of that year at the National Institute of Rural Development, Hyderabad. The Secretary of Agriculture was appreciative of the NAIP initiative and the millet technologies generated. A taskforce was formed with the co-principal investigator as one of the members. So when the government of India was willing to take up promotion of millets, the IIMR was technologically ready, thanks to development under the NAIP Millets Value Chain project. Finally, in 2011–12, a fully fledged millet programme promotion was launched through the Initiative of Nutritional Security through Intensive Millets Promotion (INSIMP) with an outlay of INR 300 crore on the project, emphasizing value chain components. About 400 beneficiary processing clusters across the millet-growing states were supplied freely to the entrepreneurs. These clusters were technologically backstopped by IIMR/NAIP.

Earlier, NIN conducted a nutritional study of the developed products on diabetic patients and school children, which shows that sorghum products are either as or more nutritious than

Fig. 6.45. (A) National seminar on millet, November 2010. (B) Global consultation on millet promotion, December 2013.

the counterparts available on-shelf in the market. In fact, promoting millets as healthy foods gained wide momentum from consumers, especially the health-conscious urban consumer. This has made it possible to sensitize the line departments of state governments and policymakers of the country.

Through various workshops, seminars, exhibitions, meetings and millet festivals, IIMR tried to engage departments such as the Ministry of Agriculture, Child and Family Welfare, Department of Agriculture and Cooperation and Human Resource Development to mainstream millets in publicly funded programmes.

Line departments of government are sensitized in state governments so that they may also adopt millet in public distribution schemes. Pilots of the Mid-Day Meal Scheme with inclusion of millet in the diet were initiated in three states of Maharashtra, Karnataka and Andhra Pradesh by the DAC, under technical guidance from IIMR and the government, who as a result have actively contemplated mainstreaming millets in publicly funded welfare programmes that target particular groups. Lastly, the government of India opted to include millets in the Public Distribution System, the Indian food security system.

6.4.2 Formation of a taskforce for promotion of millets

A joint taskforce on millets was established, made up of: DAC, ICAR, NIRD, the Central Food Technological Research Institute (CFTRI), NIN, state representatives and private companies, along with the establishment of a National Research Centre on minor millets by ICAR and the launching of a Mission Mode Programme on Millets during the 'Twelfth Plan'. Millets were introduced into PDS with a regional approach, e.g. pearl millet in Rajasthan, sorghum in Andhra Pradesh, Karnataka and Maharashtra, and finger millet in Karnataka.

6.5 Promotion of Value Added Healthy and Convenient Sorghum Product Technologies

Millets are the clear case for neglect of policy across the country as relatively, since independence, the major emphasis was given to promotion and propagation of rice and wheat, whether in terms of input subsidies or output incentives. Thus, a need arises for the promotion of millets, as nowadays the area for them is largely diverted to high value crops such as soybean, maize, cotton and sugarcane. Some of the promotion strategies taken up by IIMR under the NAIP project are discussed below.

6.5.1 Communication strategies

Creation of awareness and promotional campaigns were key interventions aggressively and innovatively pursued, as the commodity in question was generally perceived as a poor man's crop. With the two advantages of millet, that of convenience and nutritional superiority, mass awareness was created in Hyderabad and elsewhere, through fairs (Fig. 6.46) and other promotional activities.

The promotional activities were outsourced through event managers: 360 degree communication strategies were implemented (Fig. 6.47) with brand designing logos through BTL (below the line advertising, that is, all non-media adverts such as promotions, direct marketing and launches) and ATL (above the line advertising, which is advertising on television, radio, publications, and outdoor and ambient advertising) in urban markets and electronic media. An *eatrite* portal was created for this purpose and jingles were made. Mass awareness was created on sorghum as a health

Fig. 6.46. International Trade Fair, Delhi 2013, with DG, ICAR.

Fig. 6.47. Interaction with media and industrialists.

food through more than 200 roadshows (Figs 6.48, 6.49) and through wet sampling (tasting) in public parks, malls and institutes in Hyderabad as well as in hundreds of exhibitions (Fig. 6.50), imparting awareness of sorghum across 400,000 consumers through the *Jowar Rath* (Fig. 6.51). The *Jowar Rath* is a brightly coloured exhibition van that tours in Pune, Bangalore, Jabalpur, Chennai, Coimbatore and New Delhi. It advertises the taste, convenience and health benefits of sorghum and millet foods, with in-van facilities for mobile exhibitions and training. Further, through medians and billboards (Fig. 6.52) the sorghum *eatrite* products were popularized; even local road transport buses were stickered. A professional approach was also given to the labelling, packaging and branding of the various *eatrite* products, with tag lines such as 'eat *jowar* – stay healthy', 'health meets taste' and 'small grains – a big gain'.

At the same time, promotional activities were undertaken to improve practices in sorghum cultivation and to attract entrepreneurs to develop processing of sorghum. A unique initiative of *choupal haats* was launched; these were conducted in villages on a normal *haat* day (that is, a normal working day where participating farmers gather together), engineered specifically to attract farmers and rural women/consumers. A portable stall with a distinct canopy was erected in a prominent location and specific literature (booklets/pamphlets) on packages of practices and sorghum-based recipes distributed. To date, a total of about 90 *choupal haats* have reached out to around 27,000 farmers.

Aggressive campaigning to create awareness among both urban consumers and rural farmers in the project area was undertaken. Many entrepreneurs have been given promotional support for increasing their sales volumes and private sector sensitization.

The launch of the brand *eatrite*, labelled and branded as health foods based on nutritive value established by NIN studies was targeted separately for urban upmarketing (middle and higher income classes) and rural markets. IIMR promoted the products to nutritionists, doctors and dieticians (Fig. 6.47) and launched the *eatrite* website.

Fig. 6.48. Roadshow for promotion of millets.

Fig. 6.49. Free health and nutritional counselling.

Fig. 6.50. Exhibition at FICCI (International Workshop) for promotion of millets.

Fig. 6.51. Sorghum *Jowar Rath* – a mobile exhibition van for promoting *eatrite* products.

Fig. 6.52. Medians and billboards for millet products.

6.6 Conclusion

In this chapter, we have reviewed the studies carried out on consumer acceptability and market feasibility of sorghum products, and the development of entrepreneurship, communication strategies, sensitization and promotion of millets; all necessary for the successful value chain development of millets. The results of the consumer acceptability study shows that taste, texture, preparation process, appearance, swelling attribute, colour etc. were found to be the driving factors for willingness to purchase sorghum products. Entrepreneurship development has been carried out in the field and in IIMR premises to disseminate technology. IIMR has developed and standardized sorghum products that have been commercialized under the *eatrite* brand and has created a promotional campaign. The success story of millets' value chain under the NAIP project has captured the attention of high profile scientists and agricultural policymakers in India. In the next chapter we will discuss the impact of all these efforts thus far.

References

Dayakar, R.B., Patil, J.V., Nirmal Reddy, K., Soni, V.K and Srivastava, G. (2014a) *Sorghum: an emerging cash crop*, 1st edn. Cambridge University Press India, New Delhi.

Dayakar, R.B., Patil, J.V., Hymavathi, T.V., Nirmal Reddy, K. and Rajendra Prasad, M.P. (2014b) *Creation of Demand for Millet Foods through PCS Value Chain*. Final report of NAIP (ICAR). Directorate of Sorghum Research, Rajendranagar.

Live mint (2013) www.livemint.com/Opinion/VZCNX5OhwpT3UHKwqJoHbL/Bring-back-the-millets.html (accessed 1 March 2016).

7

Impact Assessment of Value Chain on Millet Foods

One of the major goals of the National Agricultural Innovation Project (NAIP) was to revive demand for millets through value addition. In examining the sustainability of value addition under this project, it is important to assess the impact on the socio-economic status of the various beneficiaries and stakeholders involved (e.g. sorghum farmers, NGOs, small-scale processors, women's groups, self-help groups (SHGs), new entrepreneurs, research institutes, policy makers and consumers). As part of the project, the lead institute, the Indian Institute of Millets Research (IIMR), trained a number of these stakeholders, empowering them with the relevant skills and enabling them to take advantage of new market opportunities emanating from branding and popularization of millets as health foods.

The positive impact of a value chain on millet foods will not only provide opportunities for higher income generation to stakeholders (i.e. farmers, primary processors and food processing entrepreneurs/companies) but will also ensure nutritional security for consumers. Most importantly, millet crops will significantly contribute to a reduction in nutritional deficiencies, and are climate-resilient crops that can mitigate the adverse impacts of climate change. Thus we must account for both tangible economic benefits as well as intangible benefits in order to measure the impact of the value chain project. Later in the chapter we also discuss the business plan for sorghum value-added products.

7.1 Impact Assessment on Various Stakeholders in Value Chain Development of Sorghum

7.1.1 Farmers – impact of end product-specific sorghum cultivation

Shifting of allocation of sorghum cultivation from marginal to better lands

The provision of on-farm technological backstopping with end product-specific cultivars/hybrids, extension services and buy-back assurance among 1500 farmers in Adilabad Nanded (*kharif*) and Parbhani (*rabi*) has changed their mindset, with their opting to allocate better lands for sorghum cultivation over the past 4 years (2008–09 to 2011–12) owing to higher returns from its cultivation. It has resulted in an increase in productivity of *kharif* sorghum of 48% and of *rabi* sorghum of 37% over the baseline (Fig. 7.1), and an increase in net return in *kharif* of 541% (INR 35,091/ha) and in *rabi* of 436% (INR 16,667/ha) over the baseline. The value of sorghum grain increased from INR 800/t at the beginning to INR 1250/t at the end of the project (Fig. 7.2). A mini demonstration processing plant (Fig. 7.3) comprising a dehuller, parboiling unit and flaking unit machine was commissioned in Borisawant village, in the Parbhani district, Maharashtra, and purchased under the NAIP for exploring

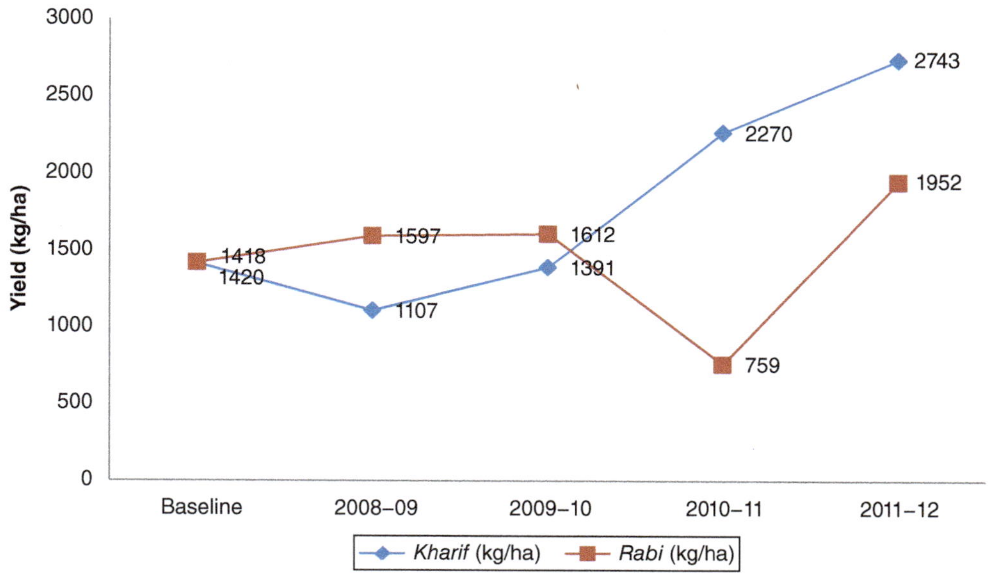

Fig. 7.1. Yield of sorghum in farmers' fields under the NAIP subproject.

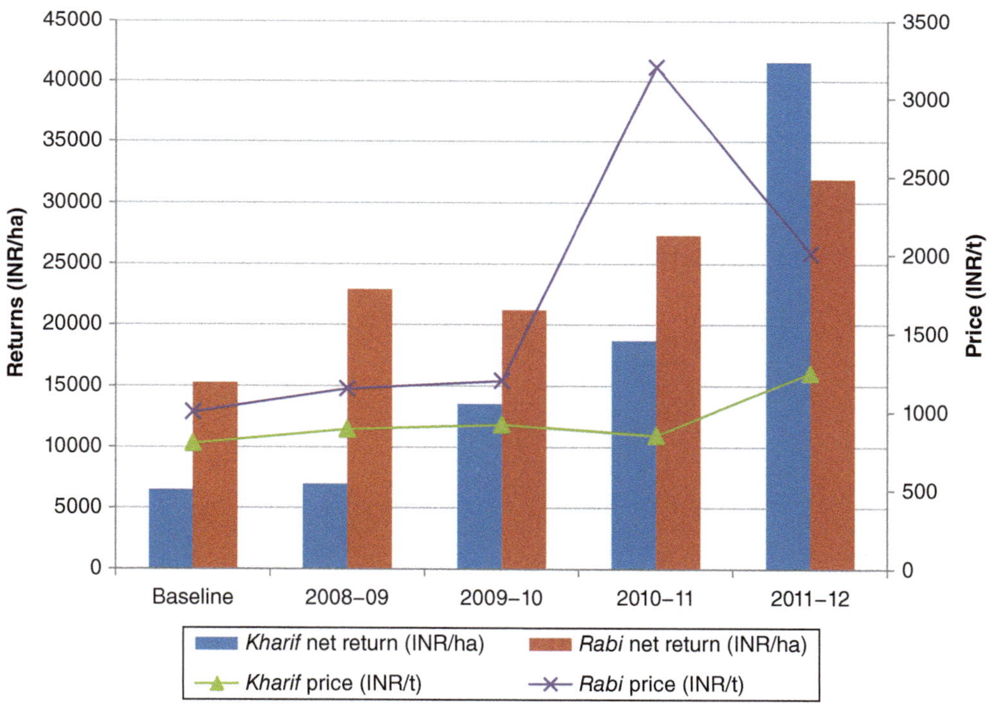

Fig. 7.2. Net return and price of sorghum in farmers' fields under the NAIP subproject.

commercial opportunities and feasibility at the farm level (Fig. 7.3). This helped to spread and popularize sorghum products in many other *choupals* in the district. Farmers engaged in value-added sorghum business at the farm gate and realized handsome profits. A farmers' group was formed in Tadkalas village to undertake value-addition business in sorghum.

Before the project, the participating farmers sold the entire surplus in their local market. However, under intensive cultivation their produce was either procured by ITC (ABD) Ltd at the ideal price or the farmers themselves converted it into flakes or *rawa* at the sorghum processing demo unit in Borisawant. Their incomes doubled to INR 2000–3000/qt excluding by-product value, whereas previously they would sell at a maximum of INR 1000/qt as grain. The economics of up-scaling plans for sorghum processing at the entrepreneurial farmer level is shown in Table 7.1.

So farm-level value addition efforts have resulted in manifold increases in the net income of farmers. Farm-level consumption of sorghum products also increased due to creating awareness of sorghum nutrition and processing (convenience foods).

Due to the increased awareness of sorghum's nutritional merits, a supply chain management model was developed among the farmers and private and public sectors, with this public–private partnership (PPP) model proving successful. There was an increase in demand for sorghum cultivation. The farmers shifted from subsistence to commercial cultivation as the net return was high and there was demand for value-added products of sorghum grain in the

Fig. 7.3. Sorghum mini processing unit.

Table 7.1. Economics of up-scaling plans at the entrepreneurial farmer level.

	Cost of production at various capacities (INR/kg)			
Efficiency levels	50 kg/day	100 kg/day	200 kg/day	500 kg/day
50% recovery	33.32	25.34	22.65	21.56
60% recovery	28.67	22.12	19.88	18.97
70% recovery	25.44	19.82	17.89	17.11

market. Increase in the yield through backstopping raised income in line with the increase in produce, a result of direct linkages between farmers and the market.

7.1.2 Processors – processing diversification and product development

Around 30 machines were retrofitted to make them suitable for sorghum processing, e.g. destoner and grader, dehuller, hammer mill, *chakki* mill, roaster, edge runner, parboiling unit, roller flaking unit, semolina making machine, cold extruder, dough kneader, ribbon blender, sifter, planetary mixer, automatic cookie cutting machine, popping machine, grinder, automatic sealing machine, foot operated *roti* making machine, automatic *roti* making machine and rotary oven. About 500 processing clusters/beneficiaries of primary/secondary processing machinery technologies across millet-growing states under the Directorate of Agriculture and Cooperation (DAC)'s Initiative for Nutritional Security through the Intensive Millets Promotion (INSIMP) programme were technologically backstopped by IIMR.

About 50 processing technologies (ready-to-eat (RTE)/-cook (RTC) sorghum foods) were developed: pearl millet flour, pearl millet-based multigrain flour, pearl millet semolina, pearl millet-based RTE extruded snacks, sorghum flour, sorghum-based products such as coarse, medium and fine semolina, parboiled semolina, designer *rawa*, multigrain flour (four types), multigrain *roti* (two types), flakes, vermicelli, pasta, gluten-free pasta (three varieties; soy, caseinates, flax seed incorporated) biscuits (17 varieties; salty, sweet, groundnut, coconut, gluten-free salty and sweet biscuits (two types), trans fat free salty, sweet (two types), groundnut and coconut and low calorie salty, sweet, groundnut and coconut), RTE extruded snacks, gluten-free crunches (two types), *lassi* (sweet and salty), instant foods such as *idli, dhokla, upma, dosa , pongal*, vermicelli *kheer* mix, flakes *kheer* mix, and *masala* sorghum flakes, pops (two types, *aam chat masala, pudina* masala), *peda*, bran soup, bran *peda*. More than 30 sorghum recipes were developed and standardized, e.g. *idli, dosa, pongal, upma, roti*, multigrain *roti, bisibela bath, sankati, ambali, samosa, khichidi*, vegetable *bath, tawa roti, sweet pongal, kesari, laddu, peda*, vermicelli *kheer*, flakes *kheer*, semolina *kheer, chuduwa, muruku, chekkalu, punugulu, chegodi, vada*, crispies, sorghum cake, fur cake and bread. Project partner ANGRAU, Hyderabad, developed ten more product technologies.

Nine sorghum-based products were successfully commercialized on a pilot scale under IIMR's brand *eatrite*, and marketed through retail outlets in Hyderabad: pure sorghum flour, sorghum-based multigrain flour, coarse, medium and fine semolina, flakes, pasta, vermicelli and biscuits. 'ANGRAU foods', a trademark, is also registered by our partner ANGRAU to commercialize five products: sorghum crispies, multigrain biscuits, gluten-free pasta, vermicelli and hot extruder products. The processed products not only make cooking convenient for consumers but are also economically feasible for commercial purposes. The economic analysis of the processed products developed in IIMR are shown in Table 7.2; the return over investment is quite high (except for in *jowar atta* and *jowar* biscuits where it is slightly low) showing that it is economical to produce the value-added products of millets.

IIMR, Hyderabad, established partners for commercialization of *eatrite* sorghum products, through MoUs (Memoranda of Understanding) with potential entrepreneurs for technology transfer and marketing of sorghum foods and MoAs (Memoranda of Agreement) for marketing of products under the project. An MoU was even signed with a multinational company, i.e. Britannia Industries Ltd (for the first time in the history of ICAR) and IIMR to work jointly on R&D for bakery and biscuit products. Furthermore, 15 MoUs and 3 MoAs were signed with private entrepreneurs under various business plans for the commercialization of sorghum processed foods in Hyderabad (Table 7.3).

The revenue generated under NAIP is given in Table 7.4. The total revenue collected under NAIP was INR 50.0 lakhs. In addition, it is estimated that ANGRAU had accumulated a revenue of INR 30.0 lakhs through the sale of ANGRAU foods (part of the value chain efforts on sorghum and millets).

Through the development of RTE/RTC food for sorghum, the inconveniences in sorghum food preparation were removed and shelf life was enhanced. Furthermore, as the nutritional superiority of millet over many other cereals was also established through the nutritional studies at NIN, a choice of nutritional food products to the consumers was provided. Moreover, this has

Table 7.2. Economics of value-added sorghum products (per SKU).

Sorghum product	Cost of production (INR)	SKU	Basic price (INR)	Profit (INR)	Return over investment	MRP (INR)
Jowar atta	35	1 kg	37.51	2.51	7.17	57
Multigrain *atta*	27.81	1 kg	35	7.19	25.85	54
Jowar rawa/idli rawa/ khichidi rawa	23.01	1/2 kg	28.5	5.49	23.86	44
Jowar flakes	11.84	1/4 kg	25	13.16	111.15	32
Jowar vermicelli	13.36	180 g	17.48	4.12	30.84	23
Jowar pasta	5.5	85 g	13.68	8.18	148.73	18
Jowar biscuits	10.8	100 g	11.4	0.6	5.56	18

Note: SKU, stock-keeping unit (indicating quantity of the product per packet).
Also, at IIMR, Hyderabad, more than 4000 farmers, SHGs, small-scale processors, women's groups and rural entrepreneurs were trained in sorghum processing through Entrepreneurship Development Programmes (EDPs).

Table 7.3. List of partners commercializing *eatrite* sorghum products.

Organization	Location	Purpose
MoUs		
Britannia Industries	Bangaluru	R&D
Hope Blessing Pvt Ltd	Delhi	Marketing
ITC (Foods) Ltd	Bangaluru	Licensing
Bhavani SS Industries, Mahal Bagayat	Bijapur	
Bhagyanagar Foods	Hyderabad	Licensing
Vegan Enterprises	Hyderabad	Licensing
GUV Foods	Hyderabad	Licensing
Kottaram Foods	Bangaluru	Licensing
Nova Traders	Hyderabad	Marketing
Sri Venkateswara Enterprises	Hyderabad	Marketing
Chandu Rotis	Hyderabad	Marketing
Matheses Industries	Hyderabad	Technology licensing
Madhav Kalyan Foods	Rajahmundry	Marketing
ISA Millets	Hyderabad	
Fountainhead Foods Private Ltd	Secunderabad	
MoAs		
Heritage Fresh Ltd	Hyderabad	Marketing
Balaji Grand Bazar	Hyderabad	Marketing
Satya Supermarket	Hyderabad	Marketing

Table 7.4. Revenue generated under NAIP initiative (up to 2012).

Source of revenue	Total amount (INR)	Responsible partner
Royalty fees from entrepreneurs (IIMR, ANGRAU)	500,000	IIMR, ANGRAU
Counter sale in IIMR	560,000	IIMR
Consignments of heritage and ITC	640,000	IIMR
Sales in exhibitions, trade fairs etc.	300,000	IIMR
R&D	3,000,000	IIMR
Total	5,000,000	

Note: An additional INR 50.0 lakh has been generated as of September 2015.

attracted entrepreneurs to take up sorghum in their business, branding it as a health food that adds more nutritional value to the daily diet. There has been an increase in consumer satisfaction owing to more options being made available. The production of RTE/RTC millets expands the basket of consumable cereal food, thus increasing the nutritional security of the country, as millets are a compliant crop for the majority of the farmers on rainfed areas. This will also help the farmers to generate revenues for their livelihood, as demand for millet grains for value-added products increases.

7.1.3 Entrepreneurs – entrepreneurship development

More than 4000 rural entrepreneurs, 1000 urban entrepreneurs, 3000 government servants, SHGs and NGOs and 300 entrepreneurs engaged in millets have been trained in sorghum processing through EDPs. Many rural industries (more than 100 processing clusters) have been established. Through these training programmes, 15 MoUs have been established with the IIMR. Out of these, six are running profitably. As discussed in detail in Chapter 6, entrepreneurial enthusiasm in the sorghum food business has increased following first-hand experience, provision of information and generation of awareness of its nutritional merits.

7.1.4 Policymakers

Due to the successful sorghum value chain model under NAIP, there has been a shift in policymakers' attitudes. Promotion has been outsourced and assertively conducted through activities such as roadshows in public parks and malls, exhibitions, health awareness programmes, social network campaigns, media coverage, video films and booklets, the internet and seminars. Through this, IIMR has created awareness of millet's nutritional superiority, sensitizing policymakers. In the advent of climate change and the escalating problem of lifestyle diseases, policymakers are increasingly aware that millets are going to be crops of the future.

In view of the successes from the NAIP subproject and the promising future in sorghum processed foods, the government of India allocated INR 300 crores in 2011 under INSIMP to the project Rashtriya Krishi Vikas Yojana. This was set up to promote millets as nutricereals, replicating the NAIP value chain initiative. As such, more than 200 processing clusters have been set up across the country. The establishment of 'Centres of Excellence' on sorghum value addition and food processing at IIMR, Hyderabad, has been a direct impact of the NAIP value chain initiative. Apart from its routine R&D activities on product development and standardization, IIMR has been serving as a nodal centre for showcasing technologies, developing marketing linkages, creating awareness through popularization measures and developing entrepreneurship.

7.1.5 Consumers

Consumer acceptability, feasibility and sustainability of newly developed sorghum RTE and RTC products under the IIMR brand name *eatrite* was evaluated, both in the traditional sorghum-consuming cities of Hyderabad and Pune, and the non-traditional sorghum-consuming city of Delhi during 2012 (DSR, 2011). Details of the study are given in Chapter 6. Results show that the willingness to purchase sorghum products in the targeted cities as a whole are: sorghum multigrain *atta*, 90%; sorghum vermicelli, 85%; sorghum-rich biscuits, 93%; sorghum *rawa/suji*, 80%; sorghum flakes, 77%. In Hyderabad, 90% of respondents are willing to purchase sorghum products. Pune was similar to this, whereas in Delhi it is 70–80%, except for multigrain *atta* where it reached 92%. Taste, texture, preparation process, appearance, swelling attribute and colour were found to be the driving factors for willingness to purchase the products. In general, the value addition of millets provides a wider choice of nutritional food products to the consumer. Consumer satisfaction increases, owing to more options being made available; a major deviation from traditional inconvenient *roti* made from sorghum *atta*.

7.1.6 Conclusion

The positive impacts of the value chain described above have been achieved by modification of the existing value chain (Fig. 7.4), which was very simple. The value chain map that was proposed by IIMR and approved by a project monitoring cell (Fig. 7.5) was then implemented (Fig. 7.6).

7.2 Business Plan for Sorghum Value-Added Product Technologies

Under the aegis of NAIP, IIMR has developed many post-harvest technologies for efficient and enhanced use of sorghum for the benefit of the consumers and various stakeholders. Three products were identified by IIMR as being most suitable for entrepreneurial activity for which a business plan, prepared by the National Academy of Agricultural Research Management – Business Planning & Development (NAARM-BPD), has been created. The products are: (i) sorghum-based multigrain *atta*, (ii) sorghum-based biscuits and (iii) sorghum-based flakes. All of these technologies have been developed and standardized as value-added sorghum products to provide healthy foods at an affordable price to the customer.

The business plan included: (i) the opportunity to develop value-added sorghum products, (ii) seasonality and quality of sorghum, (iii) backward integration and supply chain management, (iv) prospective areas (districts) where backward linkages as well as production can be taken up and (v) structure of the organization and phases of development. These are discussed below in detail.

7.2.1 Opportunity for developing value-added sorghum products

Over the years, there has been a drastic change in farming, production and consumption of sorghum in rural and urban India. Since sorghum is termed a poor man's food, the effect has been most visible in rural areas. Generally, demand for such crops is higher where the farming is cost effective, where there is a highly marketable surplus, and where the grains have a long shelf life and are readily processed, which creates employment through food processing. Previously, sorghum farmers had little interest in cultivation, as there was neither support from government policies (e.g. no subsidies such as there were for rice and wheat, and not included in

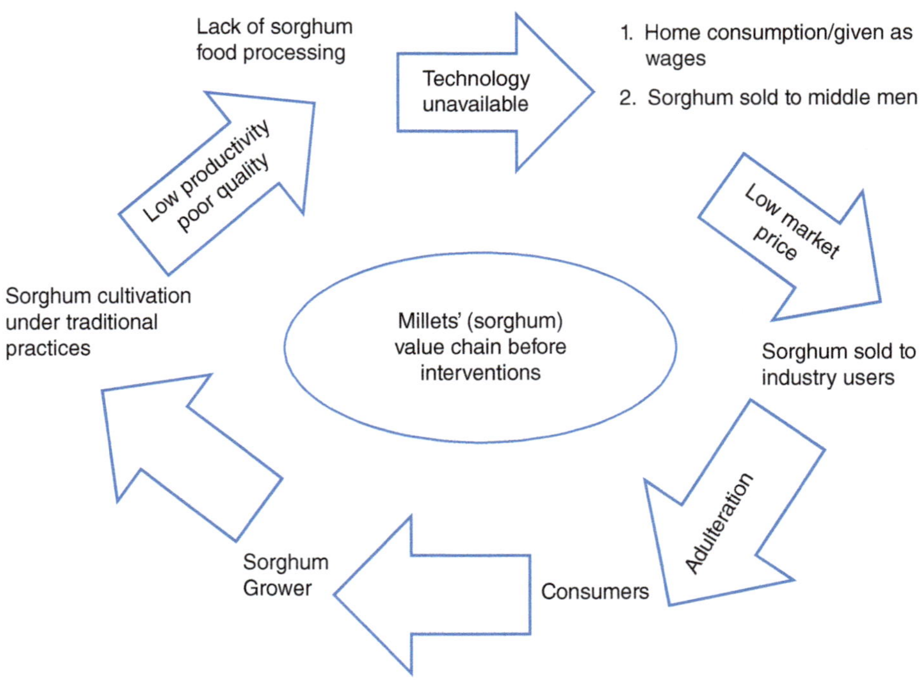

Fig. 7.4. Millets' value chain before interventions.

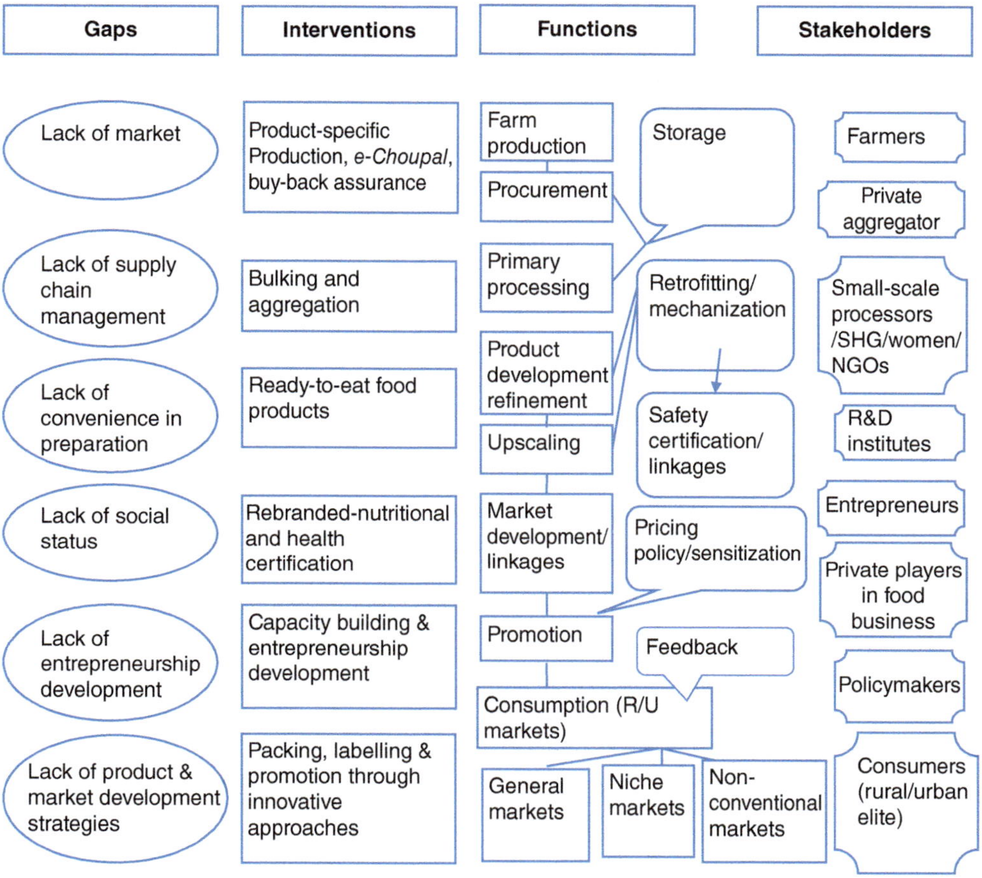

Fig. 7.5. Millets' value chain proposed by the implementing institute and approved by a project monitoring cell.

India's Public Distribution System) nor incentives in terms of higher remuneration for its cultivation. Finding a market for the cultivated grain was therefore difficult for the farmers, making them divert resources to other cereals and cash crops. The sorghum grain-keeping quality was also very low. *Rabi*-season grain was of good quality but the yields were very low, while the *kharif* sorghum gave high yields but the quality was very poor and was not suitable for consumption. Primary and secondary processing of sorghum was a difficult process that made rural and urban populations shift to using other cereals. These are reasons for the decline in sorghum cultivation in India.

A gap analysis, carried out by IIMR (Dayakar *et al.*, 2015) highlighted the inconveniences associated with the existing methods of millet food preparation, which were time consuming and involved drudgery. This is primarily due to lack of processing interventions in millet zones. Also, the absence of processing machinery suitable to mill and process sorghum compared to rice, wheat and maize rendered its cultivation non-remunerative. Also, the nutritional merits of sorghum over other fine cereals were not exploited due to lack of awareness among the consumers. The shift in preference from millet to other foods in urban populations can be addressed by innovation in secondary processing methods. The IIMR's value chain approach described in this publication has produced excellent RTC and RTE sorghum value-added products, which have gained great demand in urban markets where there is growing concern about health-related problems. Some of the products commercialized by IIMR include: sorghum-rich multi grain *atta*, sorghum *rawa* (fine and coarse *rawa* for *upma*,

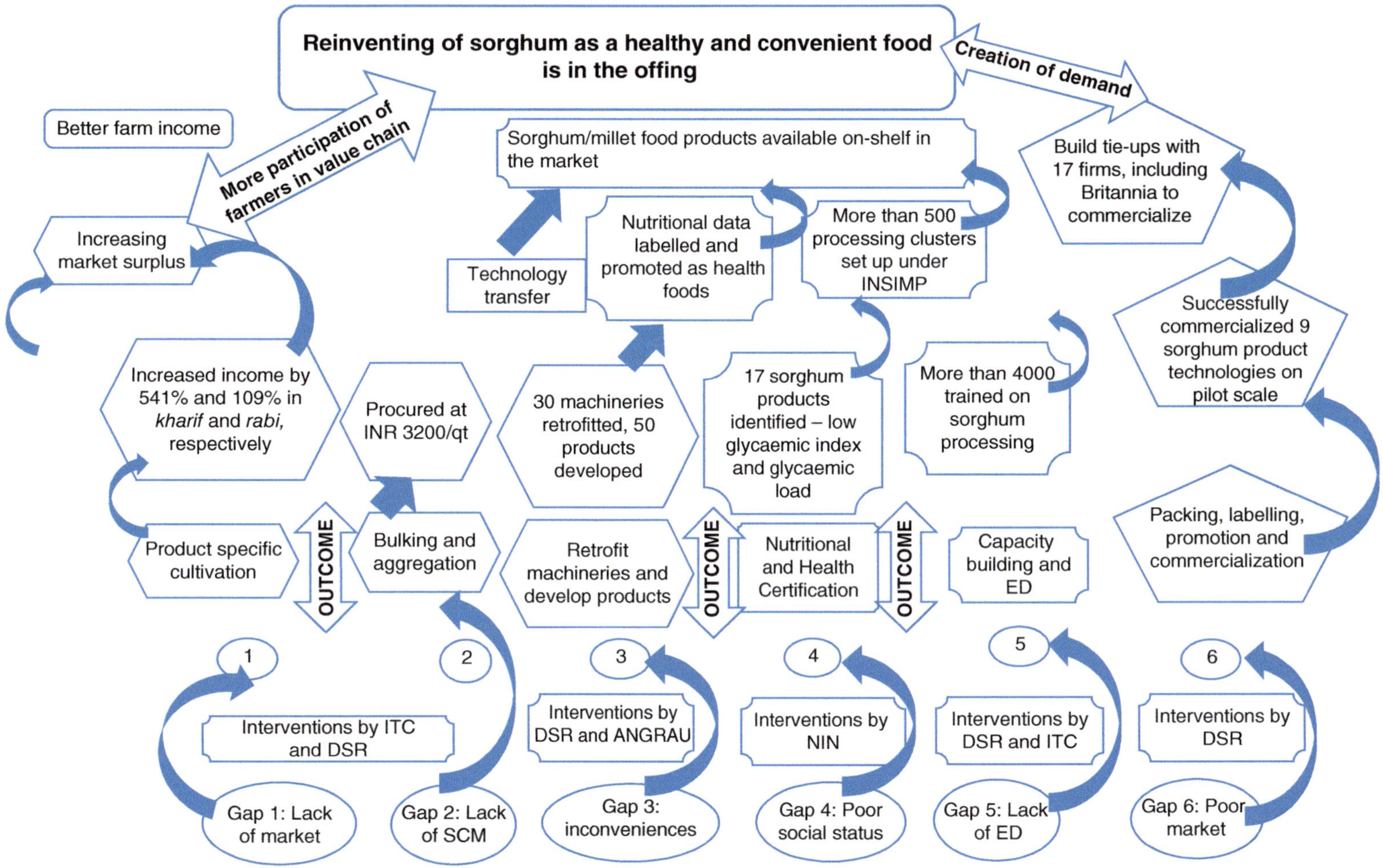

Fig. 7.6. Millets' value chain actually implemented.
Note: SCM, supply chain management; ED, entrepreneurship development.

dosa, *idli* etc.), and extruded products and instant mixes (*dosa* mix, *idli* mix etc.). These products are now being successfully commercialized on a pilot scale in the city of Hyderabad.

IIMR is aggressively promoting these millet technologies for replication and horizontal expansion activities, targeting the private sector either through start-up entrepreneurs or through big players in the form of technology transfer-driven processes. Thus, a detailed business plan is very much required to provide understanding on the potential for investing in the business.

7.2.2 Seasonality vs quality of sorghum

Although the *kharif* sorghum cultivars produce higher grain and fodder yields, they are prone to mould due to excess or untimely rain at harvest, usually during October. This affects the grain quality and makes it unfit for human consumption. Thus the price of *kharif* grain is lowered due to its poor quality and the grain is increasingly diverted for non-food industrial uses (at least three times in every 5 years). Thus *kharif* sorghum production is mainly determined by the fodder requirement from the crop. The introduction of hybrids has initiated a major change in Indian sorghum farming specifically in *kharif* season. Although its productivity is comparatively low compared to rain-bound *kharif* grain, the *rabi* sorghum grain is far superior in quality, due to its maturity during the colder months. However, of late some of the *rabi* sorghum grain is cultivated under protective irrigation, yielding almost double the quantities. M35-1 was popular for a long time, with excellent grain quality traits, though yields are not so high. Recently released *rabi* varieties – Phule series, CSV 14R and CSV 18 – have been well received by farmers. In general, due to its better quality, the profit realized by *rabi* grain is almost double that of *kharif* grain.

There has been an increase in the use of machinery post-NAIP, after the advent of the retrofitting process, especially for dehulling, and the supply of machinery through INSIMP. These prototypes are used by R&D institutes and small-scale entrepreneurs. The dehullers decorticate the outer layer, and although it may lead to partial loss of minerals and vitamins, such grains can be used for direct food consumption and are amenable for product development and used for preparation of various RTC/RTE sorghum products, which are convenient and have improved taste. The market price of dehulled grains will be higher than that of moulded grains, with the black colour from infestation removed by the process of dehulling.

7.2.3 Backward integration and supply chain management

Commercialization of sorghum as processed products has the inherent problem of (i) continuous supply of raw material throughout the year (sorghum being a seasonal crop) and (ii) the requirement of identity-preserved grain, free from adulteration (the crop often being a mix of farmer's varieties with the prescribed cultivar). Thus, quality and quantity are important determinants. Currently, things are not fully developed as far as supply chain management is concerned. However, tested models are available and are being used by well-renowned agencies, e.g. ITC (ABD) Ltd is working with crops such as soybean, which are low value crops and have low marketed surpluses owing to their smallholdings, while they have also tested the model feasibility in millets on a pilot scale. ITC, under their *e-Choupal* system, extended backward integration in sorghum cultivation in the Parbhani and Nanded districts over 5 years and undertook buyback procurement at market price, bulking, aggregation and storage functions. They supplied the produce to processors, thus linking farmers with the rest of the value chain. IIMR attempted to tackle the problem of uniform quality of a particular variety of grain through provision of seeds of improved cultivars suitable for specific end products. The contractual arrangements between farmers and processors will ensure such uniform quality for an unhindered, reliable and efficient production system.

7.2.4 Prospective districts where backward linkages as well as production can be taken up

According to the recent data derived from the Directorate of Economics and Statistics (DES,

2011), ten districts each in Maharashtra and Karnataka and five districts in Andhra Pradesh during *rabi* season have high potential for sustainable take-up of sorghum food processing through linkages with the farmers. Tables 7.5 and 7.6 show the 3-year average in production and yield, i.e. 2007–08 to 2009–10 (in Karnataka) and 2008–09 to 2010–11 (in Andhra Pradesh and Maharashtra). The *kharif* production potential in these areas is comparatively low; most of the districts produced <1 lakh t since they cultivate mostly for fodder purposes and not for human consumption.

7.2.5 Structure of organization and phases of development

Organization, structure and manpower

To implement and run a plant for the production of sorghum-based products, entrepreneurs or enterprises need to have technical and soft skills. The organogram in Fig. 7.7 represents the organizational structure needed for running a successful business model. The company's structure should include a Chief Executive Officer (CEO); the entrepreneur could play this role, being the key person taking all the decisions related to day-to-day activities. The CEO should have people in the key departments of sales and marketing, procurement and production, and supply chain and finance. He should have managers (one each for each division) to manage the day-to-day activities in a smooth manner. There should be three sales executives, one production executive, one procurement executive, and one warehouse and supply chain management (SCM) executive. Two contractual or daily wage labourers would also be required, for loading and unloading of the raw material and finished goods.

Such manpower needs to have the technical and necessary soft skills, as the finished products (e.g. multigrain *atta*, biscuits and flakes) involve skilled manufacturing techniques, procurement requires plenty of decision making and has various procurement options (especially to keep the procurement cost under control and for effectively utilizing the raw material for the production of the final products), and communication skills are required to extol the benefits of the products to the end consumer. Also, a plant that will be running 25 days/month needs to have backup manpower in order to ensure smooth functioning of the plant as well as smooth running of the normal sales operation.

Phase-wise production and product selling plan

A business plan discussed in detail with the head of finance would clarify to the entrepreneur or enterprise (if setting up their business plan for the first time) the need to produce multiple products. Taking on only one technology will take much time and effort to reach the break-even point, so multiple technologies and multiple finished products are required in order to utilize the efficiency of the plant and machines, and also to cover the cost of selling and marketing the goods to the consumer. However, if the entrepreneur or enterprise is already in a similar business and looking for diversification of their product mix, they could take on a single technology. Multigrain flour and biscuit production can be performed together as both require the same basic raw material (sorghum flour) and the production processes are similar to a certain extent, but sorghum flakes have a different manufacturing process and require a separate machine.

Phase-wise implementation

The entrepreneur can take up the production of multigrain flour and biscuits as the first two products in phase one. Once the production and marketing channel is established for these two products, the entrepreneur can add sorghum flakes at phase two. This way, the cost of the set up will be split into two parts and also returns from the first phase will give him confidence to diversify. Set distribution channels will be used for sales of the products, so adding a third product will have only an incremental cost, which can be covered through sales of the other products. Sales of the products should be discussed in advance to set up the most efficient and effective distribution channels. Once the product is ready for sale and put in the identified channel, it should be followed by a newspaper advertisement and paper insert to attract the customer.

Table 7.5. Major *rabi* sorghum-producing districts of Andhra Pradesh, Maharashtra and Karnataka. (Source: DES, 2011.)

Karnataka				Maharashtra				Andhra Pradesh			
District	Area	Production	Yield	District	Area	Production	Yield	District	Area	Production	Yield
Bagalkot	1.16	1.32	1.16	Ahmednagar	5.23	3.32	0.64	Adilabad	0.21	0.17	0.77
Belgaum	1.30	1.30	1.01	Beed	1.93	1.58	0.82	Guntur	0.04	0.27	6.33
Bellary	0.24	0.30	1.29	Jalna	1.38	1.24	0.90	Kurnool	0.56	1.10	1.90
Bijapur	2.06	1.96	0.96	Jalgaon	0.48	0.73	1.53	Mahabubnagar	0.10	0.09	0.87
Dharwad	0.37	0.25	0.70	Osmanabad	2.74	2.18	0.80	Medak	0.16	0.18	1.07
Gadag	0.60	0.47	0.79	Parbhani	1.68	1.19	0.70				
Gulbarga	2.74	2.61	0.96	Pune	4.34	3.07	0.71				
Haveri	0.28	0.25	0.89	Sangli	1.80	1.05	0.58				
Koppal	0.46	0.27	0.60	Satara	1.44	1.19	0.83				
Raichur	1.05	1.34	1.40	Solapur	6.62	3.96	0.60				

Notes: Area is lakh ha; production is lakh t; yield, t/ha. Values are 3-year averages, i.e. 2007–08 to 2009–10 (Karnataka) and 2008–09 to 2010–11 (Andhra Pradesh and Maharashtra).

Table 7.6. Major *kharif* sorghum-producing districts of Andhra Pradesh, Maharashtra and Karnataka. (Source: DES, 2011.)

Karnataka				Maharashtra				Andhra Pradesh			
District	Area	Production	Yield	District	Area	Production	Yield	District	Area	Production	Yield
Belgaum	0.30	0.48	1.64	Amravati	0.46	0.47	1.02	Adilabad	0.25	0.38	1.53
Bellary	0.30	0.34	1.18	Beed	0.34	0.30	0.90	Kurnool	0.08	0.23	2.95
Bidar	0.56	0.70	1.28	Buldhana	0.50	0.79	1.59	Mahaboobnagar	0.42	0.31	0.77
Chamarajanagar	0.19	0.12	0.64	Dhule	0.23	0.32	1.35	Medak	0.21	0.21	0.96
Davangere	0.20	0.42	2.18	Hingoli	0.39	0.44	1.12	Rangareddy	0.16	0.17	1.01
Haveri	0.11	0.15	1.46	Jalgaon	0.80	1.73	2.09				
Koppal	0.12	0.13	1.14	Latur	1.22	1.32	1.08				
				Nandurbar	0.28	0.44	1.60				
				Nanded	1.21	1.10	0.91				
				Osmanabad	0.62	0.37	0.61				
				Parbhani	0.62	0.68	1.10				
				Sangli	0.73	0.71	0.92				
				Satara	0.46	0.77	1.68				

Notes: Area is lakh ha; production is lakh t; yield, t/ha. Values are 3-year averages, i.e. 2007–08 to 2009–10 (Karnataka) and 2008–09 to 2010–11 (Andhra Pradesh and Maharashtra).

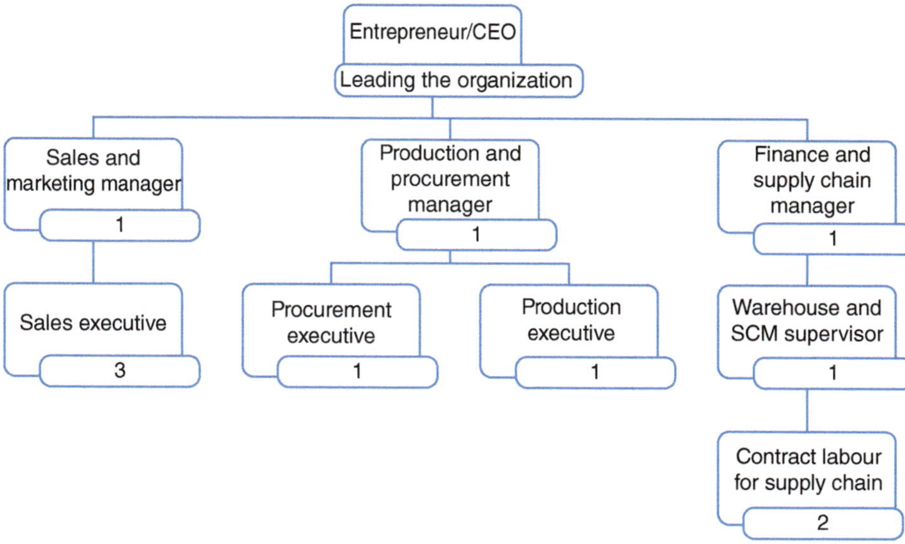

Fig. 7.7. Organizational structure of a model firm. Values indicate the number of staff in each department required.

7.2.6 Product launch

The entrepreneur should start the promotional activity with the setting up of the manufacturing plant. The sales and marketing team should start functioning on the ground and begin discussions with customers, channels, distributors and retailers, and create the initial buzz about the product. Banners, posters and flyers should be placed in key locations where they grab the attention of the customers and the channel interested in selling those products. The health benefits of the products should be highlighted, creating temptation in the mind of the customer to try the product.

Implementation schedule

The firm has to plan its activity in detail followed by the weekly monitoring of the implementation process. The following points are critical and need to be planned well in advance:

1. Identify the site for the construction of the plant for manufacturing of the products.
2. Identification of the channels for the raw material procurement.
3. If procurement of the raw material is planned to be directly from farmers, procurement planning should be in line with the manufacturing unit functioning.
4. Building of the manufacturing plant (time), should be scheduled properly with the procurement of the raw material and the manufacturing plan.
5. Storage of the grain: whether to outsource the storage or to have an inbuilt storage facility. If an inbuilt storage facility is needed, the erection of the storage building should be in line with the procurement and manufacturing plan.

The schedule followed for the implementation of the above discussion is briefly summarized in Table 7.7.

7.2.7 Market and competitor analysis

Advantage of sorghum-based products

- Complete family product: for the individual, family, young, aged, kids, health conscious and traditional.
- Healthy product: wholesome nutrition for all members of the family, no restriction for diabetic/old/ill members, energy powerhouse and perfect diet management.
- Perfectly substitutable: no variation in terms of preparation, appearance and taste of regular *atta*.
- Convenience: no extra effort required and less time consuming.
- Addresses both health and taste.

Table 7.7. Schedule for implementation of the plan for setting up the business.

Particulars	Activities	Time to complete	Estimated time to start from the date of manufacturing
Pre-manufacturing phase	Site for the plant, including all documentation	10 days	5 months
	Site for the storage (modalities (own storage/hiring) to be completed)	10 days	5 months
	Building and plant erection and trial run on purchase of machineries	4 months	5 months
	Site/place identification of raw material procurement	7 days	5 months
	Contracts with farmers if direct procurement is planned or contracts with suppliers	5 months before manufacturing	5 months
	Identification and contract of the selling channel	14 days	1 month
	Identification of the target market and locality	14 days	1 month
	Margin discussion and finalize contracts	14 days	1 month
	Hiring of the staff	1 month	2 months (as per need)
Manufacturing phase	Manufacturing of the product	Daily basis	8–10 h, depending on the product
	Promotional activity	15 days before manufacturing and should happen at regular intervals	15 days
	Newspaper advertisement	3 days before product reaches the market	1 week before actual selling starts
	Product placement in market	2–3 days	1 week before manufacturing starts
Post-manufacturing phase	Retailer outlet promotional activity	Regular interval	With the placement of the product in the market
	Mass campaign	Regular interval	With the placement of the product in the market

- Available products are either positioned as 'healthy' (cholesterol management and diabetic (Saffola)), gluten-free, diabetic) or 'tasty' (e.g. Ashirvaad, Pillsbury, Annapurna).
- Available biscuits are either positioned as 'healthy' (digestive, light, nutritious, diabetic), 'tasty' (Good Day, Hide & Seek, Butter Bite), or are loose (*nankhatai*, freshly baked).

Competitiveness of sorghum-based multigrain atta

The pure sorghum flours are positioned in the consumer sector at a lower price level, while the multigrain *atta* is relatively highly priced compared to regular *atta*. On the other hand, when we see the products which are positioned as healthy, national brands like Saffola are highly priced when compared to the regional brands or labels within a similar sector. After a consumer perception analysis, the study proposes to position the sorghum multigrain atta in the healthy sector with an equivalent price of nationally branded products, as presented in Fig. 7.8.

Currently, the multigrain *atta* in the market is priced at INR 50/kg. Sorghum-based multigrain *atta* is also priced at INR 50/kg, and

Fig. 7.8. Competitive analysis of sorghum multigrain *atta.*

at INR 230 for the 5 kg packs. Currently, most of the other multigrain *atta* has wheat as a basic ingredient; the sorghum-based multigrain *atta* has a low glycaemic index and glycaemic load, making it a healthier choice for diabetic patients.

Competitiveness of sorghum-based biscuits

Biscuits currently available in the market are either positioned in the regular or healthy consumer sectors. The products shown in Fig. 7.9 explain the positioning of the various biscuits and pricing, and it is recommended that sorghum biscuits should be positioned in the healthy sector with pricing the same as other biscuits in the higher price position. In the market, regular biscuits of both brand and non-brand are available. In the healthy sector, biscuits mostly have *ragi* (finger millet) as a base, so this is the first of its kind to have sorghum as its base ingredient. The sorghum-based biscuit is priced at INR 17/100 g and INR 38/250 g, which is between the price range of the regular biscuit and the high-value biscuit.

Competitiveness of sorghum-based flakes

As shown in Fig. 7.10, within the healthy flakes sector, currently corn flakes are the only products available in the market. The private-label cornflakes are sold at a lower price compared to national-level brands. Similarly, some organic sorghum flakes are also available in the market, but currently sold to the niche sector. The studies recommend that sorghum flakes should be placed in the healthy sector and their price should be higher compared to corn flakes.

7.2.8 State-wise potential markets for launching sorghum products

The state-wise potential area/markets for launching sorghum products are tabulated in Table 7.8.

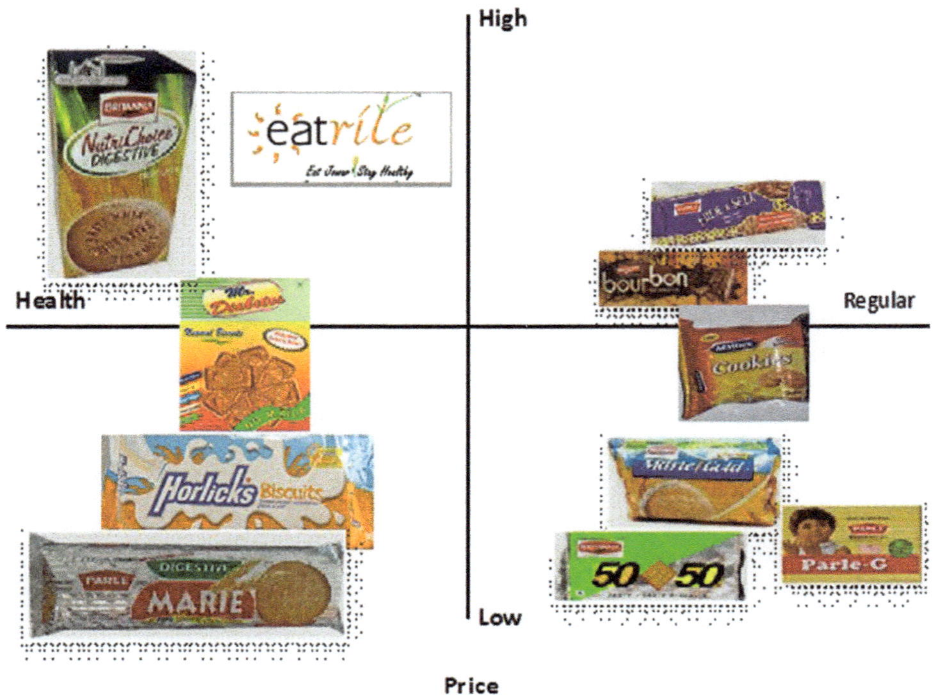

Fig. 7.9. Competitive analysis of sorghum-based biscuits.

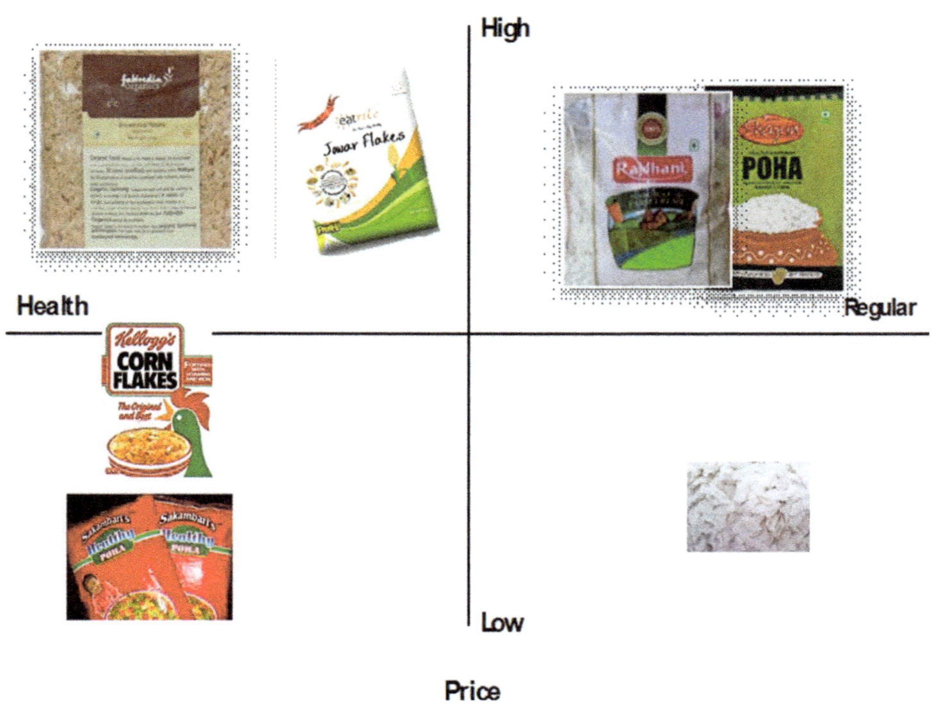

Fig. 7.10. Competitive analysis of sorghum flakes.

Table 7.8. State-wise potential markets.

State	Areas	Supply options	Market entry	Effort required
Andhra Pradesh	Vijayawada, Vizag, Kurnool,	Central manufacturing	Easy with low marketing cost	Consumer acceptance study with launching strategy
Karnataka	Bangalore, Dharwad, Gulbarga, Bidar, Hubli	Central manufacturing	Easy with low marketing cost	Consumer acceptance study with launching strategy
Rajasthan	Jaipur, Ajmer, Jodhpur, Pali	Outsourcing/ tie-up	Easy, moderate marketing cost	Opportunity analysis, market survey, pre-launch consumer study
Gujarat	Bharuch, Vadodara	Outsourcing/ tie-up	Easy, moderate marketing cost	Opportunity analysis, market survey, pre-launch consumer study
Tamil Nadu	Chennai, Coimbatore	Central manufacturing	Easy with low marketing cost	Opportunity analysis, market survey, pre-launch consumer study

7.2.9 Industry size estimation

Multigrain atta industry

- The market size of multigrain *atta* is very small compared to the regular *atta* market.
- It is estimated to be around 15,000–25,000 t per annum (on average 1–1.5% of the total packed *atta* industry of India).
- Mostly, organized players dominate the segment and national brands are available in the market.
- Few regional players are also there in the market.
- Multigrain *atta* is sold at a premium (brands Ashirvaad, Pillsbury and Saffola).
- Growth of this particular segment is estimated to be 3–4% annually.

Biscuit industry

- The size of the biscuit industry is estimated to be around 25–30 lakh t per annum.
- About 55–65% of the total volume is catered for by the organized players. 'Glucose' biscuits constitute 45–50%, followed by 'Marie' to the tune of 10–15% in the total biscuits consumption basket.
- Share of cookies is estimated at around 3–4% of the total market size.
- Growth of the biscuit sector overall has been estimated at around 12–15% per annum.

Rice flake industry

- Rice flake market size is estimated to be around 800,000–1,000,000 t per annum.
- The unorganized players dominate the segment.
- No branded players exist in this segment.
- It is sold under private labels in most urban areas.
- Growth in consumption is estimated to be 5–7% annually.

7.2.10 SWOT analyses of sorghum food products

A 'strengths, weaknesses, opportunities and threats' (SWOT) analysis of sorghum food products is essential for establishing its uniqueness amongst the available products of the same category in the market. A number of points define sorghum as a healthy food product because of its nutritional content. However, it has a few weaknesses such as its shelf life and lack of commercial business model, but its potential to emerge as a healthy food product is an added advantage (Table 7.9).

7.2.11 Product-wise SWOT analysis

Sorghum multigrain atta

A SWOT analysis of sorghum-based multigrain *atta* (Table 7.10) shows that people have

Table 7.9. SWOT analysis of sorghum food products.

Strengths	Weaknesses
• Awareness of sorghum as a healthy and nutritious food is favoured. • Appealing response to quality certification from NIN. • Strong research and development facility. • Only brand (IIMR's brand *eatrite*) in the market offering a versatile product range based on sorghum. • Easily adoptable in terms of appearance, preparation and taste when compared to traditional products. • Taste of the product is liked by the respondents. • People are willing to consider the product if *eatrite* is launched in the market.	• Infant brand in the market. • Lack of awareness among non-traditional respondents. • Lack of commercial business operation, sales/ distribution network and presence in the market. • Immature economy of scale. • Lack of communication and advertisement about the product. • Shelf life of the product is short (1 month). • The typical sorghum smell is disliked by non-traditional consumers. • Sorghum presents challenges in preparation/ cooking for urban consumers. • Inferior appearance attributes compared with traditional product.
Opportunities	**Threats**
• Growing consumer eagerness to try new products with health benefits. • Specialty consumers (diabetic, obese) are eager for a favourable result. • Able to substitute for traditional products giving variety and choice. • Health food market size is around INR 10,000 crores and expected to grow by a compound annual growth rate (CAGR) of 23% (Tata Stategic Management Group (TSMG)). • Increasing disposable incomes, diminishing culinary skills and the rising need for convenience are the driving factors for the growth of sorghum products in traditional areas.	• Initial hesitation of retailers to keep the products in non-traditional areas. • Concerns about consistency and quality are prominent in consumers' minds, as the supplier is not a branded 'fast moving consumer goods' (FMCG) one. • Sustainability in the short term is high due to marketing set-up cost. • Increasing raw material price and uncertainly in the production of sorghum. • Negative health related perceptions (non-edible varieties for some seasons, dental disease issues). • The health product segment size is very small and many consumers fall in the 'late majority' category.

developed a liking for it. The main health benefit of *atta*, its gluten-free content, makes it suitable for coeliac patients. There are also weaknesses related to the product, such as its smell and shelf life, but the opportunity is great as there are very few players in the market who are selling this product. Also, the consumer's continuous search for better health food items provides them with an opportunity to tap the current multigrain *atta* market.

Sorghum-based biscuits

A SWOT analysis of sorghum-based biscuits (Table 7.11) shows that currently very few food companies have them in their product portfolio. Also, with the increase in awareness of the content of regular biscuits, people are trying to have more balanced and nutritional types in their day-to-day life. The opportunities for this segment are great as it is still at a very early stage.

The sorghum-based biscuit, being new to the market, can capture a large segment of the nutritious biscuit sector as its nutritional content is high compared to other millets and cereal crops.

7.2.12 Production plan

Table 7.12 depicts the production plan for sorghum multigrain *atta*, biscuits and flakes. It is considered that the plant will run for 8 h on a

Table 7.10. SWOT analysis of sorghum multigrain *atta*.

Strengths	Weaknesses
• The product is very much liked by most of the respondents who have tasted it. • More than 90% of the survey respondents expressed their willingness to purchase the product. • The health benefits of sorghum *atta* are the major acceptance criteria in traditional areas. • The respondents felt no significant difference between sorghum *atta* and wheat *atta*. • Sorghum *atta* is an ideal product for diet plans and weight management. • It is suitable for all family members, and there is no restriction for its consumption by diabetics, the ill, children or the aged.	• Shelf life: shelf life of sorghum *atta* is 3 months. • Smell: typically the sorghum smell in the *atta* during preparation is not liked by a few consumers (especially new consumers). • Swelling: swelling attribute of *chapathi* was not liked by the Delhi consumers. • Stickiness and texture of sorghum *atta* is another issue to be addressed. • No existing business experience, supply chain is still in formation stage. • No professional manufacturing, marketing and sales department. • Economies of scale, learning curves, expertise in commercial operations and packaging are yet to be fine-tuned.
Opportunities	**Threats**
• Sorghum-based multigrain *atta* is the only product available in the *atta* market that is rich in sorghum. • Existing players in the multigrain *atta* segment of the market are very few. • Multigrain *atta* consumers are not brand loyal, preferring to seek more health benefits from *atta*. • Increasing health consciousness, disposable incomes, diminishing culinary skills and the rising need for convenience could be the driving factor for the growth of sorghum products in urban traditional sorghum-consuming areas.	• The size of the multigrain *atta* industry is very small, and dominated by the national brands. • Growth of this segment is also very low (3–4% per annum). • The higher price of sorghum seed would in turn increase the price of sorghum-based multigrain *atta*. • The cost of marketing and distribution is very high for a new product. • The use of sorghum as a food product is restricted to traditional consuming areas.

daily basis. The length of the working month is assumed to be 25 days. Financial analysis has been performed for 4 years and production will be increased on a yearly basis. Packaging size has also been identified, accounting for the movement of the product from the shelf (Table 7.13).

7.2.13 Financial plan

The financial implications for constructing the plant for production of the three sorghum products are tabulated in Table 7.14. It is advisable that the multi-grain *atta* and biscuits plant should be considered together, as the machinery is common up to a certain point in the early processing of raw sorghum. There are also commonalities in other materials and in the preparation of manufacturing inputs. Flakes require altogether different machines and set-up. However, if an entrepreneur (early start-up) is planning for these technologies, he should go for all three together – this will not only help him in reducing or sharing the cost in terms of putting up manpower but will also provide variety to the consumer, leading to enhancement of sales. The fixed cost in the calculation has been stated separately for all the products. However, while calculating the IRR (internal rate return) and NPV (net present value), we have taken the fixed cost of the biscuit and flakes together, as the machineries will be the same in the case of biscuits and in the case of multigrain *atta*. (The IRR and NPV are used to calculate project feasibility and the attractiveness of the investment.)

The financial plan makes the assumption that the plant should have internal storage in

Table 7.11. SWOT analysis of sorghum-based biscuits.

Strengths	Weaknesses
• Sorghum biscuit is very much liked by most of the respondents. • More than 90% of the survey respondents showed their willingness to purchase the product. • The health benefits associated with sorghum biscuit are the major acceptance criteria in traditional areas. • Sorghum biscuit is suitable for all family members and there is no restriction for its consumption by diabetics, the ill, children or the aged. • It is an ideal product for diet plans and weight management. • Taste-wise, sorghum biscuits barely differ from existing cookies.	• Shelf-life: the product shelf life is very short (6 months). • Smell: typically, the sorghum smell of the biscuit is disliked by some consumers. • Crispiness: coconut biscuits were less crispy while normal biscuits were hard while chewing. • The sorghum biscuit gumminess (while chewing, biscuits are broken into particles, becoming ready to swallow) is less as compared to normal cookies. • Fullness of taste after eating and the feeling in the mouth is disliked by some consumers. • No existing business experience; the supply chain is still at the formation stage. • No professional manufacturing, marketing, or sales departments. • Lack of production and operational advantages, e.g. economies of scale, learning curves, expertise in commercial operations, packaging.
Opportunities	**Threats**
• The sorghum-based biscuit is the only product available on the market that is rich in sorghum. • The product can compete with other cookies targeted at the health conscious consumer segment. • Health biscuit consumers are not brand loyal as they prefer to seek more health benefits. • Increasing health consciousness, disposable incomes, diminishing culinary skills and the rising need for convenience could be the driving factor in the growth of sorghum products in urban-centric traditional areas.	• The share of 'cookies' in the total biscuit market is very small (5–7 % of total). • Healthy cookies are a niche product and the market is dominated by organized players or brands. • The growth of the cookie sector is also lower (3–4% yearly; CAGR, 2008–13) as compared to biscuits overall (12–15%; data monitor, 2013). • The higher price of sorghum seed would in turn increase the price of sorghum biscuits. • The cost of marketing and distribution is very high for a new product. • Usage of sorghum as a food product is restricted to traditional consuming areas.

order to have greater control over the raw material to be used for production. It is also advised that the production plant should be near to the area where sorghum and other inputs are easily available in order to have a smooth supply of the production inputs.

Based on the first year of operations, projections for the next 3 years have been made in terms of total quantity of each product to be produced on an annual basis. Table 7.15 provides a suggested estimate of quantities to be produced on a year-on-year basis in order to achieve the break-even point and recover the initial investment in the first 4 years of operations. Based on the production plan for all 4 years, IRR and NPV have been calculated as 9% and INR 48.4 lakh, respectively (Table 7.16).

It is also important to highlight that the quantities on which the first year calculation is based are close to the break-even point of the variable cost. So this means that quantities below that will lead to a loss on the operational front. It was also identified during the financial calculation that single products or single technologies based on sorghum will not give a viable income. Therefore, it is suggested that a combination of products should be considered before starting a fresh venture. However, if the

Table 7.12. Daily, monthly and annual production quantity of sorghum-based multigrain *atta*, sorghum-based biscuits and sorghum flakes.

Particulars	Daily production (t)	Monthly production (t)	Annual production (t)
Multigrain *atta*	1.2	30	360
Biscuits	0.36	9	108
Flakes	0.8	20	240

Table 7.13. Pack sizes for different products.

Particulars	Pack sizes	Total annual SKU quantity
Multigrain *atta*	1 kg	270,000
	5 kg	18,000
Biscuits	0.1 kg or 100 g	810,000
	0.25 kg or 250 g	108,000
Flakes	1 kg	180,000
	5 kg	12,000

Table 7.14. Finance for the first year of operations (all costs in INR lakh).

Particulars	Multigrain *atta*	Biscuits	Flakes	Total
Building and plant, including plant machineries and other legal and set-up costs	82.8	77.8	99.8	166.5
Variable cost				
Cost of raw materials	68.8	52	96	216.8
Operational cost, which includes electricity, depreciation of the machineries, packaging of the products etc.	12.66	13.75	9.93	36.34
Marketing cost	18	18	18	54
Manpower cost	7.62	7.62	7.62	22.86
Channel margin (@ 25%), transportation and logistics (@5%) and taxes (@ 4%)	39.9	33.72	48.94	122.56
Total variable cost	146.39	125.09	180.49	452.56
Total sales (It is assumed that 10% of the total SKU will be sales returns and 10% of the smaller SKU will be used in promotional activities. Therefore, the actual quantity for sales will be 80% of the smaller SKU and 90% of the larger SKU)	148.5	131.22	195.3	475.02
Net profit before tax	2.1	6.34	14.8	23.24
Net profit after tax (@30%)	1.47	4.44	10.36	16.27

Table 7.15. Annual production quantity (t) for the 3 years following the first year of operation.

Particular	2nd year	3rd year	4th year
Multigrain *atta*	420	420	420
Biscuits	120	120	120
Flakes	300	360	420

Table 7.16. Internal rate of return (IRR) and net present value (NPV) of the project.

Targets			Fixed cost (INR)	Year	Revenue (INR)	Cost (variable) (INR)	Revenue minus cost (INR)
Multigrain *atta*	Biscuit	Flakes	166.5				–166.5
1200	360	800		1	475.0	451.6	23.3
1400	400	1000		2	563.0	513.0	50.0
1400	400	1200		3	612.0	548.2	63.8
1400	400	1400		4	660.8	583.0	77.8
						IRR	9%
						NPV (INR lakh)	48.40

entrepreneur is already in this business and is diversifying their product mix, they could opt for the single product or technology.

7.2.14 Marketing plan

It is suggested that in the first year the entrepreneur should opt for intensive marketing strategies in order to establish the brand. The marketing plan for the product will have both strategies, i.e. above the line (ATL) and below the line (BTL – a form of sales promotion that uses outlets such as direct mail, special discounts and point of sale offers, as opposed to the mass media used in ATL). A combination of mass campaigns, roadshows, newspaper insertions, cable TV and newspaper advertisements should be undertaken in order to create awareness of the brand to the consumer, leading to fast movement of the product from the retailer's shelves.

It is also suggested that these activities should be conducted at regular intervals, i.e. on a monthly basis, in order to grab the attention of consumers. Apart from the above-mentioned activities; entrepreneurs should also use items such as banners, posters, T-shirts and caps in order to promote the brands in the market. Banners and posters should be put up in key locations that receive maximum customer footfall. The sales and marketing team should wear T-shirts and caps to create a buzz in the market. T-shirts and caps should also be given out during mass campaigns and roadshows, so that the brands reach the consumer household and create word-of-mouth publicity.

The bundling of products can also be considered during product promotion activities. In Table 7.17, a few combinations have been suggested which the entrepreneur can consider for their bundling plan during the promotional activities. Suggested activities to be implemented to establish the brand in the market and attract consumers are given in Fig 7.11.

Activities in terms of BTL and ATL that should be carried out on a monthly basis in the initial year are listed in Table 7.18. This is essential in the initial period in order to establish the brand in the market. Once people start recognizing the brand, momentum is given to product sales. Also, the likelihood of tie-ups with big retail chains will increase.

7.2.15 Alternative strategies for selling the products

The business plan also suggests an alternative marketing or sales plan. The entrepreneur or enterprise can become a contract manufacturer for the big players in the market. They can ensure supply of the products for a particular region with a timely supply commitment. This may also facilitate the development of a cluster-based sorghum production centre in the major sorghum-producing state, leading to an increase in production area, ultimately providing dryland farmers with an opportunity to improve their livelihoods.

It is proposed that IIMR should initiate this alliance between entrepreneurs or enterprises and bigger players that have greater experience and expertise in the marketing and selling of the products. The entrepreneurs can play a role in

Table 7.17. Suggested combinations for bundling three products.

Product purchased	Quantity	Free gifts (product) with purchased product
Multigrain *atta*	5 kg pack	100 g pack of biscuits
Flakes	1 kg pack	100 g pack of biscuits
Flakes	5 kg pack	250 g pack of biscuits
Multigrain *atta* plus flakes	5 kg of *atta* plus 5 kg of flakes	1 kg pack of multigrain *atta*

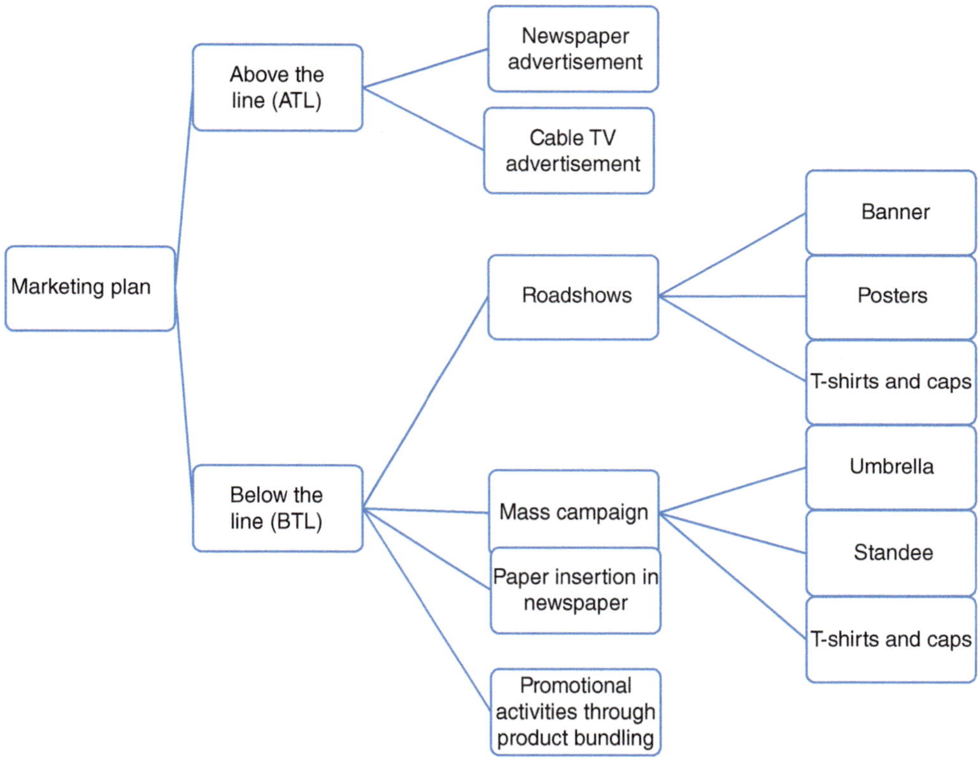

Fig. 7.11. Marketing plan for millet products.

Table 7.18. Brief activity under BTL and ATL.

Marketing strategy	Activities per month or per quarter year
Below the line (BTL)	
Roadshows in the most popular places and roads on which footfall of customers is high	3/month
Mass campaign with the help of umbrella and standees in key locations for the market	2/month
Paper insert	3/month
Above the line (ATL)	
Newspaper advertisement	1/quarter
Cable television advertisement	1/month

doing the backend manufacturing for those bigger players. The companies that will be buying the products can implement their quality standard for maintaining the quality of the products. Suggested roles for the stakeholders involved in the partnership are given in Table 7.19.

7.2.16 Risk factors

Availability and accessibility of the sorghum

The availability of sorghum for the production of a value-added and nutritional product is a critical factor. The plant needs to have very strong backend integration in terms of procurement of the grain. It is advisable that the plant should be established near the production area. Also, there should be integration with the farmers under the contract farming agreement in order to have smooth and continuous functioning of the plant. The procurement planning and mapping of the required quantity should be planned well ahead, before the start of the season.

The price fluctuation in sorghum is another critical factor. If crop production goes down, the price of sorghum grain may hit the range of INR 4400/qt. Therefore, the backward integration in terms of procurement will lead to better control over the input cost of production.

Entrepreneurial skills and capabilities

It is important to highlight an entrepreneur's skills and capabilities as these products are premium products in the market and have a much lower market share in terms of regular *atta* or biscuits available in the market. The entrepreneur should have very strong business acumen and should be able to lead the business from the front. He should have an understanding of the market channels and of the various ways through which he can procure the sorghum while keeping his operations unaffected due to price variation. Based on analysis of data, the following conclusions emerge:

1. First time entrepreneurs or enterprises need two or more sorghum-based food products to start their business. This will ensure that the business model is more economically viable. However, if an existing entrepreneur or enterprise wants to diversify their existing business they can go for any number of sorghum-based products so long as the processing of the raw material into the final product is the same.
2. Data analysis indicates the minimum quantity of each food product that needs to be manufactured on a monthly or yearly basis. These quantities will ensure the recovery of investment at the end of the fourth year of operations.
3. Key districts for potential sorghum cultivation should be explored for the setting up of the manufacturing plant. This will ease the difficulty in raw material procurement for the manufacturer and make it more cost effective.
4. It is clear that sorghum-based food products are relatively new to the market. Hence it is essential that entrepreneurs develop competent skills

Table 7.19. Suggested roles for the different players involved in the partnership.

Particulars	Entrepreneur or enterprise	IIMR	Large Indian/multinational company
Procurement and production	Manage and concentrate on the production of quality sorghum-based products.	Can help entrepreneur with backward integration with farmers and provide technology for the production of sorghum-based products.	Provide technical guidance on setting up the plants and help in implementation and usage of production quality standards.
Marketing and sales	Focus on quality production with attractive packing.	Technical help in highlighting the key qualities of the products.	Through their established distribution network, provide necessary momentum for sales.

for sales and marketing of these food products. These products also need focused marketing activities in terms of ATL or BTL. This is essential to develop the brand in the marketplace, leading to customer preference towards the purchasing of these food products.

5. As an alternative strategy to the sole entrepreneur mode, a tripartite partnership between the entrepreneur, IIMR and a big player (who is already in business) is also suggested. The entrepreneur will have the role of manufacturing standard quality food products and IIMR can provide backward integration for the raw material procurement. Forward integration will be provided by the big players through development of a supply chain for the food products. The entrepreneur can act as a contract manufacturer for the big companies. The company can provide technical guidance to the entrepreneur for the manufacturing of quality food products and play the role of marketer, a field in which they are experts.

7.2.17 Social impact

The products identified have a huge social impact on society. Sorghum-based products are more nutritional in comparison to wheat- or rice-based products. Their gluten-free content makes them suitable for the health conscious and coeliacs, and they are also recommended for people who have diabetes. The products provide children with healthy food that helps with balanced growth and the development of their bodies. Manufacture of these products can also lead to an increase in the area under cultivation, providing dryland farmers with a better price for a product whose potential is still not recognized by the market. The farming family engaged in sorghum production in dryland areas will also earn good income as the products will fetch premium prices from the market. This will also help in the development of the sorghum-based product industry and can provide employment to rural youth by engaging them in the process of production, procurement and manufacturing of products and their sales and marketing.

References

Dayakar, R.B., Patil, J.V., Chavan, U.D., Kachui, N. and Shiva Kumar, P. (2015) *Emerging Competitiveness of Sorghum in India: an Intensive On-Farms Analysis of Major Rabi Sorghum Regions.* ICAR-IIMR, Rajendranagr, India.

DES (2011) Directorate of Economics and Statistics, Department of Agriculture and Cooperation, Ministry of Agriculture, Government of India.

DSR (2011) *Report on Market Research and Strategy on Jowar Food Products: Consumer Acceptance, Profiling, Segmentation, Positioning and Communication Strategies.* Directorate of Sorghum Research, Rajendranagar, Hyderabad.

8

Value Chain of Millets: Lessons to Learn and Prospects for the Future

8.1 Constraints and Challenges in the Value Chain of Millets

As discussed in previous chapters, millet consumption has declined in the recent past as a result of both demand- and supply-led factors. Some of the crucial gaps in the value chain pertained to: low marketed surpluses, inconvenience and more time required in preparation, lack of skill in the development of convenience millet products, short shelf life, lack of processing machineries and product technologies, lack of participation in processing, inadequate awareness of nutritional benefits, lack of policy support and low market penetration. Millet (in contrast to rice, wheat and maize) utilization was constrained by the lack of processing diversification and value-addition technologies. A reduction in these value chain gaps has been attempted through demand-creating interventions.

The value chain development of sorghum was a pilot attempt to revive demand through various interventions, from on-farm production through to end-user utilization. Although it became a successful model with inherent potential for replicability in other parts of the country, it has not been altogether free from constraints and challenges. Here we elaborate on these so as to learn from the experience and benefit future attempts for replication.

8.1.1 Production constraints

Millets require less water for cultivation and are known for their climate resilience. But during the National Agriculture Innovation Project (NAIP) subproject, there were at least three out of four project years that had severe drought, the result being that the targeted sorghum yields were not attained at the participating farmer beneficiary level – other crops grown by them had >20% reduction in yields. However, the increase in the price of sorghum has partly compensated farmers, realizing positive net returns, while most of the other crops in the area have had negative returns over the past year. If millet is to be commercialized, the supply chain will have to be strengthened to ensure continual supply of identity-preserved grain, which needs to be bulked and supplied in large quantities throughout the year. This would ensure that value-added processing would remain unhindered by any reduction in grain production.

8.1.2 On-farm technological backstopping

Technological backstopping through product-specific improved cultivars in public–private partnership (PPP) mode has resulted in an increase in sorghum yields and incomes of the targeted sorghum farmers. It has also created sustainable links between farmers and markets through facilitating buy-back of marketable surplus of farmers' produce. Involvement of the private sector helps not only the farmers in obtaining market price but also the processors in receiving the required quality and quantity of raw material. However, it has been difficult to orient all the farmers up to the desired level and to change their mindset, as they can be reluctant to follow the recommended practices. Only a limited number of farmers were covered under this project because of its pilot-scale nature. Some of the participating farmers did not use the inputs provided to them for the sorghum, instead they sold them off to fellow farmers.

8.1.3 Clustering of farmers

There was a gap following the project period and some of the participating clustered farmers (i.e. those groups of farmers with similar objectives) did not want to continue to grow sorghum. It was difficult to cluster and select all the farmers who were willing to follow the package of practices for achieving the specific quality of end-product grain for processing. Sometimes, farmers would sell adulterated or mixed grain sorghum, which posed low quality grain issues, and such grain is not fully suitable for direct processing into various value-added products. Occasionally, beneficiary farmers could not keep up their commitment for the buy-back agreement made with ITC Ltd to supply identity-preserved grain at market price.

8.1.4 Primary and secondary processing of sorghum

Creation of demand and value addition of sorghum through processing is the main objective of the NAIP subproject. Primary and secondary processing was taken up by IIMR (the Indian Institute of Millets Research) retrofitting of 30 machines that had been suitable for processing rice and wheat. However, after retrofitting the machines, there were problems faced with the lack of servicing availability; very few engineers and manufacturing workshops were available for fabrication and upscaling of standard machinery adapted for millet processing. This remains a grey area when it comes to upscaling of the prototypes.

The recovery of processed products was low, which is a major concern in future upscaling. Quality of sorghum grain is essential for food processing. As some of the participating farmers gave low quality adulterated sorghum grain in the buy-back scheme, there was no uniformity in grain size, and this posed a problem during the dehulling process, with the grain becoming pulverized instead of dehulled. The percentage of by-products is more than the required standard, which causes huge wastage and loss.

8.1.5 Shelf life remains a challenge

The short shelf life issue was addressed and, through interventions, shelf life was improved from a baseline of 45 days (for the flour) to 6 months (for vermicelli and pasta). However, the shelf life of the flour could not be enhanced beyond 90 days. This is bound to pose a constraint, especially when targeted for a pan-Indian supply; there is a possibility of losing at least a month in the supply-chain operation itself. Alternative processing methods are to be attempted to increase shelf life beyond 90 days for powders such as *atta* and *suji*.

8.1.6 Policy

Through policy sensitization, the NAIP subproject has made policymakers aware of the benefits of millets and their being included in major government programmes. Mass consumption of millet in India needs to be generated through the government's Public Distribution System (PDS), but there is a lack of procurement from the government side. Furthermore, the customary announcement of Minimum Support Price from time to time is of little use as it does not cover the cost of cultivation.

8.1.7 Inadequate promotional awareness on health aspects

Sorghum and other millets are 'healthy cereals', containing excellent sources of complex carbohydrates (65–75% available carbohydrates), proteins (7–12%), micronutrients and phytochemicals with nutraceutical properties, and dietary fibre (15–20%). However, while concerted efforts have been made in Hyderabad (where the pilot attempt has attained the desired objective), they are inadequate given the potential to exploit millets across the country. Generation of awareness is the collective responsibility of all the stakeholders, be it government, private sectors, nutritionists, doctors, researchers or NGOs.

8.1.8 Conclusion

As discussed above, despite it emerging as a successful model, there are many constraints and challenges at every step of the value chain of sorghum. To overcome these constraints, sustained efforts on development by both public departments and the private sector – especially on promotion of value-added technologies – would go a long way. Subsidized processing machinery for entrepreneurs and self-help groups (SHGs) should be made available to encourage value addition at the farm level. Further, increased and wholehearted participation by both policymakers and private players would help in improving prospects for the poor dryland millet farmers.

8.2 Scope for Replication, Upscaling, Horizontal Expansion and Export Avenues

Interventions through the millets' value chain project have been successful in an overall context; through an integrated approach and active participation by all stakeholders in the value chain, they have all benefited, from farmers to consumers. Some of the positive aspects of the millet value chain interventions are: the changing of mindset of the farmers from subsistence farming to product-specific cultivation for a better price; the quality and quantity assurance that end product-specific sorghum cultivation created for the processors/entrepreneurs in running their business; the increase in interest and participation from various rural and urban groups, NGOs and new entrepreneurs from both traditional and non-traditional sorghum consuming regions; and the advocacy of the policymakers for a need to promote millets in the country. Here, in light of the project's successes and limitations, we discuss the scope for replication of the millet value chain intervention within India and beyond, as well as its potential for upscaling, horizontal expansion and millet export potential.

8.2.1 Upscaling and replication

Upscaling challenges are numerous, although post-NAIP, the issues are being addressed in the country in an assertive manner: by R&D, private sectors, NGOs and policymakers, especially regarding development of machinery suitable for various processing technologies, primary processing, product development, marketing and entrepreneurship development. Government support is also continued through the Directorate of Agriculture and Cooperation's National Food Security Mission. Three Centres of Excellence (CoEs) for sorghum, pearl millet and small millets have been established in the country. Pilot-scale studies are being attempted in some states to assess the feasibility of the mainstreaming of millets in the Mid-day Meal Scheme. The private sector has created multigrain formulations with millets as major ingredients, which are likely to enhance millet utilization. What has been achieved thus far is only the beginning, and requires consolidating through sustained integrated efforts.

But concerning the questions of sustainability and replicability of the model for millet promotion, policy changes on the part of the responsible central/state governments are required. These should include: the inclusion of millets in the PDS, procurement of millets, subsidies for millet cultivation, special funds for R&D and promotional and awareness programmes on millet promotion and their nutritional importance, encouraging start-up entrepreneurs for value addition in millets and subsidies for processing machinery development. Of course, the role of the private sector is equally important, especially with increased health awareness among the urban population.

The value chain project, which focused on sorghum, is replicable not only with other millets, but also geographically. With the various components of value chain replicable beyond India, millet interventions are possible in other countries; especially relevant as lifestyle diseases have become an issue irrespective of country, leading to consumers searching for alternative cereals to provide carbohydrate in their diet. As discussed, with its rich nutrient composition, high dietary fibre content and slow-releasing carbohydrates, millet could provide the answer to millions across the continents. Africa could be the first to adopt the model, which already has a somewhat disjointed value chain on sorghum, although there have already been some efforts to strengthen this. Asia, specifically Middle Eastern countries, could also take it on, having the advantage of being an integral part of the Indian continent. The USA is not averse to such value chains that are nutri-centred and millet has strong potential in a nation where wheat and maize is increasingly consumed. The USA has many migrants settled from India and, indeed, replication is possible wherever Indian communities are settled across the continents. And while the pilot-scale value chain intervention may not have proved to be entirely successful in India, a number of the components are worth emulating to enable the success of millet promotion for the nutritional security of consumers.

However, dissemination of the developed technologies and concepts needs to be introduced at the national level. For as long as the technologies do not reach the primary stakeholders – farmers in particular – the production scenario cannot be improved, which is crucial for promotion and commercialization of millet-based food products. The wider the area under technological backstopping with improved end product-specific millet cultivation, the broader will be the prospect of millet in the country's food basket.

Entrepreneurs never used to include millet in their business due to lack of raw material and unavailability of processing machinery. The meagre marketable surplus of the farmers is due to low remuneration of the crops in comparison to other fine cereals. While under NAIP, the IIMR could only disseminate the technologies to the participating farmers and partner entrepreneurs; through INSIMP (Initiative for Nutritional Security through Intensive Millets Promotion), training programmes are being organized for farmers' groups, NGOs, Krishi Vigyan Kendras (a grass-root level institution designed and devoted to impart need-based and skill-oriented vocational training to the practising farmers, in-service extension personnel and to those who wish to go in for self-employment through 'learning by doing'), entrepreneurs and government servants in all major millet-producing states. This training will enhance capacity building on the various aspects of value addition in millets. Besides IIMR scientists, invitations to experts from partner institutes, private consultants and retired scientists add more value to the training.

8.2.2 Horizontal expansion within India

The main objective of millet value chain development is to revive/create demand for sorghum and other millets. This demand cannot be created unless these grains are available in convenient ready-to-eat (RTE)/ready-to-cook (RTC) forms so as to be favoured by consumers. Of late, consumers have started opting for convenient, healthy and nutritious foods in the market, so with millet being a nutri-cereal, there is scope to expand its market in the food sector if it is available in a convenient form. Once again, reviving their production and demand will contribute enormously to the nutritional and health security of the country.

With successful commercialization of sorghum products at the pilot scale in the twin cities of Hyderabad and Secunderabad through Heritage retail stores, IIMR has further spread the marketing operations of their *eatrite* products in Mumbai through Big Bazaar's Future Retail stores. Being a government agency, the institute's main objective is to promote millets across the country by persuading more and more potential firms to include or exclusively take up sorghum and millets in their business. The aggressive promotional and health awareness programmes distributed through roadshows, exhibitions, seminars and workshops, and advertisements over the past 5 years has created much awareness, especially in Hyderabad. Taking advantage of the growing awareness among consumers and policymakers, around 15 firms have signed Memoranda of Understanding (MoUs) with IIMR, and Future Retail Ltd group (Big Bazaar) have signed an MoU to market millet products not only in Mumbai but also to expand horizontally across the country, especially to NCR Delhi.

Since millet, especially sorghum, products are new to consumers in non-traditional areas, firstly, commercialization needs to go side-by-side with their promotion as health foods. Secondly, the products need to be available on the shelves of as many stores as possible in the targeted areas. Thirdly, they need to be sold by a reputable firm/s for positive perception and word-of-mouth promotion. In view of the above, IIMR has developed a protocol to shortlist the potential entrepreneurs based on their credibility. Technology incubation will allow them to produce the food products using the facility provided by the lead centre for market testing so that once it is successful, the firm itself could beneficially start their own production and market it under its own brand, while IIMR retain the intellectual property rights to the technologies.

Biofortification of millet products for enhanced nutrition and increased shelf life lies solely on the R&D side of the institutes. The past achievements and failures in developing novel sorghum technologies show that millets have great prospects in the RTE and RTC food sector. Also, the organoleptic studies of sorghum extruded products, sorghum cakes, sorghum *lassi*, and sorghum snacks show high sensory value for these products, which has widened the scope for their commercialization in the near future.

8.2.3 Export avenues

Millet is grown in India mainly for subsistence consumption and feeding livestock. According to the data portal indexmundi.com, exports of India's grain millets have fluctuated greatly over the years. Between 2004 and 2014 (Table 8.1), exports in sorghum were lowest in 2005 at 4000 t, while in 2012 they hovered at 22,8000 t and in 2014 they stood at 50,000 t. Domestic consumption, on the other hand, has come down to 5 million t in 2014 from 7 million t in 2004. In the case of other millets, almost all the production, including the stock, is used for domestic consumption. The sorghum is traded as a commodity for meeting the feed requirements of importing countries. Despite having a large area under sorghum, the quantity and value of exports is much lower compared to other countries. Conducive policy framework and institutional support for promotion of exports should be taken up by the related public departments.

However, the value of sorghum exports as a share of the value of total cereal exports from India was <1%, which makes it almost negligible (Table 8.1). Enhancing infrastructural facilities and logistic support for handling the grain and other value added products for exports will make sorghum export competitive. Of late there is a renewed interest in projecting millets as a value added export product that will not only enhance export competition but also make it profitable through enhanced utilization.

The Agricultural and Processed Food Products Export Development Authority (APEDA) was established by the government of India, entrusted with responsibility for export promotion and development of various agro products. APEDA has been actively engaged in the development of markets besides upgradation of infrastructure and quality to promote the export of agro products. In its endeavour to promote agro exports, APEDA provides financial assistance to the registered exporters under its schemes for Market Development, Infrastructure Development, Quality Development and Transport Assistance. APEDA has approved its support to finance a project proposal by IIMR to develop domestic markets for export purposes. This might give a new approach for millet exports, with processed and RTE millet products for the first time potentially exported from India to the Middle East, China, the USA and African countries.

8.3 The Future of Demand Creation for Millets: a Sustainable Approach

The NAIP subproject has been sustained beyond its time period. It has kickstarted a development programme, INSIMP, formulated to increase promotion and production of millets. Through INSIMP, the government of India allocated INR 3000 million in 2011 under Rashtriya Krishi Vikas Yojana (RKVY) to promote millets as nutri-cereals. As such, more than 200 processing clusters have been set up across the country for the benefit of stakeholders, technologically backstopped by IIMR. Now, with the setting up of CoEs on sorghum, IIMR has been entrusted with responsibility for showcasing technology, recipe development, entrepreneurship development,

Table 8.1. Domestic consumption and export of India's millets, 2004–2014.

	Production ('000 t)		Production growth rate (%)		Export ('000 t)		Export growth rate (%)		Domestic consumption ('000 t)		Domestic consumption growth rate (%)		% share of value of sorghum export to total value of cereal export from India
Year	Sorghum	Other millets	Sorghum	Other millets	Sorghum	Other millets	Sorghum	Other millets	Sorghum	Other millets	Sorghum	Other millets	
2004	7,240	10,840	8.38	–24.46	24	0	–42.86	N.A.	7,200	10,840	7.46	–22.85	0.24
2005	7,630	10,500	5.39	–3.14	4	0	–83.33	N.A.	7,500	10,900	4.17	0.55	0.33
2006	7,150	10,340	–6.29	–1.52	9	0	125	N.A.	7,100	10,300	–5.33	–5.50	0.18
2007	7,930	12,410	10.91	20.02	94	0	944.44	N.A.	2,900	12,400	11.27	20.39	0.29
2008	7,250	11,370	–8.58	–8.38	52	0	–44.68	N.A.	7,200	11,300	–8.86	–8.87	0.72
2009	6,700	8,780	–7.59	–22.78	127	0	144.23	N.A.	6,600	8,800	–8.33	–22.12	0.76
2010	7,000	13,290	4.48	51.37	27	0	–78.74	N.A.	6,800	12,600	3.03	43.18	1.31
2011	6,030	12,800	–13.86	–3.69	134	0	396.3	N.A.	6,000	13,000	–11.76	3.17	0.30
2012	5,300	10,760	–12.11	–15.94	228	0	70.15	N.A.	5,150	11,100	–14.17	–14.62	0.54
2013	5,250	11,520	–0.94	7.06	75	0	–67.11	N.A.	5,200	11,500	0.97	3.60	–
2014	5,000	9,500	–4.76	–17.53	50	0	–33.33	N.A.	4,900	9,700	–5.77	–15.65	–

Sources: indexmundi.com, FAO Statistics (2015).

creating awareness, establishing market linkages etc. by the Directorate of Agriculture and Cooperation (DAC), the arm of the government of India responsible for producing, promoting and popularizing various millet-based products. Annually, at least 12 Entrepreneurship Development Programmes (EDPs), organized by three CoEs, are mandated by DAC on sorghum, pearl millet and small millets. Furthermore, an EDP has included all the potential stakeholders in the value addition and commercialization of millets. Thus they became the backbone for continuing some of the activities of the value chain on sorghum/millets. (In 2014–15 INSIMP was subsumed within the National Food Security Mission (NFSM) programme – its objectives are now covered under the 'coarse cereals component'.)

ICAR's platform research on secondary agriculture and its Agri-Business Incubation (ABI) programme are devised to continue EDPs and commercialization processes that were initiated and successful under NAIP. Supply chain management or backward integration is one important area where more attention is needed to continue the progress made on the pilot scale with the help of ITC (ABD) Ltd. Although some private companies have initiated some effort in providing this function, plenty more needs to be done in this area to make it sustainable.

With the positive impact of on-farm sorghum cultivation being visible over the past 8 years (2008–09 to 2013–14), the revamped image of sorghum cultivation with improved commercial prospects has encouraged even more farmers in the targeted districts to link up with the INSIMP programme. All participating farmers were given entrepreneurship development training to involve themselves in on-farm value addition. Under INSIMP, processing clusters have been provided at no cost to the farmer to encourage farm-level value addition and enable higher net returns. As mentioned above, INSIMP's objectives are now covered under NFSM, if not in totality.

Other departments of the government such as the Ministry of Health, Women and Child Welfare, the Ministry of Education, the Ministry of Food Processing and the Ministry of Commerce are supporting the promotion of millets for health and nutrition. APEDA is also targeting sorghum value added products for export purposes. There is a renewed interest in the funding of pilots for the inclusion of millets in the Mid-day Meal Scheme for children in the states of Telangana, Maharashtra and Karnataka, and scaling-up plans are being made. Tamil Nadu has also developed its own 'millets mission' for promotion and awareness of millets.

At the processing level, development and standardization of other millet products will be continued at IIMR under the NFSM project and certified by the National Institute of Nutrition (NIN) after critical examination of the products. The retrofitting and fabrication of machines to produce various millet products will continue according to the needs of each product development, and will be supplied to SHGs or rural entrepreneurs with NFSM support. Commercialization will continue under the IIMR brand *eatrite* and ANGRAU foods either directly by IIMR/ANGRAU or through those companies that have entered MoUs with IIMR/ANGRAU. Many state agricultural universities (SAUs) have included value addition programmes in their core remit in millet-growing states, addressing various applied issues such as product development and shelf life. The Central Food Technological Research Institute (CFTRI), NIN and IIMR are addressing other R&D issues such as biofortification, bioavailability of minerals, breeding of cultivars for product specificity, functional foods, machinery development for upscaling existing ones, product development, processing diversification, and various biochemical issues related to sorghum and other millets.

As already mentioned, 15 MoUs have been signed with entrepreneurs for the production and marketing of sorghum food products, but many more are in the queue to sign MoUs with IIMR for the same purpose. The entry of big players in the food processing business into the nutri-choice sector, such as Britannia Industries Ltd, ITC Ltd and Bambino, is expected to generate demand for sorghum cultivation in the long term due to the high volumes of supply demanded by these companies. Many private labels have also been established in the market to undertake business in value addition of millets. Brand franchisees of *eatrite* could extend operations across the country but, apart from *eatrite*, there are more than 50 brands that have arisen in the initial phase of commercialization that are struggling hard to attain break-even.

Through the Hyderabad pilot, IIMR has built a wide network of government agencies, entrepreneurs, wholesalers and retailers to help market sorghum products. However, horizontal expansion is envisaged to target markets across the country even in non-traditional sorghum consuming urban metros such as Delhi, Bengaluru, Chennai and Pune.

To conclude, the future strategies for creating demand and enhancing supply of millets in India are listed below, based on recommendations made in policy paper NAAS (2013), to which the author of this book has contributed substantially.

Strategies for creating demand and future promotion

1. The production and consumption of millets must be augmented with appropriate policy initiatives as a continuation of their efforts under INSIMP and NFSM.
2. Consortium-mode research may be pursued for validating the advantages of millets as health and functional foods.
3. Traditional, non-traditional, ready-to-use convenience foods and foods that can be used for complementary feeding programmes may be developed with proven nutrient content and bioavailability mapping.
4. Millet-based complementary foods such as *khichdi, upma, roti* etc. should be introduced in feeding programmes such as the Mid-day Meal Scheme and Integrated Child Development Services.
5. R&D on millets as fodder and forage for livestock feed security may be strengthened.
6. Commercialize and promote millet-based processed RTE snacks and convenience foods through more PPPs.
7. Awareness regarding nutritional, health and environmental advantages may be created through known communication strategies.

Strategies for enhancing supply

1. Development of varieties/hybrids of sorghum and other millets with better recovery capacity on reversal of dry spell for harsh environments and drought-prone areas.
2. Exploration of zero tillage for millets under rice fallows, particularly for southern states of India.
3. Development of varieties/hybrids resistant/ tolerant to salt/high temperature. Strengthening breeding programmes through conventional breeding, marker-assisted breeding as well as biotechnology for biofortification and other traits, such as varieties with better root architecture.
4. Validation of high productive technology under real farming situations.
5. Effective deployment of trait-specific germplasm available in gene banks for genetic enhancement.
6. Evolving strategies for better seed production with public, private and NGO partnerships and establishment of seed villages.
7. Research into better post-harvest management to enhance the shelf life of millets and prevent wastage.
8. R&D of an integrated toolkit for farm mechanization. The Central Institute of Agricultural Engineering in Bhopal has developed a millet mill suitable for small millets. This should be tested in a millet catchment area. Retrofitting of machinery used for rice/wheat/maize for millet foods processing.
9. Markets and entrepreneurship development through modern and innovative approaches.
10. Generation of scientific data to substantiate the claim of conservation of biological resources, low water consumption, agro-climatic limitations and high nutrition value of millets and their derivatives.
11. Promote production and consumption of millets through mixed/relay cropping with legumes and vegetables in homestead gardens.
12. R&D work to generate evidence-based information on phytochemicals with neutraceutical characteristics and authenticate their health potential, including anti-diabetes, anti-inflammatory and hypo-cholesterolemic properties, through clinical trials and nutritional studies. Functional foods for diabetic and obese consumers based on millet will have a good market. Measurements of the glycaemic index should be performed using specified WHO–FAO protocol. Such studies must be extended to the best preparations/recipes for millets with functional properties, through proper clinical trials.
13. Studies to examine the bioavailability of micronutrients from different preparations of millets.
14. Breeding to improve the levels of lysine and tryptophan and also screening the germplasm for specific end uses, such as milling, popping, malting and producing vermicelli noodles.
15. Strengthening backward integration and linking farmers with markets through PPP initiatives.

References

FAO Statistics (2015) Food and Agriculture Organization of the United Nations FAOSTAT Database. Rome, Italy. Available at: faostat3.fao.org (accessed 7 November 2015).

IndexMundi. Available at www.indexmundi.com/agriculture/?commodity=sorghum (accessed 7 November 2015).

NAAS (2013) *Role of Millets in Nutritional Security of India*. Policy Paper No. 66, National Academy of Agricultural Sciences, New Delhi.

Appendices

Appendix 1. Areas of sorghum, pearl, finger and small millets by state, India.
A. Area of sorghum (lakh ha).

		Quinquennial mean area (lakh ha)						
State	Season	1961–66	1971–76	1981–86	1991–96	2001–06	2006–11	2013–14 (annual)
Andhra Pradesh	*Kharif*	12.53	12.88	9.62	4.70	2.44	1.45	1.04
	Rabi	13.27	13.23	10.29	5.18	3.27	1.90	1.23
	Total	25.80	26.11	19.91	9.88	5.71	3.34	2.27 (3.65)
Bihar	*Kharif*	0.09	0.08	0.07	0.05	0.04	0.03	1.2 (1.93)
Gujarat	*Kharif*	11.65	8.92	7.77	3.03	1.11	0.93	0.74
	Rabi	1.99	1.49	1.51	0.97	0.53	0.50	0.54
	Total	13.64	10.41	9.28	4.00	1.64	1.43	1.28 (2.06)
Haryana	*Kharif*	2.70	1.79	1.31	1.10	1.00	0.80	0.72 (1.16)
Karnataka	*Kharif*	13.32	9.99	8.63	5.97	3.14	2.59	1.68
	Rabi	16.09	11.10	13.93	15.24	13.77	11.00	9.71
	Total	29.41	21.09	22.56	21.21	16.91	13.59	11.39 (18.33)
Madhya Pradesh	*Kharif*	20.13	20.43	20.31	12.41	6.61	4.91	2.86
	Rabi	0.24	0.20	0.26	0.11	0.03	0.02	0.00
	Total	20.37	20.63	20.57	12.52	6.64	4.93	2.53 (4.07)
Maharashtra	*Kharif*	25.78	25.62	29.23	24.61	16.18	11.25	6.18
	Rabi	35.31	33.33	36.67	32.35	31.58	30.90	22.44
	Total	61.09	58.95	65.90	56.96	47.75	42.15	28.62 (46.06)
Orissa	*Kharif*	0.06	0.21	0.33	0.22	0.11	0.09	0.08 (0.13)
Rajasthan	*Kharif*	11.29	9.10	9.45	6.83	6.10	6.62	5.79

Continued

Appendix 1. Continued.

State	Season	Quinquennial mean area (lakh ha)						
		1961–66	1971–76	1981–86	1991–96	2001–06	2006–11	2013–14 (annual)
Tamil Nadu	*Kharif*	5.32	5.51	5.72	4.23	2.35	1.97	2.25 (9.32)
	Rabi	2.21	1.55	1.24	0.64	1.11	0.67	1.22
	Total	7.53	7.06	6.96	4.87	3.46	2.64	3.47 (5.58)
Uttar Pradesh	*Kharif*	8.72	6.96	6.37	4.56	2.76	2.06	1.65 (2.66)
All India	*Kharif*	112.44	101.67	98.99	67.88	42.08	32.88	24.26
	Rabi	69.11	60.90	63.90	54.49	50.29	44.99	37.88
	Total	181.55	162.57	162.89	122.37	92.37	77.87	62.14 (100)

B. Area of pearl millet.

State	Season	Quinquennial mean area (lakh ha)						
		1961–66	1971–76	1981–86	1991–96	2001–06	2006–11	2013–14 (annual)
Andhra Pradesh	*Kharif*	6.04	5.62	4.56	1.55	0.92	0.49	–
	Rabi	–				0.10	0.13	–
	Total	6.04	5.62	4.56	1.55	0.98	0.61	0.49 (0.63)
Bihar	*Kharif*	0.18	0.15	0.12	0.08	0.02	0.05	0.03 (0.04)
Gujarat	*Kharif*	14.47	17.74	13.91	11.85	8.08	6.06	6.94
	Summer	–				1.51	2.15	
	Total	14.47	17.74	13.91	11.85	9.59	8.21	6.94 (8.88)
Haryana	*Kharif*	7.81	9.33	7.75	5.74	5.77	6.21	4.04
Karnataka	*Kharif*	4.85	4.68	5.36	3.56	3.40	3.39	2.75
	Rabi	–				0.01	0.01	–
	Total	4.85	4.68	5.36	3.56	3.42	3.40	2.80 (3.59)
Madhya Pradesh	*Kharif*	1.89	2.04	1.70	1.50	1.79	1.74	1.89 (2.42)
Maharashtra	*Kharif*	16.78	16.82	16.82	18.08	14.47	11.34	7.62 (9.76)
Punjab	*Kharif*	1.85	1.47	0.47	0.11	0.07	0.04	0.08 (0.10)
Rajasthan	*Kharif*	45.21	48.01	47.77	46.40	47.46	51.58	44.34 (56.77)
Tamil Nadu	*Kharif*	4.63	4.29	3.27	2.13	1.13	0.57	0.54 (0.69)
Uttar Pradesh	*Kharif*	10.11	10.21	9.56	8.01	8.45	8.71	9.19 (11.77)
All India	Total	114.27	120.76	111.66	99.25	93.39	92.70	78.10

C. Area of finger millet (*ragi*).

State	Quinquennial mean area (lakh ha)						
	1961–66	1971–76	1981–86	1991–96	2001–06	2006–11	2013–14 (annual)
Andhra Pradesh	3.41	2.76	2.39	1.38	0.73	0.50	0.44 (3.89)
Bihar	1.84	1.58	1.43	0.89	0.17	0.12	0.07 (0.62)
Chhattisgarh					0.11	0.09	0.06 (0.53)
Gujarat	0.67	0.52	0.46	0.24	0.23	0.18	0.14 (1.24)
Jharkhand					0.21	0.12	0.12 (1.06)
Karnataka	11.07	9.89	10.99	10.19	9.10	7.18	6.71 (59.32)
Madhya Pradesh	0.16	0.20	0.19	0.15	0.01	0.08	0.04 (0.35)
Maharashtra	2.27	2.06	2.26	1.76	1.45	7.67	1.26 (11.14)
Odisha	0.77	1.85	2.86	1.13	0.73	0.64	0.56 (4.95)
Tamil Nadu	3.30	2.87	2.16	1.47	1.17	0.87	1.19 (10.52)
Uttar Pradesh	1.74	2.27	1.59	1.41	0.01	0.00	–
Uttarakhand					1.48	1.32	1.12 (9.90)
West Bengal	0.10	0.11	0.15	0.13	0.13	0.12	0.10 (0.88)
All India	25.55	24.42	24.74	18.91	15.63	13.00	11.31 (100)

D. Area of small millets.

State	Quinquennial mean area (lakh ha)						
	1961–66	1971–76	1981–86	1991–96	2001–06	2006–11	2013–14 (annual)
Andhra Pradesh	32.12	7.15	4.64	1.08	0.49	0.31	0.28 (4.11)
Bihar	2.23	1.91	1.16	0.51	0.05	0.06	0.028 (0.41)
Chattisgarh					2.49	1.79	1.12 (16.42)
Gujarat	1.82	1.34	1.20	0.32	0.19	0.49	0.38 (5.57)
Jharkhand					0.25	0.25	–
Karnataka	4.37	4.47	3.21	1.11	0.61	0.32	0.24 (3.52)
Madhya Pradesh	14.00	16.07	14.19	10.23	4.03	2.97	2.27 (33.28)
Maharashtra	2.00	2.02	1.80	1.19	0.87	0.71	0.57 (8.36)
Orissa	0.52	1.86	1.48	0.65	0.30	0.18	0.19 (2.79)
Rajasthan	0.77	0.65	0.42	0.25	0.18	0.16	0.13 (1.90)
Tamil Nadu	4.99	3.89	2.77	1.18	0.58	0.36	0.32 (4.69)
Uttar Pradesh	5.62	4.58	2.85	1.47	0.25	0.13	0.08 (1.17)
Uttarakhand						0.734	0.64 (9.38)
All India	46.77	44.89	34.59	18.82	11.73	9.17	6.82 (100)

Appendix 2. Genotypes identified with corneous endosperm suitable for coarse *rawa.*

Entry no.	Genotype	Accession no.	Grain size (µm)	Grain weight (g)	Endosperm texture	Suitable product
3	EB-13	IC-332472	3.35	3.4372	3	Coarse *rawa*
14	GGUB-5	IC-319851	3.35	3.3933	3	Coarse *rawa*
18	GGUB-9	IC-319855	3.35	3.9459	3	Coarse *rawa*
71	EC-19	IC-345721	3.35	3.5197	3	Coarse *rawa*
105	EP-6	IC-305887	4	5.0232	3	Coarse *rawa*
125	EP-36	IC-305917	3.35	3.8656	3	Coarse *rawa*
135	EP-101	IC-345190	3.35	3.9693	3	Coarse *rawa*
137	EP-109	IC-345198	3.35	3.8073	3	Coarse *rawa*
196	SPV-1253		3.35	3.4479	3	Coarse *rawa*
233	SPV-1383		3.35	4.2140	3	Coarse *rawa*
316	SPV-1808		3.35	3.7954	3	Coarse *rawa*

Appendix 3. Genotypes identified with floury endosperm.

Entry no.	Genotype	Accession no.	Grain size (µm)	Grain weight (g)	Endosperm texture	Suitable product
30	GGUB-40	IC-319883	3.35	3.8732	7	Flour
32	ELG-2	To receive	3.35	3.5316	7	Flour
62	EC-4	IC-345706	3.35	3.4512	7	Flour
94	E-71	IC-333429	3.35	3.535	7	Flour
101	EP-2	IC-305883	3.35	3.9987	7	Flour
102	EP-3	IC-305884	3.35	3.5408	7	Flour
103	EP-4	IC-305885	3.35	3.9712	7	Flour
119	EP-29	IC-305910	3.35	4.2553	9	Flour
120	EP-30	IC-305911	3.35	3.6476	7	Flour
126	EP-38	IC-305919	3.35	3.6788	7	Flour
127	EP-39	IC-305920	3.35	3.8823	7	Flour
130	EP-44	IC-305925	3.35	4.0704	7	Flour
199	SPV-1259		3.35	3.6499	7	Flour
258	SPV-1626		3.35	4.2248	7	Flour
299	SPV-1759		3.35	3.6345	7	Flour
313	SPV-1802		3.35	4.0184	7	Flour
321	DSV-5		3.35	3.35	9	Flour
322	CSV-216 R		3.35	5.0346	7	Flour

Appendix 4. Suitable genotypes identified for *rawa* and flakes.

Entry no.	Genotype	Accession no.	Grain size (µm)	Grain weight (g)	Endosperm texture	Suitable product
66	EC-33	IC-345735	3.35	4.068	5	*Rawa*, flakes
107	EP-8	IC-305889	3.35	4.3192	5	*Rawa*, flakes
110	EP-15	IC-305896	4	4.6398	5	*Rawa*, flakes
111	EP-18	IC-305899	3.35	4.3035	5	*Rawa*, flakes
112	EP-19	IC-305900	3.35	4.1802	5	*Rawa*, flakes
113	EP-20	IC -305901	4	4.6142	5	*Rawa*, flakes
114	EP-21	IC-305902	3.35	4.2335	5	*Rawa*, flakes
115	EP-25	IC-305906	4	4.2104	5	*Rawa*, flakes
117	EP-27	IC-305908	3.35	4.0669	5	*Rawa*, flakes
121	EP-31	IC-305912	4	4.2977	5	*Rawa*, flakes
132	EP-98	IC-345187	3.35	4.0545	5	*Rawa*, flakes
133	EP-99	IC-345188	3.35	5.9968	5	*Rawa*, flakes
134	EP-100	IC-345189	4	4.73	5	*Rawa*, flakes
173	SPV-727		3.35	3.9570	5	*Rawa*, flakes
186	SPV-1155		3.35	4.0341	5	*Rawa*, flakes
303	SPV-1763		3.35	4.0145	5	*Rawa*, flakes
304	SPV-1764		3.35	3.9019	5	*Rawa*, flakes
305	SPV-1766		3.35	4.3030	5	*Rawa*, flakes
310	SPV-1787		3.35	3.9087	5	*Rawa*, flakes
317	SPV-1809		3.35	4.0787	5	*Rawa*, flakes
318	CSV-22R		3.35	5.0126	5	*Rawa*, flakes
319	PKV Kranti		3.35	3.9668	5	*Rawa*, flakes
320	DAGADI		3.35	4.3885	5	*Rawa*, flakes
323	CSV-18 R		3.35	4.9495	5	*Rawa*, flakes
324	M35-1		3.35	4.2682	5	*Rawa*, flakes

Appendix 5. Proximate and biochemical composition of released parental lines.

Genotype	Fat (%)	Starch (%)	Protein (%)	Phytic acid (mg/100g)	Protein digestibility (%)	Amylose (%)
2077 B	2.18	54.60	8.99	4.4	45.78	9.79
2219 B	4.27	51.70	10.83	3.9	49.13	9.46
296 B	2.63	66.75	9.19	5.1	47.03	10.28
27 B	2.65	68.00	8.58	4.5	49.59	10.18
AKMS 14 B	2.77	64.95	10.91	4.4	54.57	9.97
7B	2.65	76.25	8.13	3.7	47.26	10.77
IMS 9B	2.43	80.75	9.77	4.6	29.45	9.42
C 43	2.95	65.85	10.83	3.5	59.43	10.48
CS 3541	2.75	62.70	10.51	3.6	31.77	15.19
RS 29	3.16	60.85	9.84	2.4	61.98	16.17
RS 627	2.77	62.40	9.22	2.6	55.83	12.88
RS 673	2.82	60.10	9.22	2.8	52.5	12.75
MR 750	2.62	63.30	9.72	3.0	54.84	15.17
AKR 73	2.89	63.60	8.71	3.2	56.56	12.87
AKR 150	3.00	67.40	7.65	3.2	46.42	13.45
AKR 354	3.18	67.65	9.47	3.6	69.12	11.19
INDORE 12	3.10	61.45	10.58	3.0	51.77	11.25

Appendix 6. Proximate and biochemical composition of released varieties/hybrids.

Genotype	Fat (%)	Protein (%)	Starch (%)	Phytic acid (mg/100g)	Amylose (%)	Protein digestibility (%)
CSH 9	2.90	9.15	65.10	3.00	10.77	52.42
CSH 14	2.49	9.62	69.20	3.80	10.90	40
CSH 16	2.79	9.78	66.10	2.40	1098	46.2
CSH 17	2.51	10.19	60.20	2.40	9.20	42.34
CSH 18	2.95	9.57	65.90	2.60	9.56	49.56
CSV 13	2.75	8.70	61.00	2.40	10.75	41.96
CSV 15	2.48	9.27	67.40	3.20	9.68	44.52
CSV 17	2.97	9.82	62.30	3.60	9.77	58.22
CSV 19SS	2.91	9.49	62.55	3.80	11.04	47.81
SPV 1616	3.04	8.94	64.20	2.80	9.72	42.18
NSV 13	2.79	9.11	72.50	3.70	9.79	44.69
RSSGV 3	2.17	13.80	63.80	3.90	10.85	50.32
ICSV 745	2.31	10.77	65.90	4.00	8.35	44.49
SSV 84	2.98	10.65	63.80	4.40	9.96	40.05
APK 1	2.42	10.01	62.70	3.60	12.55	54.53
BSR 1	2.00	9.83	62.80	3.90	13.28	40.32
GJ 9	1.94	8.72	72.50	5.10	12.45	49.62
GJ 35	2.27	9.30	66.80	4.40	18.98	44.50
GJ 37	2.41	9.05	68.20	4.40	12.45	41.10
GJ 38	2.22	7.98	65.20	4.60	12.45	42.00
GJ 39	3.16	9.08	65.80	5.10	19.01	46.15
GJ 40	2.51	9.83	66.40	5.20	18.98	67.10
JJ 741	2.90	11.21	62.80	6.20	14.46	62.20
JJ 938	2.85	9.68	63.20	6.30	15.66	40.76
JJ 1022	2.76	11.44	63.20	6.20	18.8	40.25
JJ 1041	2.51	10.03	70.80	5.90	16.0	44.60
PVK 400	2.82	9.85	63.60	6.30	12.3	56.09
PVK 801	3.39	8.51	59.20	5.90	14.77	41.80
PVK 809	3.05	9.17	63.80	6.70	16.67	45.53
PSV 1	2.38	7.46	66.90	5.90	15.19	50.0
PSV 2	2.71	11.04	63.80	5.80	15.22	47.04
PSB 3	2.81	9.34	61.80	4.40	20.96	54.68
PSB 9	2.65	8.99	57.20	3.20	20.15	45.04
PSR 23	2.63	10.05	60.80	3.60	19.74	44.59
PSR 34	2.76	12.12	60.90	3.50	16.53	41.93
PVR 453	2.31	10.00	54.70	4.70	15.52	51.37
P Dagadi	3.07	11.25	68.00	4.80	18.34	41.68
P Amrutha	2.66	9.40	62.80	5.30	15.11	50.41
K8	2.49	11.14	65.90	4.60	15.77	59.95
MAN T1	3.50	8.38	64.00	3.60	15.63	47.89
CO26	2.69	8.30	58.60	4.30	18.43	49.28
CO(S)28	2.30	9.56	64.90	4.90	17.45	56.28
SPH 1290	2.74	8.39	63.45	5.30	16.35	46.77

Appendix 7. Biochemical composition of germplasm lines.

Entry	Protein (%)	Protein digestibility (%)	Phytic acid (mg/100g)
EA-10	9.34	24.17	5.0
EA-11	10.38	49.31	6.7
EA-6	10.83	59.52	5.0
EC-1	10.06	58.64	6.1
EC-11	11.08	61.41	5.4
EC-12	13.48	53.76	6.9
EC-21	13.54	50.00	6.3
EC-33	12.39	65.80	8.1
EC-34	13.16	65.90	3.0
EP-1	12.52	64.47	2.8
EP-102	14.34	64.87	3.4
EP-103	14.63	54.43	2.8
EP-104	11.27	45.00	1.4
EP-105	12.92	65.72	3.8
EP-106	11.19	43.71	3.8
EP-107	13.86	22.74	5.0
EP-11	10.73	50.87	5.0
EP-114	11.27	56.42	5.7
EP-115	13.43	61.79	5.2
EP-117	11.21	67.72	7.3
EP-12	11.27	65.11	6.3
EP-120	11.35	64.49	4.4
EP-124	14.07	41.25	4.2
EP-127	11.59	66.77	4.2
EP-13	14.07	47.22	5.2
EP-138	13.83	48.19	4.6
EP-14	12.71	48.90	5.0
EP-16	13.11	33.41	5.2
EP-17	13.80	58.09	7.1
EP-22	13.40	59.95	4.0
EP-23	13.51	47.05	4.7
EP-24	13.40	54.07	5.2
EP-37	13.27	66.28	4.6
EP-41	11.52	49.49	5.0
EP-42	12.82	70.83	5.4
EP-45	11.67	71.34	5.2
EP-46	8.40	67.37	6.7
EP-52	10.79	63.01	13.2
EP-54	9.80	72.57	13.0
EP-55	12.58	45.28	12.0
EP-57	9.72	69.64	14.8
EP-59	10.84	69.56	14.2
EP-64	10.95	69.6	6.1
EP-65	12.95	43.65	6.3
EP-68	10.30	34.69	6.3
EP-78	11.88	48.93	5.6
EP-80	14.12	44.63	6.7
EP-81	11.83	61.83	6.5
EP-82	13.16	65.51	7.9
EP-84	10.28	26.81	9.3
EP-87	12.04	35.53	8.7

Continued

Appendix 7. Continued.

Entry	Protein (%)	Protein digestibility (%)	Phytic acid (mg/100g)
EP-9	10.01	32.67	9.7
EP-91	11.35	40.17	11.4
EP-92	14.28	43.7	9.9
EP-93	11.86	49.23	15.6
EP-94	11.11	33.25	12.0
EP-95	11.08	56.77	12.6
EP-97	12.02	38.90	12.2
PEC-15	13.25	26.15	4.8
PEC-2	12.90	48.46	6.9
PEC-22	12.58	47.55	6.5
PEC-26	12.6	37.77	7.5
PEC-5	13.67	47.07	7.3
PEC-7	11.14	37.99	8.3
SEVS-2	12.53	53.37	9.9
SEVS-20	12.85	37.50	8.9
SEVS-3	13.06	22.44	8.3

Appendix 8. Sorghum recipes, using (A) dehulled sorghum, (B) sorghum flour, (C) sorghum *rawa* and (D) sorghum flakes.

(A) Dehulled sorghum recipes

Sorghum *dosa*: Soak the sorghum grain (dehulled) and black gram *dal* (2:1 ratio) together in water for 4 h. Grind the soaked sorghum grain and black gram *dal* into a fine batter. Add salt to taste and then leave to ferment (8 h). Apply a teaspoon of oil to the pre-heated *dosa* making *tawa* and pour the batter on it; spread with a scoop into a thin round shape. Fry until a crisp *dosa* is obtained. Serve hot with chutney.

Jowar annam: Cook pearled or dehulled sorghum in boiling water until the grains become smooth and palatable. This can also be prepared in a pressure cooker for 5–10 min.

(B) Sorghum flour recipes

Sorghum *roti*: Add hot water to the sieved sorghum flour. Knead into a smooth soft dough. Make a round ball (c. 50 g) and spread it into a round shape on a polythene sheet using a rolling pin or by hand pressing with your palm. Bake the *roti* on both sides of a preheated *tawa*. Serve hot with any curry or *dal*.

Ambali: Mix sorghum flour with lukewarm water carefully to avoid the formation of lumps. Add rice starch soup and salt to the above mixture and cook for 15–20 min. Allow it to cool and then drink, or else ferment overnight in an earthen pot for consumption the next day.

Samosa: Take one cup of *jowar* flour and one cup of wheat flour. Mix well and add the required amount of water to make a dough. Make small *chapathi* balls and spread the dough with a rolling pin and then cut into halves. Mash boiled potatoes, and then add chopped onions, green chillies and curry leaves. Add seasoning to the chopped onions, green chillies, curry leaves and boiled mashed potatoes, boiled peas and salt to taste. Fill the above mixture in each half fold of the dough in a triangular shape and deep fry.

Vegetable *tawa roti*: Take 25g of *jowar* flour, chopped onion leaves, onions, capsicum, carrot, cabbage, green chilli paste, sesame seeds, cumin seeds and salt, and add the required amount of water to make a *chapathi* dough. Make small balls, flatten them into round shapes and then fry in oil on both sides.

(C) Sorghum *rawa* recipes

Upma: Roast 1 cup sorghum *rawa* until it turns brown. Season with mustard seeds, bengal gram *dal*, onion, green chillies, carrot, tomato and curry leaves. Add 3 cups of water, salt to taste, and boil. Add roasted *rawa* slowly. Cook well until it becomes soft at low flame and serve hot.

Sankati: Add sieved sorghum *rawa* to the boiled water. Cook at low flame for 5 minutes, then add sorghum flour slowly and mix thoroughly so that lumps are not formed. After 10 minutes of cooking, add some cooked rice, mix well and cook again for 15–20 minutes. Roll the mixture into small balls.

Kesari: Roast sorghum *rawa* and dry fruits separately in a little oil/ghee until light brown. In a pan, boil milk with a pinch of saffron. Add the fried *rawa* and sugar to the boiled milk and allow it to cool for 15 minutes. Garnish it with dry fruits and a little ghee before serving hot.

Khichidi: Soak ½ cup of green *moong dal* and 1 cup of sorghum *rawa* for 15 minutes. Season with mustard seeds, onion, green chillies, ginger garlic paste, tomatoes, curry leaves and turmeric powder. Add sufficient water and salt, cook at low flame until it is cooked properly and serve hot.

(D) Sorghum flake recipes

Chuduwa: Fry the flakes in oil and put to one side. Heat the oil and add black gram *dal*, fried *channadal*, cumin seeds, groundnuts, red chillies and fry until brown, then add a pinch of turmeric powder and salt to taste. Then fry all these ingredients together. Add fried flakes to the above ingredients and mix thoroughly.

Pongal: Cook ½ cup of *moong dal* until it becomes soft. Add ½ cup of flakes, 2 cups of milk and cook until very soft. Add salt and season with ground pepper, cumin, cashew nuts, curry leaves and serve hot.

***Pongal* (sweet):** Cook ½ cup *moong dal* until it becomes soft. Add ½ cup of flakes, 2 cups of milk and cook until it turns very soft. Add 1 cup of jaggery and cardamom powder. Heat some ghee, fry cashew nuts and dry coconut and add to sweet *pongal* and serve hot.

Peda: Roast the flakes and grind to a fine powder. Add an equal quantity of sugar to the flakes powder and mix well. Add ghee slowly to the flakes powder and roll into small balls. Decorate with cashew or *badam*.

Appendix 9. Details of cluster analysis of sample *jowar*-based products. (A) Survey details. (B) sample kit distributed to the consumers. Results for (C) *jowar* multigrain sorghum *atta* and (D) sorghum biscuits. (E) Profile of the important clusters identified for sorghum biscuits. (F) Results for sorghum vermicelli. (G) Profile of the important clusters identified for sorghum vermicelli. (H) Results for *jowar suji*/*rawa*/semolina (*uppit*/*upma*). (I) Profile of the important clusters identified for sorghum *rawa*. (J) Results for *jowarsuji*/*rawa*/semolina (*uppit*/*upma*). (K) Results for *jowar* flakes. (L) Profile of the important clusters identified for sorghum flakes.

(A) Details of survey for cluster analysis.

As per the stated objective, the study required a primary survey to be conducted among the households in Hyderabad, New Delhi and Pune. The consumers were provided with a sample kit (containing five *jowar*-based products such as multigrain *atta*, *jowar* vermicelli, *jowar* flakes, *jowarsuji*/*rawa*/semolina, and *jowar*-rich biscuits) and were advised to evaluate these products on parameters like packaging, physical appearance, and other cooking and consumption parameters such as the product attributes before and after cooking, and experiences encountered during preparation and consumption. The feedback of the consumers on various factors was captured through a structured questionnaire. The consumers provided their acceptability rating on the product attributes, including packaging, appearance, colour, smell, texture, preparation process, taste, etc., in both raw as well as cooked forms using a five-point hedonic scale (5, liked very much; 4, liked moderately; 3, neither liked nor disliked; 2, disliked moderately; 1, disliked very much). Consumers rated prior awareness on *jowar* and its benefits and their decision to purchase or not on a binary (yes/no) scale.

Sampling plan: A combination of random sampling and partial judgment sampling methods was adopted for the selection of the respondents because of the following two criteria:

1. The targeted respondents belonged to households/families. Other segments of population such as bachelors/caterers were also covered, though at a minor level.

2. Certain segments were deliberately left out such as slums, below poverty line (BPL) populations, and the highly affluent class so as to avoid skewing the sample.

Respondents were classified into four groups/segments, based mainly on socio-economic status comprising income level, education, food habits, etc. The criteria for the sample selection included a higher-income group, higher middle-income group, middle-income group, minority groups, and food habits (vegetarian /non-vegetarian).

Sample area selection: The areas within each city were identified based on the availability of the target groups of respondents in the majority. Identification and selection of area were done by analysing the available secondary data, expert view, literature review based on the reports of the municipal corporation of the area, and the prevailing rent in the concerned city. The households were chosen randomly within the defined areas/locations.

Sample size: The sample size was determined to achieve a statistically significant result at 95% confidence level and 5% confidence interval. On this assumption, from any infinite population (above 50,000), 380 samples are required to achieve the desired result and accuracy level. Hence, about 400 respondents from each city were selected for the interview.

Data collection: The questionnaire was divided into two sections:

- Section I: Information pertaining to the demography, socio-economic conditions, and basic consumption data were collected while giving the sample to the respondents.
- Section II: Feedback on various attributes of the products such as packaging, appearance in raw form, preparation process, and consumption were captured while revisiting the respondents.

Questionnaire testing: The questionnaire was tested before starting the survey to learn about the likely response from the consumers and the ability of the questionnaire to meet the end results. The questionnaire testing was performed using a controlled sampling technique. Around 15 samples were distributed to consumers in Bengaluru with a profile similar to that of the respondents under study. After testing the questionnaire, necessary modifications were incorporated to finalize it.

Field survey plan: On the basis of the sampling plan and questionnaire, the survey work was conducted in the cities of Pune, Hyderabad and New Delhi. The interviewers were provided with a sample kit, questionnaire, an authorization letter from the Indian Institute of Millets Research (IIMR), and a recipe book. They were advised to follow set guidelines while conducting interviews.

The guidelines underscored that no forceful interviews should be conducted. All the respondents should participate at their own will and there should be no coercion or monetary/award-based incentive of any kind. Once the objective of the survey was explained to the respondent, he/she should be asked whether he/she would like to participate and the interview should proceed only on affirmation. Interviews should be sourced through lawful and generally accepted ethical means. There ought to be no transgression of any legal or moral codes during the survey. The interviewer should be courteous at all times. All these guidelines were followed during the survey.

(B) Sample kit distributed to the consumers.

Products	*n*	Approx. weight (g)
Jowar multigrain *atta*	1	200
Jowar vermicelli	1	100
Jowar-rich biscuits	2	25
Jowar flakes/*poha*	1	150
Jowar suji/*rawa*/semolina	1	100
Jowar pasta (only in New Delhi)	1	100

(C) Results for multigrain sorghum *atta*.

Factor name, associated variable and Cronbach's alpha	Factor load	Eigenvalue	Variance explained (%)
Factor 1: Organoleptic (0.944)		9.191	27.85
Softness, texture, taste, smell and overall properties of *chapathi*	0.784		
Preparation (softness of dough, ease of rolling, swelling, smell)	0.735		
Appearance, texture and colour of raw form *atta*	0.699		
Factor 2: Prior awareness (0.926)		5.169	15.66
Jowar is a source of energy and a digestive product	0.887		
It provides wholesome nutrition to the family	0.879		
It is a perfect blend of modern and health food	0.867		
It is good for the sick and the aged	0.771		
Factor 3: Comparison with traditional products (0.768)		2.143	6.5
Comparison of taste with that of wheat *atta*	0.819		
Comparison of preparation process with that of wheat *atta*	0.789		
Comparison of appearance with that of wheat *atta*	0.776		
Factor 4: Socio-economic and value for money		1.703	7.0
Income	0.797		
Willingness to pay the price	0.531		
Expenditure on food items	−0.847		
Food habit/ preferences	0.705		
Factor 5: Demography		1.340	5.0
Family size	0.826		
Number of children in the family	0.783		
Background of the family head	0.700		0.91906364
Kaiser's measure of sampling adequacy: (overall MSA)			
Bartlett's test of sphericity (Sig.)			0.000

(D) Results for sorghum biscuits.

Factor name, associated variable and Cronbach's alpha	Factor load	Eigenvalue	Variance explained (%)
Factor 1: Organoleptic (0.896)		5.91	21
Sweetness, crispiness, colour and overall properties of biscuits	0.790		
Taste, filling, thickness	0.752		
Appearance, taste in mouth	0.743		
Smell of *jowar*	0.406		
Factor 2: Prior awareness (0.913)	0.891	4.08	15
Jowar is a source of energy and is easy to digest			
It provides wholesome nutrition to the family	0.881		
It is a perfect blend of modern and health food	0.796		
Factor 3: Comparison with traditional products (0.748)		1.99	7
Comparison of overall properties with other cookies	0.760		
Smell and taste	0.697		
Crispiness and colour of the biscuits	0.610		
Factor 4: Socio-economic		1.66	6
Expenditure on food items	0.843		
Income	0.875		
Factor 5: Myth associated with *jowar* and income		1.43	5
Jowar products are not suitable for all family members	0.784		
Occupation of the female members	0.436		

Continued

Continued.

Factor name, associated variable and Cronbach's alpha	Factor load	Eigenvalue	Variance explained (%)
Factor 6: Demography		1.27	4.5
Family size	0.834		
Number of children in the family	0.752		0.867
Kaiser's measure of sampling adequacy: (overall MSA)			
Bartlett's test of sphericity (Sig.)			0.000

(E) Profile of the important clusters identified for sorghum biscuits.

Parameters		Good cluster		Avg. cluster			
		N	%	N	%	Chi-square	*P*-value
Family size (total family members)	<3	196	20	28	17	4.33	0.2279
	4	372	37	58	35		
	5	215	21	30	18		
	>5	222	22	48	29		
No. of children	0	470	47	97	47	8.98	0.0292
	1	260	26	35	26		
	2	217	22	26	22		
	>2	58	6	6	6		
Food habit	Veg	736	73	107	65	4.47	0.0344
	Non-veg	269	27	57	35		
Occupation of female	Housewife	209	21	33	20	0.0390	0.8423
	Working	796	79	131	80		
Occupation of male	Service	627	63	98	60	2.07	0.3551
	Other	378	37	49	30		
Monthly household income (INR)	<10 K	81	8	16	10	22.84	<0.0001
	11–20 K	482	48	76	46		
	21–30 K	205	20	55	34		
	>30 K	237	24	17	10		
Food expenditure (% of income)	Up to 15	165	16	13	8	18.31	0.0004
	16–25	268	27	55	34		
	26–35	206	20	50	30		
	>35	366	36	46	28		
Quantity consumed (g/month)	≤100	44	4	1	1	16.66	0.0008
	110–500	309	31	74	45		
	500–100	420	42	56	34		
	>1000	232	23	33	20		
Buying preference	Branded	928	92	130	79	28.02	<0.0001
	Unbranded	77	8	34	21		
Place of majority of purchases	Kiranas/ Retail	800	80	133	81	0.4094	<0.8149
	Supermarket	141	14	20	12		
	Other	64	6	11	7		
Price level (INR/100 g)	4–5	313	31	45	27	28.87	<0.0001
	6–9	272	27	22	13		
	10	385	38	96	59		
	11–15	35	3	1	1		

(F) Results for sorghum vermicelli.

Factor name, associated variable and Cronbach's alpha	Factor load	Eigenvalue	Variance explained (%)
Factor 1: Organoleptic (0.950)		8.82	28
Overall experience with cooked and raw vermicelli	0.837		
Appearance, colour, texture and stickiness of cooked vermicelli	0.776		
Colour of raw vermicelli and smell of cooked vermicelli	0.766		
Preparation time, thickness and appearance of raw vermicelli	0.745		
Factor 2: Prior awareness (0.914)		4.02	13
Jowar is a source of energy and is a digestive product	0.881		
It provides wholesome nutrition to the family	0.867		
It is a perfect blend of modern and health food	0.818		
Factor 3: Comparison with traditional products (0.699)		1.81	6
Comparison of appearance with that of wheat vermicelli	0.752		
Comparison of taste with that of wheat vermicelli	0.714		
Comparison of the preparation process with that of wheat vermicelli	0.657		
Factor 4: Socio-economic and value for money		1.65	5
Income	0.876		
Expenditure on food items	0.823		
Willingness to purchase	–0.360		
Factor 5: Myth associated with *jowar* and income		1.36	4.5
Jowar products are not suitable for all family members	0.784		
Occupation of the female member	0.446		
Factor 6: Demography		1.28	4.0
Family size	0.848		
Number of children in the family	0.736		
Background of the family head	0.232		0.917
Kaiser's measure of sampling adequacy: (overall MSA)			
Bartlett's test of sphericity (Sig.)			0.000

(G) Profile of the important clusters identified for sorghum vermicelli.

		Good cluster		Avg. cluster			
Parameters		N	%	N	%	Chi-square	*P*-value
Family size (total family members)	< 3	108	18	112	20	2.94	0.4016
	4	229	39	199	35		
	5	121	21	122	21		
	>5	126	22	141	25		
No. of children	0	233	40	328	57	42.52	<0.0001
	1	159	27	128	22		
	2	144	25	102	18		
	>2	48	8	16	3		
Food habit	Non-veg	439	75	392	68	6.76	0.009
	Veg	145	25	182	32		
Occupation of female	Housewife	490	84	426	74	16.44	<0.0001
	Working	94	16	148	26		
Occupation of male	Service	360	61	355	62	0.0194	0.990
	Other	224	39	219	38		

Continued

Continued.

Parameters		Good cluster N	Good cluster %	Avg. cluster N	Avg. cluster %	Chi-square	*P*-value
Monthly household income (INR)	<10 K	35	6	60	10	23.57	<0.0001
	11–20 K	272	47	284	49		
	21–30 K	120	21	136	24		
	>30 K	157	27	94	16		
Food expenditure (% of income)	Up to 15	99	17	74	13	16.23	0.001
	16–25	158	27	165	29		
	26–35	101	17	148	26		
	>35	226	39	187	33		
Quantity consumed (g/month)	≤100	24	4	8	1	23.57	<0.0001
	110–500	304	52	357	62		
	500–100	199	34	140	24		
	>1000	57	10	69	12		
Buying preference	Branded	440	75	407	71	2.90	0.088
	Unbranded	144	25	167	29		
Place of majority of purchases	Kiranas/ Retail	450	77.0	411	71.60	21.56	<0.8149
	Supermarket	116	19.8	107	18.64		
	Other	18	3.08	56	9.75		
Price level (INR/200 g)	5–9	118	20	100	17	18.45	0.001
	10	189	32	252	44		
	11–15	173	30	150	26		
	16–20	77	13	57	10		
	21–35	27	5	15	3		

(H) Results for sorghum *suji*/*rawa*/semolina (*uppit*/*upma*).

Factor name, associated variable and Cronbach's alpha	Factor load	Eigenvalue	Variance explained (%)
Factor 1: Organoleptic (0.946)		9.38	29
Taste, overall qualities and softness of *uppit*	0.810		
Appearance, water absorption and texture of *uppit*	0.750		
Colour and *rawa* and *uppit*	0.736		
Smell of *rawa* and *uppit*	0.716		
Texture of *rawa* and ease of roasting	0.702		
Factor 2: Prior awareness (0.720)		2.58	8
Jowar product is a high energy source and easy to digest	0.781		
It is perfect blend of of modern and health food	0.661		
It provides wholesome nutrition to the family	0.371		
Factor 3: Comparison with traditional products (0.697)		2.45	7.5
Comparison of taste with that of wheat *rawa*	0.688		
Comparison of preparation process with that of wheat *rawa*	0.667		
Comparison of appearance with that of wheat *rawa*	0.651		
Factor 4: Socio-economic and value for money		1.57	5
Income	0.810		
Price of the product	0.621		
Expenditure on food items	−0.771		
Factor 5: Myth associated with *jowar* and income	0.759	1.50	4.75
Jowar products are not suitable for all family members			
Occupation of the female members	0.453	1.40	4.0
Factor 6: Demography	0.810		
Family size			
Number of children in the family	0.743		0.892
Kaiser's measure of sampling adequacy: (overall MSA)			
Bartlett's test of sphericity (Sig.)			0.000

(I) Profile of the important clusters identified for sorghum *rawa*.

Parameters		Good cluster		Avg. cluster			
		N	%	N	%	Chi-square	*P*-value
Family size (total family members)	<3	28	25	26	13	6.29	0.098
	4	42	37	79	41		
	5	22	19	46	24		
	>5	22	19	43	22		
No. of children	0	95	83	109	56	24.11	<0.0001
	1	11	10	46	24		
	2	8	7	36	19		
	>2	0	0	3	2		
Food habit	Veg	55	48	109	56	1.82	0.1775
	Non-veg	59	52	85	44		
Occupation of female	Housewife	15	13	32	16	0.6183	0.4317
	Working	99	87	162	84		
Occupation of male	Service	64	56	125	64	3.53	0.1710
	Other	50	44	69	36		
Monthly household income (INR)	<10 K	13	11	14	7	3.92	0.2706
	11–20 K	61	54	118	61		
	21–30 K	25	22	46	24		
	>30 K	15	13	16	8		
Food expenditure (% of income)	Up to 15	8	7	10	5	1.61	0.6591
	16–25	21	18	46	24		
	26–35	25	22	44	23		
	>35	60	53	94	48		
Quantity consumed (g/month)	101–500	81	71	79	41	27.92	<0.0001
	501–1000	25	22	72	37		
	>1000	8	7	43	22		
Buying preference	Branded	46	40	53	27	5.59	0.0081
	Unbranded	68	60	141	73		
Place of majority of purchases	Kiranas/ Retail	109	96	166	86	14.29	0.0008
	Supermarket	4	4	3	2		
	Other	1	1	25	13		
Price level (INR/100 g)	10–19	53	46	98	51	2.75	0.4310
	20–25	32	28	58	30		
	26–30	24	21	27	14		
	31–45	5	4	11	6		

(J) Results for sorghum *suji*/*rawa*/semolina (*uppit*/*upma*).

Factor name, associated variable and Cronbach's alpha	Factor load	Eigenvalue	Variance explained (%)
Factor 1: Organoleptic (0.940)		8.13	25
Overall qualities and appearance of *sira*	0.808		
Texture and water absorption of *sira*	0.767		
Smell, colour and softness of *sira*	0.755		
Overall qualities in raw form, ease of roasting, colour and taste of *sira*	0.736		
Texture, appearance and smell of *rawa*/*suji*	0.690		
Factor 2: Prior awareness (0.917)		4.91	15
Jowar is a source of high energy and easy to digest	0.880		
It is the perfect blend of modern and health food	0.860		

Continued

Continued.

Factor name, associated variable and Cronbach's alpha	Factor load	Eigenvalue	Variance explained (%)
It provides wholesome nutrition to the family	0.854		
Jowar products are not suitable for all family members	0.673		
Factor 3: Comparison with traditional products (0.641)		1.68	5
Comparison of appearance with that of wheat *rawa*	0.753		
Comparison of taste with that of wheat *rawa*	0.726		
Comparison of preparation process with that of wheat *rawa*	0.720		
Factor 4: Socio-economic		1.64	5
Income	0.800		
Occupation of the female members	−0.510		
Expenditure on food items	−0.713		
Factor 5: Demography		1.38	4.0
Family size	0.810		
Number of children in the family	0.743	1.12	3.5
Factor 6: Value for money and food habits			
Price of the product	0.663		
Vegetarian/non-vegetarian	0.361		0.901
Kaiser's measure of sampling adequacy: (overall MSA)			
Bartlett's test of sphericity (Sig.)			0.000

(K) Results for sorghum flakes.

Factor name, associated variable and Cronbach's alpha	Factor load	Eigenvalue	Variance explained (%)
Factor 1: Organoleptic (0.952)		8.44	28
Overall experience with raw and cooked (*poha*) flakes	0.830		
Water absorption, softness, colour and smell of cooked flakes (*poha*)	0.810		
Taste and stickiness of *poha,* and colour and smell of the *jowar* flakes	0.789		
Thickness and appearance of *jowar* flakes and appearance of *poha*	0.757		
Factor 2: Prior awareness (0.914)		4.20	14
Jowar is a source of energy and is a digestive	0.860		
It provides wholesome nutrition to the family	0.876		
It is a perfect blend of modern and health food	0.814		
Factor 3: Myth associated with *jowar* (0.732)		1.89	6
Jowar products are not suitable for all family members	0.804		
Factor 4: Comparison with traditional products (0.572)		1.53	5
Comparison of appearance with that of wheat vermicelli	0.801		
Comparison of preparation process with that of wheat vermicelli	0.752		
Comparison of taste with that of wheat vermicelli	0.566		
Factor 5: Socio-economic and value for money		1.40	4.5
Income	0.879		
Expenditure on food items	−0.829		
Occupation of the female member	−0.266	1.27	4.0
Factor 6: Demography			
Family size	0.843		
Number of children in the family	0.763		
Food habit	−0.265		0.917
Kaiser's measure of sampling adequacy: (overall MSA)			
Bartlett's test of sphericity (Sig.)			0.000

(L) Profile of the important clusters identified for sorghum flakes.

Parameters		Good cluster		Avg. cluster		Chi-square	*P*-value
		N	%	N	%		
Family size (total family members)	<3	124	18	51	22	7.37	0.0610
	4	269	39	70	30		
	5	151	22	47	20		
	>5	149	22	62	27		
No. of children	0	288	42	173	75	82.44	<.0001
	1	186	27	37	16		
	2	171	25	16	7		
	>2	48	7	4	2		
Food habit	Non-veg	530	76	117	51	54.03	<0.0001
	Veg	163	24	113	49		
Occupation of female	Housewife	531	77	191	83	4.18	0.0408
	Working	162	23	39	17		
Occupation of male	Service	431	62	128	56	5.76	0.0561
	Other	262	38	102	44		
Monthly household income (INR)	<10 K	54	8	21	9	29.51	<0.0001
	11–20 K	313	45	103	45		
	21–30 K	138	20	77	33		
	>30 K	188	27	29	13		
Food expenditure (% of income)	Up to 15	129	19	22	10	12.62	0.0055
	16–25	174	25	61	27		
	26–35	126	18	57	25		
	>35	264	38	90	39		
Quantity consumed (g/month)	≤100	213	31	91	40	12.28	0.0065
	110–500	276	40	63	27		
	500–100	163	24	59	26		
	>1000	41	6	17	7		
Buying preference	Branded	43	6	41	18	28.19	<0.0001
	Unbranded	650	94	189	82		
Place of majority of purchases	Kiranas/ Retail	538	78	208	90	18.26	0.0001
	Supermarket	114	16	16	7		
	Other	41	6	6	3		
Price level (INR/kg)	8–10	14	2	0	0	63.71	<0.0001
	11–15	115	17	7	3		
	16–20	274	40	139	60		
	21–25	147	21	63	27		
	26–30	119	17	21	9		
	31–45	24	3	0	0		

Index